Fundamentals of Environmental Site Assessment and Remediation

Fundamentals of Environmental Site Assessment and Remediation

Edited by
Yue Rong

CRC Press
Taylor & Francis Group
Boca Raton London New York

CRC Press is an imprint of the
Taylor & Francis Group, an **informa** business

CRC Press
Taylor & Francis Group
6000 Broken Sound Parkway NW, Suite 300
Boca Raton, FL 33487-2742

Printed on acid-free paper

International Standard Book Number-13: 978-1-138-10515-7 (Hardback)

Cover photograph provided courtesy of Civil & Environmental Consultants, Inc.

Library of Congress Cataloging-in-Publication Data

Names: Rong, Yue, 1958- editor.
Title: Fundamentals of environmental site assessment and remediation / [edited by] Yue Rong.
Description: Boca Raton : Taylor & Francis, a CRC title, part of the Taylor & Francis imprint, a member of the Taylor & Francis Group, the academic division of T&F Informa, plc, 2018. | Includes bibliographical references.
Identifiers: LCCN 2018003940| ISBN 9781138105157 (hardback) | ISBN 9781315101897 (ebook)
Subjects: LCSH: Hazardous waste sites--Evaluation. | Hazardous waste site remediation. | Hazardous wastes--Risk assessment.
Classification: LCC TD1052 .F86 2018 | DDC 628.5/5--dc23
LC record available at https://lccn.loc.gov/2018003940

Visit the Taylor & Francis Web site at
http://www.taylorandfrancis.com

and the CRC Press Web site at
http://www.crcpress.com

To all people who do environmental remediation to make the world better.

Contents

Preface

I am recommending this book, *Fundamentals of Environmental Site Assessment and Remediation*, contributed by a group of environmental professionals who have been working in this field. The contents of this book fit the interests of college students, freshman environmental professionals, and other professionals who might need fundamental knowledge about site remediation. The book can also refresh and update knowledge for experienced environmental professionals working in this field.

Concepts and technologies have been evolving in environmental site assessment and remediation. This book attempts to capture some of these aspects. It starts with relevant environmental laws and regulations that apply to site assessment and remediation, followed by a discussion of the major types of contaminants in drinking water. Human health risk assessment is presented as a concept to guide site assessment and remediation. Fundamental disciplines such as groundwater hydrology, analytical chemistry, contaminant fate and transport, mathematical modeling, environmental statistics, and microbiology are discussed in the book. Environmental forensics is ushered in for the use of many available analytical tools to conduct analysis and interpretation of data. In addition, engineering treatment such as *in situ* treatment, bio-remediation, and nano-technology is presented. To conduct site assessment and remediation requires multi-disciplinary knowledge and skills. This book contains these multi-disciplinary aspects. To be more useful, the book also incorporates a set of exercises and answers at the end of each chapter.

The chapters in the book present a live and vivid picture of the practice of site assessment and remediation by current environmental professionals. I am impressed by the varieties of the theories and applications presented in this book. I will be grateful if any chapter or any information in the book helps readers, even in a very small way. I certainly hope that readers will have the same good feelings as I do after reading this book.

Enjoy reading!

Yue Rong, PhD
Editor
Los Angeles, California

Foreword

Remediation of a hazardous waste site is a complex process that may require several years to bring to completion. The process requires contributions from people with a wide variety of skills and talents, including environmental law and public policy, toxicology, analytical chemistry, environmental chemistry, statistics, earth science, and environmental engineering. The legal and technical staff of the state and federal regulatory agencies set the expectations for the cleanup, earth scientists and chemists define the extent of contamination, engineers select and implement the remedy, and earth scientists and chemists document the attainment of the cleanup goals. At most sites, the performance of the cleanup is also closely monitored by the people who are impacted by the hazardous waste and the people who are responsible for paying for the cleanup.

Those involved in the process do not need to be experts in every applicable discipline, but to be effective, they must know enough to understand and evaluate the contributions of others. If the process is to work well, everyone involved in the process must share a common core of knowledge about environmental assessment and remediation.

This book, *Fundamentals of Environmental Site Assessment and Remediation*, provides that common core of knowledge. The core knowledge is distilled from actual experience in managing environmental assessment and remediation at a wide variety of sites over many years. Dr. Yue Rong, the editor of *Fundamentals of Environmental Site Assessment and Remediation*, is the environmental program manager at the Los Angeles Region of the California Regional Water Control Board. He has more than 25 years' experience with the board dealing with groundwater contamination and vapor intrusion. His chapter authors share his direct practical experience with environmental assessment and remediation. I was a research microbiologist for the US Environmental Protection Agency from 1979 to 2014. Since 2014, I have been the principal scientist at Scissortail Environmental Solutions, LLC. As a groundwater scientist, I have followed Dr. Yue Rong's career for decades. He has a remarkable talent for focusing on the specific information that is needed to make a decision and move a site forward, and avoiding information that does not contribute to a decision and is simply a distraction. This book reflects that talent.

The writing is directed toward the intelligent lay reader. Acronyms and professional jargon are carefully and completely explained. If formulas and equations are used, exercises are provided that allow readers to test their understanding of the subject matter. Individual chapters can allow an expert in a particular discipline to fill in the gaps and gain a working knowledge of the other disciplines. The entire book will be useful to someone who needs a general background in environmental assessment and remediation, such as people involved in real estate transactions, finance, insurance, or community relations.

Environmental Assessment and Remediation puts a heavy emphasis on California and will be most useful to someone working in that state. However, the book is applicable and useful for any state in the United States.

The first chapter reviews the applicable law and public policy. The emphasis is on the Resource Conservation and Recovery Act, the Comprehensive Environmental Response Compensation and Liability Act or Superfund, and the Clean Water Act.

The second chapter reviews the major contaminants in groundwater. It provides short descriptions of the health effects, regulatory standards, occurrence, and treatment of benzene, methyl tertiary butyl ether, trichloroethylene, tetrachloroethylene, arsenic, lead, nitrate and nitrite, perchlorate, hexavalent chromium, 1,2,3-trichloropropane, 1,4-dioxane, and perfluorooctanoic acid and perfluorooctane sulfonate.

The third chapter discusses human health risk assessment, including hazard identification, dose-response assessment, exposure assessment, and risk characterization. It presents examples for the calculation of dose. It discusses the evaluation of exposure pathways, and it discusses the application of screening levels to evaluate risk. It presents an extensive discussion of the California Underground Storage Tank "Low Threat Closure Policy."

Chapter 4 provides a succinct description of the essentials of soil science and groundwater hydrology. It describes good practice for sampling subsurface solids, groundwater, and soil gas and provides a logic to organize the site characterization data into a site conceptual model.

Chapter 5 describes the tools and techniques for site assessment and environmental forensics at sites contaminated with petroleum hydrocarbons. It discusses the use of gas chromatograms to recognize patterns that are associated with various distillates, and the changes in the patterns caused by physical and biological weathering in the environment. It discusses the use of diagnostic compounds and hydrocarbon ratios, biomarkers, and stable isotope ratios in forensics. There is an extensive discussion of techniques to determine the age of a release.

Chapter 6 is devoted to vapor intrusion. It discusses the interaction between the properties of the contaminant, the properties of the subsurface environment, the properties of the atmosphere, and the properties of the building, which controls the final concentration of contaminants in vapors in a building. It describes a process to screen a building for the risk from vapor intrusion, and it discusses options to mitigate the risk from vapor intrusion.

Chapter 7 describes conventional methods to clean up a site, including excavation, recovery of contaminants as an oily phase, pumping of contaminated groundwater, soil vacuum extraction, applying heat to the subsurface, multi-phase extraction and air sparging, the installation of vapor barriers, the injection of chemicals to oxidize or reduce the contaminants, the use of bacteria to degrade the contaminants, and passive methods such as monitored natural attenuation of the use of plants to extract or transform the contaminants.

Chapter 8 goes into greater depth on the injection of chemicals to oxidize or reduce the contaminants, with a description of the chemistry behind the use of permanganate, persulfate, hydrogen peroxide, ozone, zero-valent iron, calcium polysulfide, ferrous iron and sodium dithionite, and biologically produced iron sulfide. The chapter also discusses in greater depth the aerobic degradation and anaerobic biodegradation of contaminants, including a discussion of substrates and electron acceptors. It provides a more detailed description of thermal treatment, and it discusses surfactant flushing.

Chapter 9 describes nanotechnology with an emphasis on nano-scale zero-valent iron.

Chapter 10 deals with bioremediation. It discusses the microbial metabolism of contaminants, with an emphasis on the role of electron acceptors and electron donors in contaminant degradation. It illustrates the use of the quantitative polymerase chain reaction to determine the abundance of DNA coding for enzymes that carry out specific metabolic processes that destroy contaminants, or the abundance of DNA associated with microorganisms that are uniquely associated with the degradation of particular contaminants. It also discusses the use of 16S amplicon genetic sequencing to characterize the entire microbial community to identify the relative abundance of organisms that are known to degrade a contaminant of concern through various metabolic pathways under various geochemical conditions.

Chapter 11 discusses the applications of conventional and widely applied statistical tools to specific problems in site assessment, including techniques to deal with non-detects in the data set, evaluation of background concentrations, whether a site is in compliance with a clean-up goal, trend analysis, and geospatial analysis.

Chapter 12 reviews the primary methods used to analyze samples in the laboratory, how an analytical detection limit is determined, and the importance of understanding the difference between a method detection limit and a practical limit of quantitation.

John T. Wilson, PhD
Principal Scientist
Scissortail Environmental Solutions, LLC

Acknowledgments

The editor would like to express appreciation to Ms. Ariel St Felix and CRC Press, Taylor & Francis Group for accepting, editing, and producing this book. The editor thanks Dr. Yiping Cao for her peer review of Chapter 10, Mr. Nhan Bao, Ms. Annclisa Moc, and Ms. Kirsten Faulkner for their assistance in graphing, Ms. Celina Rong for her assistance with language, and Dr. Jeff Kuo of California State University at Fullerton and Mr. Joseph Clements of CRC Press, Taylor & Francis Group for their vision and support in publishing this book. Gratitude also goes to the contributors of each chapter, who also peer-reviewed the book chapters and made the book possible.

Editor

Dr. Yue Rong is currently the environmental program manager at the Los Angeles Regional Water Quality Control Board of the California Environmental Protection Agency, United States. He has more than 25 years' experience with the agency in dealing with site cleanup and groundwater contamination problems in the Los Angeles area of California. His expertise includes organic pollutants' fate and transport in the subsurface soil and groundwater, environmental analytical chemistry, quality assurance and quality control, environmental statistics, risk assessment, and soil and groundwater pollution assessment and remediation. He is the recipient of the Board Outstanding Achievement Award and Supervisory Performance Award and also received the 2011 Association for Environmental Health and Sciences Foundation Achievement Award. Dr. Rong is an associate editor for the peer-reviewed journal *Soil and Sediment Contamination* and for the journal *Environmental Forensics*. He was elected in 2006 and re-elected in 2008 as the president of the Southern California Chinese American Environmental Professional Association. He is the editor of a book entitled *Practical Environmental Statistics and Data Analysis*. Dr. Rong has a PhD in Environmental Health Sciences from the University of California at Los Angeles, an MS in Environmental Sciences from the University of Wisconsin, and a BS in Earth Sciences from the Beijing Normal University, China.

Contributors

Ravi Arulanantham
Geosyntec Consultants
Oakland, California

Ning-Wu Chang
Department of Toxic Substances Control
California Environmental Protection Agency
Cypress, California

Chawn Y. Jeng
Department of Toxic Substances Control
California Environmental Protection Agency
Cypress, California

Stephen Koenigsberg
CEC Civil and Environmental Consultants, Inc.
Irvine, California

Jun Lu
School of Resources and Environment
Hefei University of Technology
Hefei, China

Aaron Peacock
MicroBac Molecular
Knoxville, Tennessee

Jian Peng
Orange County Environmental Resources
Orange County, California

Yue Rong
Regional Water Quality Control Board
California Environmental Protection Agency
Los Angeles, California

Weixing Tong
Water Quality Control Board
California Environmental Protection Agency
Los Angeles, California

Michael J. Wade
Wade Research, Inc.
Marshfield, Massachusetts

Jason J. Wen
Water Resources Department
City of Lakewood, California

Zhong Xiong
China Aerospace Kaitian Environmental Tech
Changsha, China

Xihong Zhai
Department of Toxic Substances Control
California Environmental Protection Agency
Cypress, California

1 Site Remediation–Related Laws and Regulations

Ning-Wu Chang, Jian Peng, Jason J. Wen, and Yue Rong

CONTENTS

1.1 INTRODUCTION

This chapter provides an overview of legislations and regulations related to site remediation. Besides the overarching objectives of site remediation to protect human health and the environment, the implementation of site remediation relies on regulatory framework and guidelines. There are two primary laws that govern site remediation: the Resource Conservative and Recovery Act (RCRA) and the Comprehensive Environmental Response Compensation and Liability Act (CERCLA), as will be discussed further in this chapter.

The US RCRA of 1976 is the primary legislation regulating hazardous waste management. RCRA originated as a set of amendments to the Solid Waste Act of 1965. RCRA was then signed into law in 1980 and 1984 with additional amendments. The 1984 RCRA amendment is often referred to as the Hazardous and Solid Waste Amendment (HSWA). Basically, RCRA contains the "cradle to grave" concept in the control of hazardous material and the handling of hazardous waste. The other primary legislation is the CERCLA, also known as the Superfund, which addresses hazardous waste release from past activities and emergency responses. The CERCLA was first passed in 1980 and was amended in 1986 with the Superfund Amendments and Reauthorization Act (SARA) to add new requirements and new focus to the Superfund program.

The other related environmental legislations, such as the Clean Air Act (CAA), the Clean Water Act (CWA), the Safe Drinking Water Act (SDWA), the Toxic Substances Control Act (TSCA), and the National Environmental Policy Act (NEPA), are also applicable regulations governing the implementation of site remediation.

1.2 RESOURCES CONSERVATION AND RECOVERY ACT

RCRA covers hazardous waste management from a preventive perspective. It addresses the hazardous waste management responsibilities of all involved entities, including generators, transporters, and treatment, storage, and disposal facilities (TSDF). RCRA is a federal law and provides the minimum requirements. However, under RCRA, states can develop their own hazardous waste programs. The state programs must meet or exceed the requirements in the federal program, and the states must be authorized by the United States Environmental Protection Agency (EPA) to implement the state programs.

There are areas in RCRA that are relevant to site remediation: the disposal of site remediation wastes, imminent hazards, and corrective actions for RCRA-regulated facilities. They are briefly discussed in the following sections.

1.2.1 Disposal of Site Remediation Waste

For the most part, hazardous waste from site remediation must be managed in accordance with applicable RCRA requirements. First, the waste generated from the site remediation must be profiled properly to determine whether the waste from site remediation is a hazardous waste. The criteria for the classification of hazardous waste have been codified in Part 40 of the Code of Federal Regulation (40 CFR), specifically in 40 CFR 261. In general, a hazardous waste can be a characteristic waste or a listed waste under this regulation. However, under this regulation, some wastes are exempted and excluded from full RCRA regulations. The list of excluded wastes can be found in 40 CFR 261.4, and universal waste as an exempted waste can be found in 40 CFR 261.9. However, since the state program may be more stringent than RCRA, certain excluded wastes under RCRA may still be considered as hazardous wastes under the state program. In addition, the state program may name and/or identify additional wastes as hazardous wastes.

Certain special wastes are not regulated under RCRA. For example, polychlorinated biphenyls (PCBs) are regulated under TSCA, specifically under 40 CFR 761; therefore, they are not regulated under RCRA unless the toxicity limits are exceeded. Similarly, since asbestos is already regulated under CAA, CWA, and TSCA, it is not regulated as hazardous waste under RCRA.

Under RCRA, the entity (the responsible party in the case of site remediation) will be responsible for properly identifying the hazardous waste being generated. Once identified as hazardous waste, the waste must be handled in accordance with RCRA's generator, transporter, on-site or off-site TSDF regulations.

1.2.2 Imminent Hazards and Corrective Actions

Under RCRA, the responsible persons of an RCRA-regulated entity are required to take the necessary cleanup actions when situations pose imminent hazards from hazardous waste releases. RCRA's authority is usually used to strengthen the government's action through issuing orders, decrees, and agreements.

The corrective actions under RCRA are synonymous with site remediation in general. Releases of hazardous waste from an RCRA-regulated facility will require the corrective actions necessary to protect human health and the environment. Even though RCRA usually regulates the hazardous waste management within the facility boundary, such as fence-to-fence coverage, corrective actions can extend beyond the facility boundaries as needed. An RCRA-regulated facility, through permit or administrative order, is required to conduct facility investigations, corrective measure studies, and corrective actions: a similar process to CERCLA site remediation.

1.2.3 Underground Storage Tanks

In 1984 RCRA amendments, the underground storage tank program was added as Section 601. The purpose of this program is to regulate underground tanks that store regulated substances to prevent and manage releases. The regulated substances include hazardous chemical products and petroleum products. Hazardous waste storage tanks are already regulated under RCRA Subtitle C, which details performance standards under the RCRA facility permit requirement.

This RCRA amendment specifies schedules for system installation and notification, for leak and release detection, for release response and corrective action, and for standard setting for new tanks. The technical standards and corrective action requirements for the owners and operators of underground storage tanks are further codified in 40 CFR 280.

The underground storage tank program expands the regulated communities to small quantity entities such as neighborhood gasoline fueling stations. Any releases from the underground storage tank will be subject to similar RCRA corrective action requirements, and the corrective action will usually go beyond the property boundary of a gasoline station.

1.3 COMPREHENSIVE ENVIRONMENTAL RESPONSE COMPENSATION AND LIABILITY ACT

To respond to circumstances where there is an imminent hazard from past releases, CERCLA, also known as the Superfund Program, was enacted in 1980. The CERCLA program also addresses immediate response to emergencies due to hazardous releases and establishes reporting quantities for all hazardous substances. CERCLA provides broad federal authority to respond directly to releases or threatened releases of hazardous substances that may endanger public health or the environment. CERCLA provides a framework for site remediation and requires more extensive public participation in the site remediation process (USEPA, 1992). There are several areas that require further discussion.

1.3.1 National Contingency Plan

The CERCLA implementing regulations have been codified in 40 CFR 300, the National Oil and Hazardous Substances Pollution Contingency Plan (also known as the National Contingency Plan [NCP]). The NCP provides the guidelines and procedures needed to respond to releases and threatened releases of hazardous substances, pollutants, or contaminants. These guidelines and procedures are generally followed when conducting site remediation.

The NCP spells out the typical sequence of a site remediation project, which involves the following steps: preliminary assessment (PA), site investigation (SI), listing of priority list (National

Priority List [NPL] under federal program), remedial investigation (RI), feasibility study (FS), record of decision (ROD), remedial design (RD), remedial action (RA), long-term operation and maintenance, and final site close-out. During the FS, the remedial alternatives will be evaluated using the following nine criteria:

- Threshold criteria
  - Overall protection of human health and the environment
  - Compliance with applicable or relevant and appropriate requirements (ARARs)
- Balancing criteria
  - Long-term effectiveness and permanence
  - Reduction of toxicity, mobility, or volume through treatment
  - Short-term effectiveness
  - Implementability
  - Cost
- Modifying criteria
  - State acceptance
  - Community acceptance

The threshold criteria must be satisfied for an alternative to be eligible for selection. The five balancing criteria are used to weigh the relative merits of the alternatives being evaluated. The modifying criteria must be considered during remedy selection and are evaluated only after the FS has been reviewed by the state regulatory agencies and after public comments addressed through the public participation process.

One of the threshold criteria under Section 121(d) of CERCLA requires that site remedial actions attain or waive federal environmental ARARs, or more stringent state environmental ARARs, on completion of the remedial action. The 1990 NCP also requires compliance with ARARs during removal and remedial actions to the extent practicable. ARARs are identified on a site-by-site basis for all site remedial actions where CERCLA authority is the basis for cleanup. The lead agency as well as the supporting agencies must identify and communicate information about potential ARARs to each other.

ARAR includes any standard, requirement, criterion, or limitation under any federal environmental law, such as TSCA, SDWA, CAA, and RCRA, as well as any promulgated standard, requirement, criterion, or limitation under a state or facility site–related law. Applicable requirements are those substantive standards that specifically address the situation at a site. If a requirement is not legally applicable, a decision-maker must determine whether it is relevant and appropriate for the site. A requirement that is relevant and appropriate must address problems or situations similar to the remedial action and would also be suitable to the conditions of the site.

In addition, many federal and state agencies develop criteria, advisories, guidance, and proposed standards that are not legally enforceable but would be helpful in remedial action implementation or in determining the level of protectiveness for remedy selection. These criteria, advisories, guidance, and proposed standards are "to be considered" materials (TBCs). TBCs are not ARARs, and their identification and uses are not mandatory. In a situation where no ARARs would address a particular situation or ensure sufficient protectiveness, the TBC should be used to set cleanup targets. Similarly, many ARARs have general performance criteria with no specific implementation instruction that could be contained in specific guidance documents.

The requirements under CERCLA and NCP, such as the nine criteria evaluation of the alternatives, provide the procedures to document the planning, preparedness, and implementation of the site remedial activities with adequate consideration of related environmental issues. In most cases, because the site remediation will adhere to those substantive and procedural standards, the "functional equivalent" requirements of NEPA will be met. Therefore, a remediation site under federal jurisdiction does not usually need to conduct a separate NEPA evaluation. However, state-specific environmental review may still be required for non-federal lead site remediation.

In response to sustainability initiatives, site remediation is also moving toward greener and more sustainable approaches (https://clu-in.org/greenremediation/) to minimize the environmental footprint of the remedial action while still protecting human health and the environment. The evaluation of green and sustainable remedial action alternatives is conducted during the FS. The green and sustainable evaluation is conducted in addition to the NCP's nine criteria or incorporated and expanded in the evaluation of the balancing criteria.

1.3.2 Hazardous Substances and Petroleum Exclusion

The term *hazardous substance* under CERCLA is defined in CERCLA Section 101(14) to include substances listed under four other environmental statutes (USEPA, 1987). They include

- All hazardous air pollutants listed under Section 112 of CAA
- Toxic pollutants that are subject to pretreatment standards under Section 307(a) of CWA and toxic pollutants that present an imminent danger to public health when discharged to waters of the United States under Section 311 of CWA
- Wastes that are regulated as hazardous wastes under RCRA
- Any imminently hazardous chemical substance or mixture under Section 7 of the TSCA

In addition, Section 102 of CERCLA gives EPA the authority to designate additional hazardous substances not included in this list. The list of hazardous substances and their reporting requirements are provided in 40 CFR 302.4.

However, both the definition of hazardous substance and the definition of *pollutant or contaminant* under CERCLA Section 104(a)(2) exclude petroleum, including crude oil or any fraction thereof, unless specifically listed under those statutes. Accordingly, no petroleum substance, including used oil, can be a *hazardous substance* except to the extent that it is listed as a hazardous waste under RCRA or under one of the other statutes. In its memorandum, EPA further interprets the petroleum exclusion to apply to materials such as crude oil, petroleum feedstocks, and refined petroleum products, even if a specifically listed or designated hazardous substance is present in such products. However, waste oil containing listed CERCLA substances is not within the petroleum exclusion. Similarly, pesticides are not within the petroleum exclusion, even though the active ingredients of the pesticide may be contained in a petroleum distillate. When a reporting quantity of a listed pesticide is released, the release must be reported. In addition, as authorized under RCRA, the state programs may have their own definitions on petroleum exclusion. For example, in many states, there is an underground storage tank (UST) program regulating leaking USTs.

1.3.3 Responsible Party

Under CERCLA, the liability will be triggered if all of the following take place: hazardous wastes are present at a facility; there is actual or potential release of the hazardous substances; cleanup costs will be or have been incurred; and there is a liable party.

The CERCLA liability is retroactive, so any party could be found responsible for contamination that occurred before CERCLA's enactment in 1980. The CERCLA liability is also joint and several, which means that any potential responsible party (PRP) could be held accountable for the entire site remediation even if other entities were also responsible. Finally, the CERCLA liability is strict. That means that a finding of negligence is unnecessary, so a party could be found responsible even though it was operating in accordance with industry standards.

Under CERCLA, there are four classes of PRPs. In addition to present owners and operators, a past owner could be liable if the operations coincided with hazardous waste disposal. The entity that generated the hazardous substances that ended up at the site, as well as the transporters selected the site to dispose of the hazardous waste, could be found responsible. There is only a very narrow exemption of PRP under CERCLA, such as the so-called *innocent purchaser defense*. Once a PRP has been identified, it is potentially liable for all the cleanup costs, damage to natural resources, costs of certain health assessments, and injunctive relief where imminent and substantial endangerment exists.

1.3.4 Brownfields Program

A brownfield is a property whose expansion, redevelopment, or reuse may be complicated by the presence or potential presence of a hazardous substance, pollutant, or contaminant. Cleaning up and reinvesting in these properties increases local tax, job growth, use of existing infrastructure, and proper development of land yet improves and protects the environment.

The Brownfields Program empowers regulators, communities, and other stakeholders in economic redevelopment to work together in a timely manner to remediate and sustainably reuse brownfields and has changed the way contaminated property is perceived, addressed, and managed.

The EPA started the Brownfields Program in 1995. It provided funding to local governments to launch brownfield "pilot" projects and developed guidance and tools for the remediation and redevelopment of brownfield sites. In 2002, the Small Business Liability Relief and Brownfields Revitalization Act (the "Brownfields Law") codified many of EPA's practices, policies, and guidance. The Brownfields Law expanded EPA's assistance by providing new tools for the public and private sectors to promote sustainable brownfield cleanup and reuse.

One important tool of the Brownfields Law is the provision of relief of liability from CERCLA as discussed in the previous section. The liability relief under the Brownfields Law includes

- Small business liability relief for de minimis exemption and municipal solid waste (MSW) exemption, with expedited settlements based on limited ability to pay
- Clarification on release from contiguous property or that owned by someone else
- Exemption of *bona fide* prospective purchasers (and their tenants) from owner or operator liability as long as the person does not impede the performance of a response action or natural resource restoration
- For innocent landowners, the actions landowners must take to satisfy the "all appropriate inquiries" requirement of the defense

Along with the authorized funding to promote brownfields development, the liability relief enables the new owner or developer to actively participate in the redevelopment of brownfields. At the same time, site remediation can be achieved to increase the property's value and to protect human health and the environment.

1.4 CLEAN WATER ACT

The CWA of 1972, which is actually an amendment to the Federal Water Pollution Act of 1948, is arguably the most important and most influential federal law in the recent history of the United States. In a nutshell, CWA regulates point source discharges into navigable surface waters. The significance of CWA to soil and groundwater remediation projects lies in the fact that many such remediation projects involve the discharge of treated water as a point source into surface water. For projects that no longer use surface water discharge as an option (e.g., diversion of discharge to sanitary sewer or landfill), the decisions are often due to stringent requirements imposed by CWA that made this option legally or economically impractical. Therefore, it is worthwhile to examine the requirements of CWA and their implications for soil groundwater remediation projects.

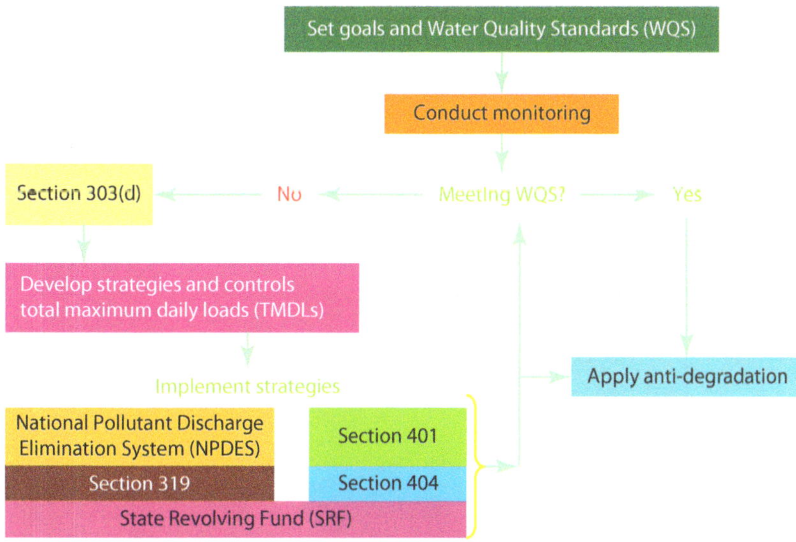

FIGURE 1.1 Schematic flowchart of the Clean Water Act (CWA).

With a goal of restoring and maintaining the chemical, physical, and biological integrity of the nation's waters, CWA uses both regulatory and non-regulatory tools to control direct pollutant discharges into the waters of the United States, to finance municipal wastewater treatment facilities, and to manage polluted runoff. Figure 1.1 summarizes the basic framework of CWA. First, the water quality standards (WQS) are established, and monitoring is conducted to determine the attainment of WQS. Based on the determination, the water body can be declared unimpaired or impaired. In the latter case, it will be put on the CWA List of Impaired Waters, that is, the CWA Section 303(d) list. Once it is on the list, total maximum daily loads (TMDLs) may be developed to bring the water quality into compliance. As part of the TMDL process, implementation measures are required, including, the National Pollutant Discharge Elimination System (NPDES) (see Section 1.4.5) permitting, CWA Section 401 programs, CWA Section 404 programs, and funding programs (e.g., CWA Section 319 Nonpoint Source Grant Program) to support the water quality improvement projects (Peng, 2011).

Analogous to CWA as the national environmental law on water quality, the Porter-Cologne Act of 1969 is the principal law that governs water quality regulation in California. It establishes a comprehensive program to protect water quality and its beneficial uses. Unlike CWA, the Porter-Cologne Act applies to both surface water and groundwater. As a state-level implementation of CWA, the Porter-Cologne Act precedes CWA but was revised to fully comply with CWA.

1.4.1 Water Quality Criteria

WQS, as defined in CWA, consist of a numeric or narrative water quality criterion (termed *water quality objectives* in California), designated uses (*beneficial uses* in California), an anti-degradation policy, and implementation procedures, as shown in Figure 1.2.

As one of the key elements in the CWA, the development of water quality criteria is the first, and perhaps the most important, step in environmental policy-making and one of the primary tools for managing water quality. While the criteria could be either numeric or narrative or both, numeric criteria provide a precise basis for deriving water quality–based effluent limitations (WQBELs) in the NPDES permits and waste load allocations (WLAs) for TMDLs to control pollutant discharges. These limitations and allocations will in turn affect many other aspects of the environmental programs and require the implementation of structural and nonstructural best management practices (BMPs) to ensure that the effluent meets discharge limits.

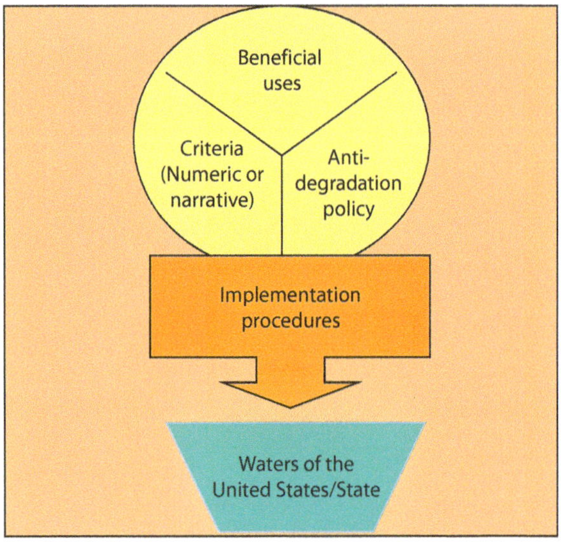

FIGURE 1.2 Components of a Water Quality Standard (WQS).

1.4.2 NTR and CTR

The National Toxics Rule (NTR) specifies the national water quality criteria for a range of toxic pollutants and has been promulgated for 14 states since 1992. The California Toxics Rule (CTR) specifies the federally promulgated water quality criteria for the State of California (USEPA, 2000). CTR is applicable to all California's inland surface waters, enclosed bays, and estuaries for all purposes and programs under CWA except for those within Indian tribal areas (where NTR applies).

1.4.3 Sediment Quality Objectives

During groundwater and soil cleanup, contaminated sediment in aqueous environments may sometimes be encountered. In some cases, the contaminated sediment needs to be removed/dredged as part of the cleanup process. In these cases, regulations for sediment quality will come into play. While there is no nationwide regulatory framework or standard for sediment quality, California's State Water Resource Control Board has adopted the Phase I Sediment Quality Objectives (SQOs) in 2009 for the protection of benthic communities in bays and estuaries from direct impacts based on a 'multiple lines of evidence' approach that integrates sediment chemistry, sediment toxicity, and benthic community information.

Under the SQO framework, the conditions of a site will be evaluated for sediment chemistry, toxicity, and impact on benthic infauna. After all three lines of evidence (LOE) are evaluated, chemistry and toxicity LOEs are combined to evaluate chemically mediated effects, and toxicity and benthic community LOEs are combined to evaluate biological effects. Finally, chemically mediated effects and biological effects are combined to evaluate the overall station-level sediment quality in five categories ranging from "unimpacted" to "clearly impacted." The classification can then be used for regulatory purposes such as 303(d) listing and TMDLs, as described in the following subsection.

1.4.4 Total Maximum Daily Loads (TMDLs)

A TMDL is the maximum amount of a pollutant that a water body can receive and still safely meet WQSs. Each TMDL must account for all sources of the pollutant, including point source WLA, nonpoint source and natural background load allocation (LA), and a margin of safety (MOS):

TMDL = LA + WLA + MOS. TMDLs allocate allowable pollutant loads for each source and identify management measures that, when implemented, will assure that WQSs are attained. Currently, there are nearly 2000 water body–pollutant combinations and over 400 TMDL projects in California. TMDLs are essentially technical documents and are not by themselves enforceable. Once formally adopted, the WLAs will be written into appropriate permits (e.g., NPDES permits, as discussed later, or other permits, such as waste discharge requirements [WDRs]) to be implemented. For groundwater and soil remediation projects that could discharge point source pollutants (including sediments) to surface waters, such as groundwater pump and treat, the allocation will be written into such permits. Meeting the permit requirements often becomes an important part of the cleanup projects themselves, since they are often fairly stringent.

1.4.5 National Pollutant Discharge Elimination System (NPDES)

As authorized by the CWA, the NPDES permit program controls water pollution by regulating point sources that discharge pollutants into waters of the United States. In fact, the NPDES program is the essential and most critical part of the CWA. The NPDES permits are a legal document between a regulatory agency and a discharger. Violating any requirements specified in the NPDES permits often results in severe legal and financial penalties. In California, the State Water Resources Control Board and nine Regional Water Quality Control Boards (RWQCBs) administer NPDES permits, with EPA providing oversight. There are two types of NPDES permits. A general permit regulates multiple entities that share similar characteristics (such as a construction general permit). An individual permit is issued to individual dischargers and imposes discharge limitations on pollutants that are specific to this single discharger. For example, all of the large wastewater treatment plants have their own individual NPDES permits due to the fact that each is unique in many different ways. For soil and groundwater remediation projects, both general and individual permits are common. Note that TMDL, powerful as it is as a regulatory tool, is not self-enforceable in California. Rather, TMDLs are executed via various permits, with NPDES permits as the primary vehicle.

An NPDES permit provides two levels of control: technology-based effluent limits (TBELs), based on the ability of dischargers in the same industrial category to treat wastewater, and WQBELs if technology-based limits are not sufficient to provide protection of the water body. For technology-based effluent limits, CWA mandates that EPA establish national technology based regulations known as *effluent guidelines* and *pretreatment standards* to reduce pollutant discharges from categories of industry discharging directly to waters or discharging indirectly through publicly owned treatment works. EPA has provided extensive guidance on TBELs (e.g., USEPA, 2009). Water quality–based limits are derived from water quality criteria, which may be obtained from a TMDL process as discussed earlier and are mostly more stringent than TBELs. The most stringent standards must be used. In practice, going beyond TBELs requires additional measures and efforts by the NPDES permittee, such as upgrades of the existing technology or replacement with a better and more efficient technology.

For soil and groundwater cleanup sites, not only do the direct discharges need to meet the NDPES permit requirements, but also the stormwater runoff needs to be managed where the bare soil is treated on the surface to meet the stormwater permits either as a construction site, as specified by the construction general permit, or as an industrial site, as specified by the industrial general permit.

1.4.6 State Implementation Plan (SIP)

The Policy for Implementation of Toxics Standards for Inland surface Waters, Enclosed Bays, and Estuaries of California, or State Implementation Plan (SIP), applies to discharges of toxic pollutants into the inland surface waters, enclosed bays, and estuaries of California (California SWRCB, 2005). The SIP was developed in accordance with the Porter-Cologne Water Quality Control Act and the CWA and is implemented through NPDES permits. The SIP establishes, among other things, implementation provisions for priority pollutant criteria in the NTR and the CTR as well as those established by the

RWQCBs; toxicity monitoring requirements; and chronic toxicity control provisions. The provisions and methodologies in the SIP are often used to calculate allocations for TMDLs and effluent limitations for NPDES permits. The goal of this policy is to establish a standardized approach for permitting discharges of toxic pollutants to inland surface waters in a manner that promotes statewide consistency. As such, this policy is a tool to be used in conjunction with watershed management approaches and, where appropriate, the development of TMDLs to ensure the achievement of WQS.

1.4.7 A Case Study in Clean Water Act Implementation

An example is given here to illustrate the significance of surface regulations for soil and groundwater cleanup operations. In the Newport Bay watershed located in central Orange County, California, naturally occurring selenium in shallow groundwater caused elevated selenium concentrations in water, sediment, and biota in the lower portion of the watershed, exceeding regulatory limits. In 1998, the lower portions of the watershed were listed on the CWA 303(d) List as impaired by selenium. In 2002, the United States EPA developed TMDLs for a number of pollutants including selenium but did not specify implementation measures. In 2004, the Santa Ana Regional Water Quality Control Board issued an NDPES general permit, "General Waste Discharge Requirements for Short-Term Groundwater-Related Discharges and de minimis Wastewater Discharges to Surface Waters within the San Diego Creek/Newport Bay Watershed, Order No. R8-2004-0021, NDPES No. CAG998002" (SARWQCB, 2004). This is a general permit regulating any groundwater discharges (well flushing; groundwater cleanup; construction dewatering, etc.) within the watershed. The discharge limits are shown in Table 1.1.

In addition to these default limitations for point source discharges, the permit specifies stringent limits on selenium, as shown in Table 1.2.

The significance of the selenium limits is that they are virtually unachievable. For most of the watershed, especially these parts that are affected by selenium-rich shallow groundwater, the selenium concentrations can be as high as 200 microgram per liter (μg/L). Current treatment technology

TABLE 1.1
Summary of Discharge Limits

Constituents	Maximum Concentration (mg/L)*
Oil and grease	15
Sulfides	0.4
Total suspended solids	75
Total residual chlorine	0.1
Total petroleum hydrocarbons	0.1

*mg/L = milligram per liter

TABLE 1.2
Discharge Limit for Selenium

Discharges to San Diego Creek and Tributaries Thereto		
Constituent	Maximum Daily Concentration Limit	Average Monthly Concentration Limit
Total recoverable selenium	8 μg/L*	4 μg/L*

*μg/L = microgram per liter

using anaerobic reduction and media filtration is practically infeasible (e.g., over a hundred million dollars would be needed for selenium treatment to reach compliance). Therefore, this permit was considered a *de facto* ban on most of the dewatering and groundwater cleanup projects. Since the issuance of this permit, all current and potential dischargers have looked at other compliance pathways stipulated in the permit, including establishing a working group to look into novel treatment technologies and/or ceasing discharges to the surface water altogether by sewer or by other means. However, after over a decade, limited progress had been made on practicable treatment technologies that could be used in the watershed. After a number of permit extensions, a new selenium TMDL was developed in August 2017 that allows a two-phase, 38-year compliance schedule. In this rare case where an NPDES permit triggered a TMDL, it is anticipated that the implementation of the selenium TMDL will require more NDPES permits and water quality–based effluent limits to regulate groundwater-related discharges.

1.5 SAFE DRINKING WATER ACT

1.5.1 Safe Drinking Water Act and Drinking Water Regulation

The SDWA is the federal law that protects public drinking water supplies throughout the nation. First enacted in 1974, SDWA has been substantially amended many times. Two major amendments were made in 1986 and 1996. The federal program consists of the establishment of drinking water standards, monitoring and reporting requirements, and public notification. The 1974 law established the current federal-state arrangement in which the primary implementation and enforcement authority for the drinking water program may be delegated to states. Forty-nine of 50 states (except Wyoming) assumed delegated primacy.

The SDWA requires the EPA to establish National Primary Drinking Water Regulations (NPDWRs) for contaminants that may cause adverse public health effects. The regulations include both mandatory levels (maximum contaminant levels [MCLs]) and non-enforceable health goals (maximum contaminant level goals [MCLGs]) for each included contaminant. MCLs have additional significance because they can be used under the Superfund law as ARARs in the cleanup of contaminated sites.

Although NPDWRs include many programs in addition to MCLs, such as monitoring requirements and public notification, the numerical standard is most applicable in remediation requirements. The drinking water standards can be organized into seven groups (including secondary standards), as shown in Table 1.3.

TABLE 1.3
Summary of Drinking Water Standards in Seven Groups

Groups	MCL/SMCL	MRDL	TT	AL	Total Number of Standards
Microorganisms	2	—	5	—	7
Disinfectants		3	—	—	3
Disinfection by-products	4	—	—	—	4
Inorganic chemicals	14	—	—	2	16
Organic chemicals	51	—	2	—	53
Radionuclides	4	—	—	—	4
Primary standards (total)	**75**	**3**	**7**	**2**	87
Secondary MCLs	15	—	—	—	15

AL: action level; MCL: maximum contaminant level; MRDL: maximum residual disinfectant level; SMCL: secondary maximum contaminant level; TT: treatment technique.

The appendix of this book includes a comprehensive table that summarizes U.S. National Drinking Water Regulations, including numeric values of MCLs, action levels (ALs), TTs (treatment technique), and MCLGs, potential health effects, and typical sources of contaminants.

Under delegated primacy, which requires a state to establish its own standards, if needed, at least as stringent as the federal standards, many states, including California, New Jersey, and Massachusetts, have established a few MCLs that are more stringent than the federal MCLs.

Under SDWA, the EPA promulgates a drinking water regulation for a contaminant if the following three criteria are met:

- The contaminant may have adverse health effects.
- It is known, or there is a substantial likelihood, that the contaminant will occur in public water systems with a frequency and at levels of public health concern.
- Its regulation presents a meaningful opportunity for health risk reduction for persons served by public water systems.

Every 5 years, the EPA must publish a list of unregulated contaminants that are known or anticipated to occur in public water systems and that may require regulation (known as a *contaminant candidate list* [CCL]). The SDWA further directs the EPA to administer a monitoring program for unregulated contaminants to facilitate the collection of occurrence data for contaminants that are not regulated but are suspected to be present in public water supplies. Every 5 years, the EPA must publish a rule (Unregulated Contaminant Monitoring Rule [UCMR]) listing no more than 30 unregulated contaminants to be monitored by public water systems. This list is based on the CCLs as well as other data. Every 5 years, the EPA is required to make a regulatory determination (whether or not to regulate) for at least five of the contaminants included on the CCL.

For each contaminant on the CCL that the EPA determines requires regulation, the EPA must set a non-enforceable MCLG at a level at which no known or anticipated adverse health effects occur and which allows an adequate margin of safety. The EPA must then set an MCL as close to the MCLG as "feasible" using best technology, treatment techniques, or other means available (taking costs into consideration). Similarly, the state may establish a similar health goal and/or guideline; for example, the Public Health Goal (PHG) in California. Once the EPA Administrator makes a determination to regulate a contaminant, the EPA must propose within 24 months and promulgate a "national primary drinking water regulation" within 18 months after the proposal.

New regulations generally become effective 3 years after promulgation. Up to two additional years may be allowed if the EPA (or a state in the case of an individual system) determines the time is needed for capital improvements. The EPA is required to review and strengthen, as appropriate, each drinking water regulation every 6 years. The SDWA regulatory processes are summarized in the flow diagram in Figure 1.3.

When developing regulations, the EPA is required to (1) use the best available, peer-reviewed science and supporting studies and data and (2) make publicly available a risk assessment document that discusses the estimated risks, uncertainties, and studies used in the assessment.

1.5.2 Drinking Water Source Protection

Drinking water may come from groundwater (aquifers), streams, rivers, and lakes. Protecting these drinking water sources is critical to sustaining safe drinking water supplies. The 1986 Amendments to the SDWA established a Wellhead Protection Program to protect groundwaters that supply drinking water wells of public water systems. The 1996 Amendments to the SDWA established a related program for states, called the Source Water Assessment Program (SWAP). The key elements of this program—protection area and zone delineation, inventory of possible contaminating activities (PCAs), and vulnerability analysis—are also elements of a Wellhead Protection Program. The intent

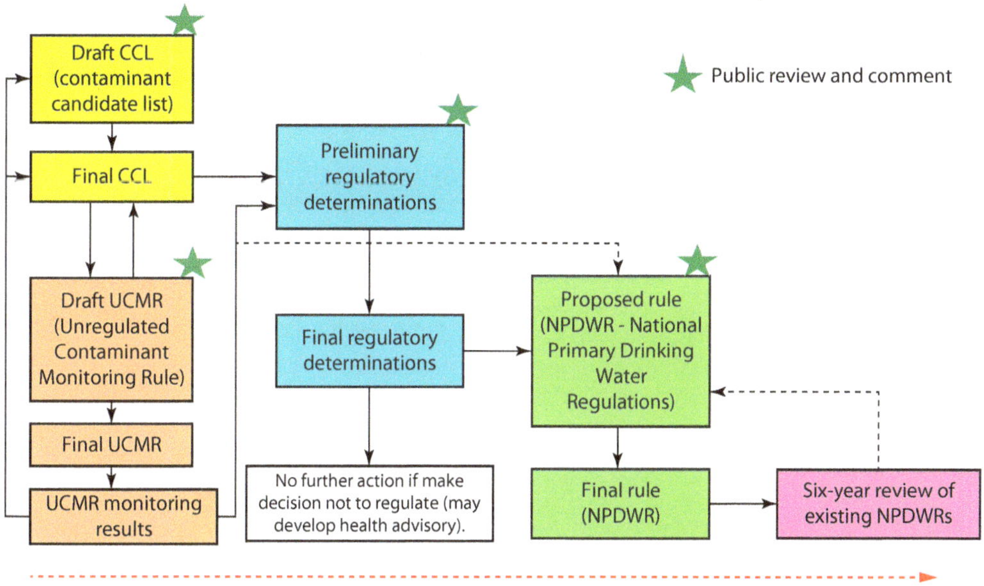

FIGURE 1.3 The process flow chart to add new contaminants to the MCL (maximum contaminant level) list.

of the 1996 SDWA amendments was to promote source water protection, with assessments being the initial step.

Source area is the capture area for a drinking water source. For a surface water source, the source area is the watershed. For a groundwater source, the source area is the recharge area and the area within delineated protection zones. Vulnerability or susceptibility evaluation determines the most significant threats to the quality of the water supply, taking into account the physical barrier effectiveness of the drinking water source. The vulnerability evaluation also considers the type and proximity to the water supply of activities that could release contaminants.

Potential sources of surface water and groundwater contaminants may include commercial, industrial, agricultural, and municipal sources and activities. More detailed information about source water protection assessment and protection can be found at www.waterboards.ca.gov/drinking_water/certlic/drinkingwater/documents/dwsapguidance/DWSAP_document.pdf.

For groundwater protection, the SDWA authorized the establishment of state underground injection control (UIC) programs and required states to prohibit any underground injection not authorized by state permit. To implement the UIC program, the EPA has established six classes of UIC wells based on similarity in the fluids injected, construction, injection depth, design, and operation techniques and issued regulations that establish performance criteria for each class. The injection wells are classified as follows:

- *Class I*: Hazardous wastes, industrial non-hazardous liquids, or municipal wastewater beneath the lowermost underground source of drinking water (USDW)
- *Class II*: Brines and other fluids associated with oil and gas production and hydrocarbons for storage
- *Class III*: Fluids associated with solution mining of minerals beneath the lowermost USDW
- *Class IV*: Hazardous or radioactive wastes into or above USDWs, which are generally banned
- *Class V*: All injection wells not covered under other classes—many of these wells inject non-hazardous fluids into or above USDWs and are typically shallow, on-site disposal systems

- *Class VI*: Carbon dioxide for long-term geologic sequestration to reduce atmospheric emissions of CO_2 from industrial sources

1.5.3 Drinking Water Standards and Remediation Goals

The goals of remediation of contaminated soil and groundwater are to clean up the contamination and restore its original function of beneficial use. Drinking water for human potable consumption is one such major beneficial use. Therefore, meeting drinking water standards has served as the basis of many remediation goals. Many toxicity data have originated from human health studies. Since humans are not the only beneficial users or the most susceptible species in the environment, many remediation standards are set much lower to meet fish and ecosystem requirements. Human health criteria used for establishing drinking water standards are designed to protect people during consumption and other potable uses. Many standards of chemical contaminants are expressed as a concentration, and their numerical values are related to body weight. Since the public health–protective concentration as a numeric standard is in general directly proportional to body weight in the risk calculation, the toxicity of such contaminants to fish and other microorganisms, which have lower weights, will be proportionally lower, which resulted in much lower remediation goals. For example, for copper, there is no drinking water MCL but only an AL at 1300 μg/L. However, copper standards for aquatic life could be several orders of magnitudes lower, at sub-parts per billion levels, since copper is very toxic to fish and aquatic life. Another example is chlorine. Water systems add chlorine at 1–3 (mg/L) for drinking water disinfection to prevent water-borne disease, but chlorine is very toxic to fish and other aquatic life. Besides body weight as a factor in deriving the standard, other factors should also be considered, such as the sensitivity of particular species or organisms.

Another major difference between drinking water standards and remediation goals is that drinking water standards are nationwide or even worldwide and are independent of location, but remedial goals or standards could be locally determined based on the beneficial uses of the water body and various other parameters such as dilution factor, temperature, pH, and so on. In the United States, drinking water standards are established and enforced by the federal EPA or at the state level. Remedial goals could be very different from site to site and are established by regional regulators based on various beneficial uses of the water bodies.

1.6 CALIFORNIA LOW-THREAT UNDERGROUND STORAGE TANK CASE CLOSURE POLICY

In addition to relevant laws and regulations, each federal and state agency may also have its own policies to implement the tasks required by these laws and regulations. Policies are evolving along with the law and regulation updates. This section gives one such example in California, which is used to evaluate leaking UST sites for case closure.

The California State Water Resources Control Board adopted a low-threat underground storage tank case closure policy (LTCP) in 2012 (California SWRCB, 2012). The policy was intended to close the low-risk sites so that more resources could be used for higher-risk sites. There are many reasons for the policy, which are discussed in the below list.

First, California has a large number (ranked number two in the nation) of leaking underground tanks as potential sources of groundwater contamination. Second, the average length of time for a case closure in California is an astounding 20 years. Third, current scientific understanding has improved in areas of natural attenuation for aromatic contaminants and risk assessment methodology. Fourth, there are fewer cases impacting drinking water sources on record. Fifth, the regulated community has long been arguing that consistent numerical standards in case closure across the state are lacking. Sixth, the California UST cleanup fund was impacted negatively during the 2008 financial crisis. With limited resources, priority should be given to the high-risk sites, and the low-risk sites should be closed in a more expedient manner.

The most distinguishing feature of this policy is that there are numerical criteria within the policy for the evaluation of a case closure. The numerical criteria were calculated based on human health risk assessment. The policy has eight general criteria, which need to be met before the site can be evaluated using the numerical criteria. To be eligible for case closure under the policy, a candidate site must satisfy all of the following criteria:

1. The unauthorized release is located within the service area of a public water system.
2. The unauthorized release consists only of petroleum.
3. The unauthorized ("primary") release from the UST system has been stopped.
4. Free product has been removed to the maximum extent practicable.
5. A conceptual site model that assesses the nature, extent, and mobility of the release has been developed.
6. Secondary source has been removed to the extent practicable.
7. Soil or groundwater has been tested for methyl tertiary butyl ether (MTBE).
8. Nuisance does not exist at the site.

After all these eight general criteria are met for a site, the following three numerical media-specific criteria can be applied:

1. Groundwater
2. Vapor intrusion to indoor air
3. Direct contact and outdoor air exposure

As an example, the groundwater numerical criteria in Scenario #2 of the five scenarios listed in the LTCP indicate that if a site satisfies the following: (1) a plume length less than 250 ft, (2) the nearest drinking water well greater than 1000 ft away, (3) no free product on-site, and (4) benzene concentration less than 3000 μg/L and MTBE concentration less than 1000 μg/L, this site will be considered to be low risk for case closure.

A more detailed discussion on each media-specific criterion (numerical criterion) can be found in Chapter 3. When a site is evaluated per all eight general criteria and three media-specific criteria and has satisfied all of them, this site is considered to be low risk and can be closed per the policy.

The entirety of the policy can be seen at www.waterboards.ca.gov/board_decisions/adopted_orders/resolutions/2012/rs2012_0016atta.pdf.

PRACTICE AND EXERCISES

Question 1: Explain what MCL and MCLG are and how they are related.

Answer 1: MCL (maximum contaminant level) is an enforceable legal drinking water standard; MCLG (maximum contaminant level goal) is an entirely health-based level and not legally enforceable. For each contaminant, MCL must be set as close to the MCLG as is "feasible" using best technology, treatment techniques, or other means available, such as costs and economic impacts.

Question 2: Why is a remediation goal for a chemical contaminant often lower than the corresponding drinking water MCL?

Answer 2: MCLs are set to protect human health, and most remediation goals are established to protect fish, aquatic life, and microorganisms. In general, the toxicity of a chemical contaminant, which is used as a basis to establish an MCL and/or a remediation goal, is dependent on body weight. Since the human body is typically much heavier than those of aquatic life, remediation goals are usually proportionally lower than MCLs for the same contaminant.

Question 3: What are the major differences between drinking water standards and remediation goals?

Answer 3: Drinking water standards are to protect human health, and remediation goals are to protect not only human but also aquatic life and the general environment; Drinking water standards are in general set nationwide, independently of location; remediation goals are typically location dependent and established to protect local beneficial uses.

To properly interpret the laws and regulations, it is important to understand the background of how the laws and regulations were derived. The U.S. Federal Register documents the rationale for the U.S. Code of Federal Regulation. In this exercise, you will read the U.S. Federal Register Volume 55, Number 46, March 8, 1990, Page 8719, related to 40CFR300.430(e)(9) regarding the detailed analysis of alternatives, using the nine criteria, during feasibility study and then summarize the background and rationale for the final rule.

The U.S. Federal Register can be accessed through https://www.gpo.gov/fdsys/search/pagedetails.action?sr=791&originalSearch=&st=Port+scanning&ps=10&na=&se=&sb=re&timeFrame=&dateBrowse=&govAuthBrowse=&collection=&historical=false&packageId=FR-1990-03-08&fromState=

Question 4: For NPDES permits, what are the differences between a general permit and an individual permit?

Answer 4: A general permit provides requirements for many permittees of a similar nature to achieve higher efficiency, such as one for all construction sites in a state. An individual permit is issued to a single permittee with unique characteristics, such a major wastewater treatment plant.

Question 5: Why is NPDES critical to the success of the CWA?

Answer 5: This is because an NPDES permit is a legal contract between a regulatory agency and a discharger. Violating any requirements in the permit will result in several penalties.

Question 6: What is the connection between a WQS and a TMDL?

Answer 6: A TMDL prescribes limits for all point source and nonpoint source discharges into a waterbody such that the WQS applicable to this water body is not exceeded.

Question 7: What are the differences between a TBEL and a WQBEL in an NPDES permit?

Answer 7: A TBEL is based on the best available technology. A WQBEL is based on achieving the WQS and is often derived from a TMDL's WLA. A WQBEL is always at least as stringent as the corresponding TBEL.

REFERENCES

California SWRCB (State Water Resources Control Board), 2012, California Low-Threat Underground Storage Tank Case Closure Policy, www.waterboards.ca.gov/board_decisions/adopted_orders/resolutions/2012/rs2012_0016atta.pdf

California SWRCB, 2005, Policy for Implementation of Toxics Standards for Inland Surface Waters, Enclosed Bays, and Estuaries of California (State Implementation Plan or SIP).

Peng, J., 2011, Statistics in environmental policy making and compliance in surface water quality in California, USA. In: *Practical Environmental Statistics and Data Analysis*, Rong, Y. (Ed.), ILM Publications, Hertfordshire, UK.

Santa Ana Regional Water Quality Control Board (SARWQCB), 2004, General Waste Discharge Requirements for Short-Term Groundwater-Related Discharges and de minimis Wastewater Discharges to Surface Waters within the San Diego Creek/Newport Bay Watershed, Order No. R8-2004-0021, NDPES No. CAG998002.

USEPA 9838.1, 1987, Scope of the CERCLA Petroleum Exclusion Under Sections 101(14) and 104(a)(2), July 31.

USEPA, 1992, CERCLA/Superfund Orientation Manual, EPA/542/R-92/005, October.

USEPA, 2000, Water Quality Standards; Establishment of Numerical Criteria for Priority Toxic Pollutants for the State of California (California Toxics Rule or CTR), 40CFR Part 131, Federal Register, Vol. 65, No. 97, May 18, 2000.

USEPA, 2009, Technical support documents for the preliminary 2010 effluent guidelines program plan, October, 2009.

FURTHER READING

California SWRCB, 2009, Water Quality Control Plan for Enclosed Bays and Estuaries, Part I. Sediment Quality (Sediment Quality Objectives or SQO).

2 Pollution Sources and Drinking Water Protection

Jason J. Wen

CONTENTS

The purpose of cleaning up polluted sites is largely to protect drinking water resources, which are mainly used for human consumption. This chapter discusses the types and sources of common water pollutants, how they are used, and how they subsequently impact drinking water resources, their drinking water standards, and their treatment methods. This knowledge is necessary for environmental professionals to conduct site assessment and remediation work.

2.1 INTRODUCTION

There are more than 51,000 community water systems (CWS) in the United States. Among these CWS, 82% are relatively small water systems, each serving 3300 or fewer individuals, and provide water to just 9% of the total population. Among the community systems, 71% rely on groundwater, and 29% rely on surface water.

Drinking water supply is either from surface water, such as rivers and lakes, or groundwater pumped from water wells. In general, surface water must be treated by a surface water treatment plant with conventional treatment processes including coagulation, sedimentation, filtration, disinfection, and other advanced treatment processes as needed, such as membrane filtration, advanced oxidation, and so on. Unlike surface water, the quality of groundwater in general is protected by deep water aquifers, and it is free from most surface contamination. Therefore, most groundwater wells have no need for additional treatment. Groundwater is pumped out, and disinfectants may be added to prevent bacteriological contamination from the distribution system. In the United States, about 77% of drinking water comes from surface water sources, including most urban and metropolitan areas; the other 23% comes from groundwater, including some urban areas and most remote rural areas. In the United States, the majority of water systems are at least partially supplied by groundwater.

Both surface water and groundwater are susceptible to pollution. Most common pollutants have an effect on surface water, including industrial, agricultural, and municipal discharges. To ensure the safety of drinking water, environmental protection agencies at both federal and local levels have eliminated waste discharges to protect source water and drinking water systems, controlled through local water supply permits, such that water purveyors must choose the best available water supplies free of contaminants or provide corresponding treatment processes to effectively remove contaminants. Under the Clean Water Act (CWA), the United States Environmental Protection Agency (USEPA) has been implementing pollutant discharge control through the National Pollutant Discharge Elimination System (NPDES). All discharges to surface water must be regulated through the NPDES permit, and most common pollutants are now under control after more than 40 years' efforts. In addition, under the Safe Drinking Water Act (SDWA), USEPA and the state regulators require drinking water purveyors to have implemented a source water protection program to ensure source water quality. They issue an operation permit only if the proposed treatment processes can treat such source water so that it meets all drinking water standards.

The SDWA requires USEPA to determine the level of contaminants in drinking water at which no adverse health effects are likely to occur with an adequate margin of safety. These non-enforceable health goals, based solely on possible health risks, are called *maximum contaminant level goals* (MCLGs). Based on MCLGs, USEPA established *maximum contaminant levels* (MCLs), which are set as close to the MCLGs as possible considering cost, benefits, and the ability of public water systems to detect and remove contaminants using suitable treatment technologies. For some contaminants, the drinking water standards adopted by various states are more stringent than the federal standards. For example, the California MCL for benzene is 1 μg/L (ppb), whereas the federal MCL is 5 μg/L.

There are five groups of contaminants typically found in drinking water supplies, as listed in Table 2.1 with their typical sources of pollution.

Although in general, microorganisms are not found in groundwater, they are always present in surface water supplies, and various disinfection treatments to remove and/or deactivate microorganisms are the key processes in a surface water treatment plant.

Most contaminants, except for disinfectants/disinfection by-products, result from pollution discharges. As a result of several decades of effort in pollution control and source water protection, most active water sources in the United States today are either free from contamination or limited to a very few contaminants from pollution.

TABLE 2.1
Common Drinking Water Contaminant Groups

Contaminant Group	Common Contaminant	Typical Source of Pollution
Microorganisms	Bacteria, *E. coli*, viruses	Human and animal fecal waste
Disinfectants/disinfection by-products	Chlorine, trihalomethanes, haloacetic acids, bromate	Disinfectants/by-products of water disinfection
Inorganic chemicals	Arsenic, chromium, lead, nitrate, perchlorate	Erosion of natural or manmade deposit; industrial discharges
Organic chemicals	Volatile organic contaminants, synthetic organic contaminants	Industrial/agriculture discharges, leaking from chemical storage tanks
Radionuclides	Alpha activity, radium, uranium, radon	Erosion of natural or manmade deposits

In the following sections, we will discuss some common contaminants from pollution typically found in drinking water sources. The focus of the discussion will include basic data on contaminants and their sources and occurrences, health issues and regulatory standards, and the best available treatment technologies for their removal. Section 2.2 will cover typical organic contaminants, including benzene, methyl tertiary butyl ether (MTBE), trichloroethylene (TCE), and tetrachloroethylene, also named perchloroethylene (PCE); Section 2.3 will discuss common inorganic contaminants, including arsenic, lead and copper, nitrate and nitrite, and perchlorate; and Section 2.4 includes a discussion of several contaminants of emerging concern (CECs), such as chromium (VI), 1,2,3-trichloropropane (1,2,3-TCP), 1,4-dioxane, and perfluoroalkyl substances (PFASs).

Numerous technical documents, publications, tools, and other resources are available from websites. The following web pages are the main references used in preparing this chapter and provide useful information for readers who would like to learn more about these topics:

- www.epa.gov/laws-regulations/summary-safe-drinking-water-act
- www.epa.gov/dwstandardsregulations
- www.cdc.gov/healthywater/drinking/public/regulations.html
- www.waterboards.ca.gov/drinking_water/certlic/drinkingwater/MCLsandPHGs.shtml
- www.oehha.ca.gov/water/public-health-goals-phgs

2.2 ORGANIC CONTAMINANTS

Common organic contaminants found in drinking water include chlorinated hydrocarbons, such as TCE and PCE, and non-chlorinated hydrocarbons associated with petroleum fuel products, such as benzene and MTBE. Since they are typical small organic compounds and have relative high vapor pressure, they are classified as volatile organic compounds (VOCs) and can be removed by either an aeration process or activated carbon adsorption. Another major group of organic contaminants is synthetic organic compounds (SOCs). SOCs are much larger organic compounds such as Atrazine and Simazine from pesticides and herbicides and widely used in agriculture activities. SOCs in general are either semi or non-volatile; therefore, most common treatment to remove SOCs is through activated carbon adsorption.

VOCs are a class of chemicals that are carbon-containing and evaporate, or volatilize, easily into air at normal air temperatures. VOCs are found in a variety of commercial, industrial, and residential products, including gasoline, solvents, cleaners and degreasers, paints, inks and dyes, and pesticides. When VOCs are found in the environment it is typically, often exclusively, the result of human activity, such as a spill or inappropriate disposal where the chemical has been allowed to soak into the ground. Once released into the environment, VOCs may be carried deeper into groundwater aquifers in the form of a non-aqueous phase liquid, and evaporating low concentrations of VOCs as it is transported through the subsurface. The USEPA estimates that volatile organic chemicals are present in one-fifth of the nation's water supplies.

Table 2.2 lists the drinking water standards for several common VOCs and their basic properties.

Since these chemicals are volatile, an aeration process, such as packed tower aeration (PTA), is very effective to remove VOCs from water, although sometimes the off gas must be treated to prevent pollutants from being released to ambient air. Granular activated carbon (GAC) is also a widely used technology to remove VOCs.

2.2.1 Benzene and BTEX

Most original underground storage tanks (USTs) for petroleum products were installed before the 1980s and are made of bare steel, which is likely to corrode over time, thus allowing UST contents to leak into the environment. Leaking petroleum products will contaminate the soil underneath and often groundwater. Chemicals can also be released into the environment from various other sources,

TABLE 2.2
Chemical Properties of Common Organic Contaminants in Drinking Water

Parameters	Benzene	MTBE	TCE	PCE
Chemical formula	C_6H_6	$(CH_3)_3C(OCH_3)$	C_2HCl_3	C_2Cl_4
Molecular weight (g/mol)	78.11	88.15	131.39	165.83
Density (g/mL)	0.88	0.75	1.46	1.62
Solubility (g/L)	1.8 (25°C)	51 (25°C)	1.07 (20°C)	0.15 (25°C)
Log octanol-water (log Kow)	2.14	1.24	2.61	3.4
Vapor pressure (mm Hg)	95.2 (25°C)	245–251 (25°C)	50 (20°C)	9.9775 (13.8°C)
Henry's Law Constant (atm-m³/mol)	5.5×10^{-3}	5.87×10^{-4}	1.1×10^{-2}	1.8×10^{-2}
MCL (ppb)	1 (CA); 5 (USEPA)	13	5	5
PHG or (MCLG) (ppb)	0.15 (0)	13 (NA)	1.7 (0)	0.06 (0)
DLR (ppb)	0.5	3	0.5	0.5
Treatment technology	GAC and PTA	GAC and PTA	GAC and PTA	GAC and PTA

DLR, detection limit for purposes of reporting; GAC, granular activated carbon; PTA, packed tower aeration.
The detection limit for purposes of reporting is a parameter that is set by regulation in state of California for each reportable analyte. It is not laboratory specific, and it is independent of the analytical method used (in cases where several methods are approved). It is expected that a laboratory can achieve a reporting limit that is lower than or equal to the DLR.

including emissions from petroleum refineries, gasoline and diesel engines, and so on. The most common contaminants leaking from USTs into groundwater are benzene, toluene, ethylbenzene, and xylenes, often referred to collectively by the acronym *BTEX*, which all belong to the class of aromatic hydrocarbons with similar physical and chemical properties. BTEX together constitute up to 20% of the composition of gasoline. Apart from being used as gasoline constituents, these chemicals are extensively used in other applications as well. For example, benzene ranks in the top 20 chemicals in the United States in terms of production volume and is used to make plastics, resins, nylon, and synthetic fibers. Benzene is also used to make some types of rubbers, lubricants, dyes, detergents, drugs, and pesticides.

Many rural and urban regions depend on groundwater for their drinking water supply. In urban areas, a large number of potential contaminant sources, such as gas stations, are located in close proximity to drinking water supply wells. When released into the subsurface, these chemicals would be subject to volatilization near the surface, but they are highly mobile in the soil and may leach into groundwater due to their solubility in water. However, these chemicals are found to be biodegradable under aerobic conditions in soil and groundwater.

Potential health effects from long-term exposure to benzene include anemia, decrease in blood platelets, and increased risk of cancer; toluene may affect the nervous system and cause kidney or liver problems; ethylbenzene may cause liver or kidney problems; and xylenes may cause nervous system damage. Table 2.3 summarizes the federal drinking water standards of BTEX as MCLs and their MCLGs.

TABLE 2.3
Federal Drinking Water Standards (BTEX)

	Benzene	Toluene	Ethylbenzene	Xylenes
MCL (mg/L or ppm)	0.005	1	0.7	10
MCLG (mg/L or ppm)	0	1	0.7	10

Some states may have different MCLs; for example, in California, the MCLs for benzene, toluene, and xylenes are 0.001, 0.15, and 1.75 mg/L, respectively. Based on California State Water Resources Control Board data in 2016 using GeoTracker groundwater ambient monitoring and assessment program (GAMA), benzene has been detected in 26 public water wells above the California drinking water standard MCL at 1 µg/L (the federal MCL is 5 µg/L). Since BTEX is rather volatile, treatment to remove BTEX can be readily achieved by either aeration or GAC. Both of these technologies are widely used for wellhead treatment at groundwater well sites.

Some of the treatment options for BTEX-contaminated soils include dig and haul, soil vapor extraction, and bioventing. Excavating contaminated soil from a site for ex-situ treatment is generally very expensive and is a feasible option only when the extent of contamination is very limited. Soil vapor extraction (SVE) is most commonly employed at impacted sites. This involves extracting gasoline vapor from the subsurface via strategically placed extraction wells with a vacuum blower and treating it aboveground either by combusting the vapors in a thermal/catalytic oxidizer or by passing the contaminated vapors through GAC. Bioventing involves injecting oxygen into the unsaturated zone to enhance in-situ biodegradation of the contaminants.

Treatment options for BTEX-contaminated groundwater include pump and treat, air sparging, and bioremediation. In a pump and treat system, contaminated groundwater is pumped aboveground from strategically placed extraction wells and treated aboveground with technologies such as air stripping and GAC. Air stripping further requires off-gas treatment such as thermal oxidation or carbon adsorption. Air sparging is an in-situ groundwater treatment technology in which ambient air is injected into groundwater to supply oxygen so as to enhance bioremediation and to volatilize the contaminants. These treatment technologies are often combined to form a treatment train for greater contaminant removal efficiency. At sites where active cleanup is not deemed necessary due to low contaminant concentrations and/or lack of potential receptors, monitoring of the natural attenuation of the contaminants is conducted.

In addition, more in-situ treatment technologies have been developed in recent years. The most up-to-date cleanup technologies will be discussed in other chapters in this book. In general, treating soil and groundwater contaminated with BTEX is a long and expensive process. The current treatment technologies are effective in reducing contaminant concentrations to acceptable levels but do not eliminate them completely from soil or groundwater.

To prevent surface contamination caused by leaking USTs, water purveyors have been abandoning old shallow groundwater wells, and new wells are being drilled in much deeper aquifers with several hundred feet of sanitary seals installed.

2.2.2 MTBE

The fuel oxygenate MTBE has been used extensively as an additive in gasoline to reduce air pollution and to enhance octane ratings. MTBE ($C_5H_{12}O$) is an aliphatic ether. When it is released into the environment, MTBE's unique chemical/physical properties (see Table 2.2), together with its low potential for natural degradation, make it a significant threat to groundwater that is used as a drinking water source. The persistence and mobility of MTBE in the subsurface environment, combined with its widespread use, have contributed to its dominant presence and frequent detection in groundwater plumes and CWS wells. MTBE has made the news; in 1996, the City of Santa Monica in California learned that two of its drinking water fields, Charnock and Arcadia, were contaminated with MTBE at levels as high as 610 µg/L. This was the first major incident of water contamination, which brought MTBE to public attention.

Possible sources of MTBE releases to groundwater include leaking underground fuel tanks (LUFTs) and pipelines, tank overfilling and faulty construction at gas stations, spillage from vehicle accidents, and home owner releases. Of particular importance are underground fuel tanks and associated pipes and dispensers at gasoline stations. The LUFT sites represent major point sources of gasoline constituents and the leading cause of fuel hydrocarbon (FHC) and oxygenate groundwater contamination. Approximately 35% of CWS wells have one or more LUFT sites within a 1-km radius.

The sheer number of leaking UST sites in close proximity to CWS wells poses a daunting task for state/federal regulators and water purveyors. The standard approach for dealing with petroleum releases is not sufficient for MTBE because of its ability to move quickly and its slow degradation in the subsurface. Response time is critical for MTBE. A quick response to a release greatly increases the ability to check the spread of MTBE and to clean up the mass of the release.

The properties of MTBE that are pertinent to its subsurface fate and transport and their significance are discussed in the context of contaminant transport and remedial techniques. Due to its high water solubility relative to gasoline hydrocarbons, MTBE is more likely to be leached into groundwater or infiltration. Gasoline that has 11% MTBE (the nominal standard for air pollution uses) is determined to have the potential of creating groundwater concentrations of up to 3,990,000 μg/L and gas-phase concentrations of 85,000 mg/m^3. Due to its high volatility when dissolved in gasoline, vapor extraction is an appropriate approach for the remediation of vadose-zone MTBE contamination from gasoline releases. A comparison of vapor extraction remediation of vadose-zone MTBE contamination due to gasoline releases with remediation of benzene from gasoline releases shows MTBE to be similar to benzene, for which the technique is often used. However, the high water solubility of MTBE makes stripping less effective as a remedial technique for dissolved concentrations.

Based on available information on the health effects of MTBE, USEPA decided that there was insufficient information available to allow it to establish quantitative estimates for health risks, and as such, it would not set health advisory limits. The drinking water advisory document indicates that there is little likelihood that MTBE in drinking water will cause adverse health effects at concentrations between 20 and 40 ppb or lower. USEPA has not set a national drinking water standard for MTBE; however, some states have set their own limits. California has a primary MCL of 13 μg/L, established in 2000 at the same level as the public health goal (13 μg/L). California also established a secondary MCL of 5 μg/L addressing taste and odor concerns.

The fact that MTBE was used to improve air quality and was found in groundwater is rather a surprise and has taught us that we should take a holistic approach and consider any potential consequences when we try to solve environmental problems. To improve urban air quality, one of the 1990 Amendments to the federal Clean Air Act (CAA) required the use of reformulated alternative fuels in the most polluted areas, such as in Los Angeles. To implement the provision of CAA, starting in 1995, gasoline in the specified areas was required to contain at least 2% oxygenates, typically MTBE, to improve combustion efficiency in automobiles. Due to the MTBE contamination found in Santa Monica wells, in 1999, California banned the fuel additive MTBE and directed that it must be eliminated from gasoline by the end of 2002. To date, many of the states in the United States have banned MTBE, because it has a high potential to impact groundwater and drinking water resources.

All chemicals vary considerably in their toxic effects depending on the dose and time of exposure. For example, USEPA has found that benzene can potentially cause temporary nervous system disorders, immune system depression, and anemia at concentrations above the MCL for relatively short periods of time, while long-term exposure to benzene above the MCL can cause leukemia and chromosomal defects. While benzene has been designated as a known human carcinogen, there is inadequate evidence to state whether ethylbenzene, toluene, xylenes, and MTBE can cause cancer after a lifetime of exposure to these chemicals. Even though water with chemical concentration less than the MCL is considered safe to drink, drinking water containing more than one chemical may still not be fit for consumption, because little is known about the additive and synergistic effects of these chemicals. Completely preventing long-term exposure and exposure to high concentrations of these chemicals is the best way to avoid the possibility of adverse health effects.

2.2.3 Trichloroethylene (TCE)

TCE is a volatile, colorless, chlorinated hydrocarbon compound that has been widely used as an industrial solvent, primarily for the vapor degreasing and cold cleaning of fabricated metal parts. It is also used in textile cleaning and solvent extraction processes. Production of TCE in the United States has been estimated at about 130,000 metric tons/year. Of the total TCE used in the United States in 1982, 66% was used in vapor degreasing, 22% for export, 7% as chemical intermediates, and 5% for miscellaneous uses. Degradation products (dichloroethylene [DCE] in the form of *cis*- and *trans*-1, 2-DCE) and relative degradation rates (half-life of approximately 1–1.5 years) have been observed in several field studies. The extent of degradation varies with the amount and type of local microorganisms.

TCE released to the environment tends to partition into the atmosphere. It has been estimated that 60%–90% of the annual world production of TCE is released to the environment. The environmental degradation of TCE primarily involves atmospheric photooxidation. The key properties of TCE that affect its movement in the environment are its high vapor pressure and relatively low solubility in water. In surface waters, TCE volatilizes rapidly into the atmosphere, while degradation occurs slowly in groundwater by microbial action.

TCE concentrations in drinking water supplies from surface waters have been measured in 133 cities through federal surveys. Thirty-two percent of the treated drinking waters contained TCE concentrations ranging from 0.06 to 3.2 μg/L and averaging 0.47 μg/L. Thirty percent of all systems sampled contained a TCE level below 1 μg/L. TCE concentrations in groundwater have been measured extensively in California. Between 1984 and 2001, TCE was detected 859 times out of 15,447 water samples taken. Between 2005 and 2008, the MCL of 5 ppb for TCE was exceeded in 133 sources, and the Public Health Goal (PHG) of 1.7 ppb was exceeded in 323 sources. TCE has the most frequently exceeded California MCL for an organic chemical.

Cancer is the primary health concern from TCE exposure. Depending on the particular exposure scenario, animal studies, and tumor sites selected in the risk analysis, the range of drinking water concentrations within the *de minimis* level (10^{-6}; one excess cancer case in 1 million people exposed over a lifetime of 70 years) is from 0.1 to 64 ppb. Since the 5 ppb MCL is within that range, it meets the acceptable risk level of 10^{-4} to 10^{-6} that federal and state regulatory agencies use for establishing drinking water MCLs to protect public health.

Packed tower aeration (PTA) and granular activated carbon (GAC) are the best available technologies for TCE removal. In most cases, air stripping is more cost-effective.

2.2.4 Tetrachloroethylene (PCE)

PCE is slightly soluble in water and is readily volatile. The primary uses of PCE are as a chemical intermediate, primarily in the production of chlorofluorocarbons, and as a solvent, primarily for cleaning operations. The breakdown of the use of PCE is approximately 55% as a chemical intermediate, 25% in metal cleaning and vapor degreasing, 15% in dry-cleaning operations, and 5% for miscellaneous other uses. Numerous household products contain some level of PCE.

Drinking water exposures may occur as a result of environmental releases of PCE from leaking storage tanks, from industrial wastes, and through spills on site or during transportation. Its numerous industrial uses and its presence in consumer products may lead to the release of PCE into water. Cleaning processes have the potential to create PCE-contaminated wastewater. PCE contamination of water arises primarily from the disposal of spent sludge and the storage and disposal of solvents. Contamination of soil with PCE may result from the disposal of sludge and wastewater containing PCE, particularly from recycling operations.

PCE has been detected in most water supplies, including drinking water, groundwater, surface waters, and rainwater. USEPA has estimated that 5.3% of the U.S. population may be exposed to PCE at levels in excess of 0.5 μg/L and 0.4% of the population may be exposed at levels in excess of 5 μg/L.

Typical exposures to PCE from drinking water are not expected to result in any acute health effects due to the low levels involved. Some studies have suggested a positive relationship between working in the dry-cleaning industry and cancer. California considers PCE to be an animal carcinogen and a possible human carcinogen. Geographically, patterns of PCE contamination in California are usually in concert with use patterns, with the highest levels of PCE contamination occurring in the historically most heavily urbanized areas. In a statewide sampling survey of large water systems, approximately 70% of the PCE-contaminated wells were found in Los Angeles County.

Similarly to TCE treatment, PTA and GAC are listed as the best available technologies for PCE removal.

2.3 INORGANIC CONTAMINANTS

Inorganic contaminants are mineral-based compounds such as metals and nitrates found in water supplies. These contaminants occur naturally in some water but can also get into water through farming, chemical manufacturing, industrial discharges, and other human activities. USEPA has set legal limits (MCLs) for 15 inorganic contaminants. It is common to have trace amounts of many inorganic contaminants in water supplies. While there are numerous metals that are necessary for healthy body functions, high levels of these essential metals can be harmful to health. Many dissolved metals that contaminate water supplies can be acutely or chronically toxic. Contamination levels vary widely by geographic location and water source. In the following sections, we will discuss several common inorganic contaminants found in drinking water supplies.

2.3.1 Arsenic

Arsenic is a naturally occurring element in the earth's crust and is very widely distributed in the environment. In certain geographical areas, natural mineral deposits may contain large quantities of arsenic, which may result in higher levels of arsenic in water. Waste chemical disposal sites may also be a source of arsenic contamination of water supplies. The main commercial use of arsenic in the United States is in pesticides, mostly herbicides, and in wood preservatives.

The erosion of rocks and minerals is believed to be the primary source of naturally occurring arsenic found in drinking water supplies and in soil. Anthropogenic sources of arsenic in water include mining, nonferrous metals, especially copper, smelting, waste water, pesticides, treated wood, sewage sludge, coal-fired power plants, urban runoff, and atmospheric deposition.

Arsenic occurs in natural waters in both inorganic and organic forms. Inorganic species such as arsenite (AsO_2^-—As(III)) and arsenate (AsO_4^{-3}—As(V)) are predominant in natural waters. The arsenate/arsenite ratio is dependent on not only the source of water but also redox conditions. As(V), which is the stable oxidation state in oxygen-containing waters, can be reduced to trivalent As(III) in anoxic or reducing environments. As(III) released into oxygenated waters can also be reoxidized to As(V) within a time scale of days. In most municipal water supplies, particularly surface reservoirs, the dominant form of arsenic is As(V). In chlorinated drinking water supplies, all arsenic forms have been found to be pentavalent as a result of oxidation by free chlorine. The major form of arsenic in well waters relatively rich in arsenic also appears to be As(V). In freshwater sources, often more than 80% is As(V), while the remaining 20% or less is composed of As(III). In the environment, some conditions that may affect arsenic valence and speciation include pH, redox potential, the presence of complexing ions, such as ions of sulfur, iron, and calcium, and microbial activity.

According to analysis of the Safe Drinking Water Information System (SDWIS) data, 11,873 groundwater systems were estimated to have mean arsenic concentrations that exceeded 2 μg/L, 5252 systems exceed 5 μg/L, and 2303 systems exceeded 10 μg/L. Arsenic concentrations were projected to be much lower in surface water systems; for example, 1052, 325, and 86 systems exceeding the 2, 5, and 10 μg/L levels, respectively.

Many scientific studies conclude that long-term exposure to inorganic arsenic through drinking water is associated with relatively high risks of cancer of the lungs and bladder and, to a lesser extent, with an increased risk of cancer of the skin, liver, and kidneys. The amounts of arsenic required to cause adverse health effects depend on the chemical and physical form of the arsenic that is ingested. Inorganic forms are generally more acutely toxic than organic forms, and more water-soluble forms tend to be more toxic than those that dissolve poorly in water. Also, the oxidation state of arsenic affects its toxicity, with As(III) being more toxic than As(V). The USEPA's arsenic rule established an MCL of 10 ppb and an MCLG of zero. USEPA's upper bound (90th percentile) estimates of lifetime cancer risk at 10 ppb ranged up to 6.1×10^{-4}.

Regarding arsenic treatment, it is important to know the form of arsenic in your source water, because As(V) is more effectively removed from source waters than As(III). Consequently, if the arsenic in the source water is predominately As(III), then oxidizing As(III) to As(V) with a chemical oxidant (such as chlorine) will result in a higher arsenic removal efficiency. A chemical oxidant should be selected, as aeration is ineffective at oxidizing arsenic.

Given various regional water quality parameters, the presence of iron (Fe) will likely play a prominent role in technology selection and the ability to treat a given water source. Because of the unique role that iron can play in facilitating arsenic removal, the level of iron in the source water is a primary consideration in the selection of an optimal treatment technology.

Precipitation/co-precipitation is frequently used to treat arsenic-contaminated water and is capable of treating a wide range of influent concentrations to meet the MCL for arsenic. The effectiveness of this technology is less likely to be reduced by characteristics and contaminants other than arsenic compared with other water treatment technologies. It is also capable of treating water characteristics or contaminants other than arsenic, such as hardness or heavy metals.

The effectiveness of adsorption and ion exchange for arsenic treatment is more likely than that of precipitation/co-precipitation to be affected by characteristics and contaminants other than arsenic. However, these technologies are capable of treating arsenic to the MCL. Adsorption and ion exchange tend to be used more often when arsenic is the only contaminant to be treated.

Membrane filtration is used less frequently, because it tends to have higher costs and produce a larger volume of residuals than other arsenic treatment technologies.

There are some innovative technologies for arsenic treatment, including permeable reactive barriers, biological treatment, phytoremediation, and electrokinetic treatment, and some of these are *in situ* treatments.

2.3.2 Lead and Copper

Lead and copper are rarely present in raw water sources. They are primarily present at the customer's tap due to corrosion of lead and copper-based materials in home plumbing.

Corrosion in water systems is defined as the electrochemical interaction between a metal surface, such as pipe wall or solder, and water. During this interaction, metal is oxidized and transferred to the water or to another location on the surface as a metal ion. Depending on the materials, there are many forms of corrosion, but usually the most important ones for drinking water are (1) uniform corrosion, where the electrochemical interaction occurs along the pipe wall, resulting in a relatively uniform loss of metal across the entire surface; (2) non-uniform corrosion, where metal is lost from a localized point, causing pitting and mounding in some cases; and (3) galvanic corrosion, which results from a coupling of dissimilar metals or internally in metallic alloys.

The form of lead and copper released into the water can be dissolved, colloidal, or particulate (i.e., bound up with other compounds such as iron and aluminum). Of great importance is the *scale* that builds up naturally on the metal surface. Pipe scales can be complex and can include two types of compounds: (1) passivating films that form when pipe material and water react directly with each other; and (2) deposited scale material that forms when substances in the water (e.g., iron, manganese, aluminum, calcium) precipitate out or sorb to, and then build up on, the pipe surface. Scales can have layers and are influenced by treatment history.

The characteristics of the scale and its structure dictate the amount of lead or copper that is released into the water. If conditions favor the formation of insoluble, adherent scale (i.e., scale that adheres well to the pipe wall), the rate of metals release will be low. However, if scales do not adhere well to the pipe wall, or if they are very soluble, the release of metals may be greater. Other compounds in the water, including aluminum, iron, manganese, and calcium, can significantly influence scale formation and properties. The type of scale will also dictate how susceptible it is to releasing particulate lead following physical disturbances.

Lead can enter drinking water when service pipes that contain lead corrode, especially where the water has high acidity or low mineral content that corrodes pipes and fixtures. The most common problem is with brass or chrome-plated brass faucets and fixtures with lead solder, from which significant amounts of lead can enter into the water, especially hot water. A number of factors are involved in the extent to which lead enters the water, including the following:

- The chemistry of the water (acidity and alkalinity)
- The types and amounts of minerals in the water
- The amount of lead it comes into contact with
- The temperature of the water
- The amount of wear in the pipes
- How long the water stays in the pipes
- The presence of protective scales or coatings inside the plumbing materials

Homes built before 1986 are more likely to have lead pipes, fixtures, and solder. The Safe Drinking Water Act (SDWA) has reduced the maximum allowable lead content—that is, content that is considered *lead-free*—to be a weighted average of 0.25% calculated across the wetted surfaces of pipes, pipe fittings, plumbing fittings, and fixtures and 0.2% for solder and flux.

Considering the fact that contamination of lead and copper in water mainly results from corrosion of household plumbing, instead of setting MCLs, the EPA established action levels (ALs) for lead at 15 ppb and copper at 1.3 ppm in the Lead and Copper Rule (LCR). The EPA has set the MCLG for lead in drinking water at zero, because lead is a toxic metal that can be harmful to human health even at low exposure levels. Lead is persistent, and it can bioaccumulate in the body over time. Young children, infants, and fetuses are particularly vulnerable to lead, because the physical and behavioral effects of lead occur at lower exposure levels in children than in adults. In children, low levels of exposure have been linked to damage to the central and peripheral nervous system, learning disabilities, shorter stature, impaired hearing, and impaired formation and function of blood cells. The MCLG for copper is set at the same level as the AL (1.3 ppm). Health effects from copper exposure above the AL include gastrointestinal distress.

The Centers for Disease Control and Prevention (CDC) recommends that public health actions be initiated when the level of lead in a child's blood is 5 μg/dL or more. It is important to recognize all the ways a child can be exposed to lead. Children are exposed to lead in paint, dust, soil, air, and food, as well as drinking water. If the level of lead in a child's blood is at or above the CDC action level of 5 μg/dL, it may be due to lead exposures from a combination of sources. The EPA estimates that drinking water can make up 20% or more of a person's total exposure to lead. Infants who consume mostly mixed formula can receive 40%–60% of their exposure to lead from drinking water.

However, because lead contamination of drinking water often results from corrosion of the plumbing materials belonging to water system customers, the EPA established a treatment technique rather than an MCL for lead. A treatment technique is an enforceable procedure or level of technological performance which water systems must follow to ensure control of a contaminant. The regulation requires systems to collect tap samples from sites served by the system that are more likely to have plumbing materials containing lead. If more than 10% of tap water samples exceed the lead AL of 15 ppb, then water systems are required to take additional actions, including:

- Taking further steps to optimize their corrosion control treatment
- Educating the public about lead in drinking water and actions consumers can take to reduce their exposure to lead
- Replacing the portions of lead service lines (lines that connect distribution mains to customers) under the water system's control

The requirement of the LCR is corrosion control treatment to prevent lead and copper from contaminating drinking water. Corrosion control treatment means that utilities must make drinking water less corrosive to the materials it comes into contact with on its way to consumers' taps. While it is important to understand and control corrosion, the LCR is specifically concerned with controlling metals release (i.e., release of lead and copper) into the water. Metals release is a function of the reactions that occur between the metal ions released due to corrosion and the physical, chemical, and biological characteristics of the water and the metal surface.

Flushing the stagnant water out of a faucet in the morning before using water is a good practice to reduce lead and copper levels in the water, since longer water contact time with house plumbing such as faucets will likely increase the levels of lead and copper in the water.

2.3.3 Nitrate and Nitrite

Nitrates (NO_3^-) and nitrites (NO_2^-) are two of the nitrogen-oxygen compounds that are used by plants and animals and eventually return to the air as nitrogen gas. Bacteria in soil and plants use oxygen to change nitrite into more stable nitrate, which can be converted back to nitrite by other bacteria when oxygen is lacking. In nature, plants use nitrate as an essential nutrient. Nitrate and nitrite salts completely dissolve in water. In commerce, the majority of nitrate is used in inorganic fertilizers. Nitrate and nitrite are also used in food preservation, some pharmaceutical drugs, and the production of munitions and explosives.

The major sources of nitrates in drinking water are runoff from fertilizer use; leaking from septic tanks, sewage; and the erosion of natural deposits. Animal wastes and nitrogen-containing fertilizers increase concentrations of nitrate in the environment.

Human exposure to nitrates and nitrites results primarily from dietary ingestion, particularly from vegetables and cured meats. Exposure estimates indicate that for more than 99% of the adult population in the United States, only 1%–3% of nitrate and nitrite intake comes from drinking water. The release of nitrate and/or nitrite to soil and water at waste disposal sites could result in contamination of drinking water sources and increased uptake by plants you eat. For infants, the exposure scenarios are somewhat different. For bottle-fed infants consuming drinking water used to prepare their formula, drinking water can be a substantial exposure pathway.

Methemoglobinemia is the primary adverse health effect associated with human exposure to nitrate or nitrite. Most people are not exposed to levels that would cause adverse health effects. Infants are generally recognized as the subpopulation most susceptible to nitrate-induced methemoglobinemia.

The MCL as well as the MCLG for nitrate as nitrogen is 10 ppm, 1 ppm for nitrite-nitrogen and 10 ppm for joint nitrate/nitrite (expressed as nitrogen) in drinking water. In this case, the MCL equals the MCLG, because analytical methods or treatment technology do not pose any limitation. The calculation of these MCLs and MCLGs is based on the protection of infants from the occurrence of methemoglobinemia, the principal toxic effect observed in humans exposed to nitrate or nitrite.

The following treatment methods have proven to be effective for removing nitrate to below 10 mg/L or 10 ppm: ion exchange, reverse osmosis, and electrodialysis. These methods for the removal of nitrate are relatively expensive including both the initial capital cost and the operating costs.

In general, the reverse osmosis process can remove 85%–95% of the nitrate. Actual removal rates may vary depending on the initial quality of the water, the system pressure, and the water temperature. Ion exchange for nitrate-nitrogen removal operates on the same principle as a household water softener. In a standard water softener, calcium and magnesium ions are exchanged for sodium ions. However, for the nitrate removal process, special anion exchange resins are used that exchange chloride ions for nitrate and sulfate ions in the water as it passes through the resin.

2.3.4 Perchlorate

Perchlorate (ClO_4^-) is the most oxygenated member of a series of four anions made up of chlorine and oxygen. The anion has a charge of −1 and can form an acid or a salt in combination with H^+ or another cation such as sodium, potassium, or ammonium. Perchlorate salts are ionic and dissociate completely when dissolved in water.

Perchlorate is an oxidizing chemical used in a variety of industrial processes. Perchlorate can occur in the environment either through industrial contamination or from natural sources. Perchlorate exposure in the United States is ubiquitous, mostly from the ingestion of perchlorate in contaminated food or water. Ammonium perchlorate (NH_4ClO_4), the salt used as an oxidizer in rocket propellants, is a white, crystalline solid and is the major source of most of the perchlorate that has been detected in drinking water sources in the states of California and Nevada.

Perchlorate can apparently be formed by sunlight or lightning interacting with oxygen and chlorine in the atmosphere. As perchlorate falls to the earth in rain, it can distribute at low levels throughout the environment in both soil and water. Perchlorate is also released to the environment from its use in highway flares, fireworks and other explosives, and rocket fuel.

Drinking water sources have become contaminated with perchlorate as a consequence of soil pollution in areas where solid rocket fuel has been manufactured, used, or disposed of. Perchlorate salts are soluble in water, and once dissolved, perchlorate ion can persist in surface water and groundwater for several decades. Since 1997, as a new, more sensitive analytical method has been developed, California has reported measurements of perchlorate concentrations in thousands of drinking water sources and wells throughout the state. Between 2010 and 2013, perchlorate concentrations above 4 ppb were reported in 248 drinking water sources in California.

The primary toxic mechanism of perchlorate is a reduction in iodide uptake into the thyroid gland. If severe enough, this can lead to reduced thyroid hormone production. The function of the thyroid gland is the production of thyroid hormone. Iodide is a key component in the structure of thyroid hormone, and by blocking its uptake into the thyroid gland, perchlorate can potentially cause a decreased production of thyroid hormone. Adequate supplies of thyroid hormone are vital for a variety of physiologic processes, and even small reductions in thyroid hormone have been associated with increased cardiovascular disease risk factors, abnormal fetal brain development, and altered childhood cognition.

There is no USEPA drinking water standard for perchlorate. However, California has set an MCL of 6 ppb for perchlorate with a PHG at 1 ppb.

Perchlorate can be removed using advanced treatment techniques, including ion exchange, reverse osmosis, and biological treatment. Many of these processes have the added advantage of removing nitrate, with often co-occurs with perchlorate. Both ion exchange and reverse osmosis produce concentrated perchlorate brine, which requires special handling and disposal.

2.4 CONTAMINANTS OF EMERGING CONCERN (CECS)

CECs comprise several groups of chemicals recently detected in the water supply. Due to limited toxic and occurrence data, USEPA has not established any enforceable standards such as MCLs for CECs. However, some studies have indicated that many CECs post health concerns, and more research and monitoring are necessary. Many CECs have been included in the federal drinking water Contaminant

Candidate List (CCL), which is the first step in the regulatory process to set drinking water standards. The systematic monitoring of many CECs is carried out under the federal Unregulated Contaminant Monitoring Rule (UCMR), which provides data needed to assess the occurrence as well as the levels of CECs detected in the water supply. Several states have set up health advisory and even regulatory standards for some CECs. In this section, we will discuss a few CECs.

2.4.1 Hexavalent Chromium

Chromium is a metallic element and generally occurs in small quantities associated with other metals, particularly iron. Chromium is an industrially important metal that has the potential to contaminate drinking water sources. In aqueous systems, chromium typically exists in two different oxidation states: the trivalent form, Cr(III), and the hexavalent form, Cr(VI). The hexavalent ionic form of Cr(VI) is more water-soluble, more easily enters living cells, and is much more toxic than the trivalent ionic form Cr(III). Trivalent chromium is an essential trace element in the human diet. Hexavalent chromium is a human carcinogen, as determined by the National Toxicology Program (NTP), the International Agency for Research on Cancer (IARC), and the USEPA. In addition to naturally occurring chromium sources, contamination from the by-products of tanning, chrome plating, corrosion control, and other industrial processes may contribute to Cr(VI) in drinking water.

Hexavalent chromium has been demonstrated to be a human carcinogen when inhaled, and a large body of evidence exists that supports the carcinogenic effect of inhaled Cr(VI); however, studies on the effects of ingested Cr(VI) are very limited. Those studies found statistically significant increases in cancer rates only at higher doses of Cr(VI), although trend tests were significant.

USEPA currently has an MCL for total chromium species (Cr(III) plus Cr(VI)) of 100 μg/L in drinking water, which was raised in 1991 from its previous level of 50 μg/L. Although USEPA raised the national MCL, the state of California maintained its own total chromium MCL of 50 μg/L. In July 2011, California's Office of Environmental Health Hazard Assessment issued a non-enforceable PHG for Cr(VI) of 0.02 ppb. In 2014, California issued a Cr(VI) MCL of 10 ppb. In 2017, due to the state agency failing to properly consider the economic feasibility of complying with the MCL, a court ordered that the Cr(VI) MCL of 10 ppb in California was invalidated. However, the state is in a process to re-establish a Cr(VI) MCL.

Chromium is present in rock and soil primarily in the form of the insoluble oxide, Cr_2O_3. Chromium is generally not mobile in soil. Chromium enters environmental waters from anthropogenic sources such as electroplating factories, leather tanneries, and textile manufacturing facilities. Chromium also enters groundwater by leaching from soil. Chromium can exist in water as either Cr(III) or Cr(VI). Rivers in the United States have been found to have from <1 to 30 μg/L of chromium. U.S. lakes usually have <5 μg/L of chromium. When high levels are present, they can usually be related to sources of pollution. There were 2208 sources of drinking water reported as having chromium levels detected above 1 ppb in California in 2009. Hexavalent chromium levels in 456 sources were between 6 and 10 ppb, and 1434 sources had levels between 1 and 5 ppb.

Since 2001, numerous research projects have investigated the most effective and affordable ways to remove chromium from drinking water. Through these efforts, three technologies emerged as the most promising for the removal of Cr(VI) in drinking water: weak base anion (WBA) exchange; strong base anion (SBA) exchange; and reduction-coagulation-filtration (RCF). Each technology has its strengths and weaknesses, so utilities must evaluate the many unique aspects of individual water qualities and treatment site characteristics to determine the most appropriate treatment solution.

2.4.2 1,2,3-Trichloropropane (TCP)

1,2,3-TCP (TCP) ($C_3H_5Cl_3$) is an anthropogenic industrial chemical and pesticide by-product. It has been used as an industrial solvent and as a cleaning and degreasing agent. It has also been found as an impurity resulting from the production of soil fumigants. It is used as a chemical intermediate in the production of other chemicals. In 1977, the production of TCP in the United States was estimated in the range from 21 to 110 million pounds.

Based on the formation of multiple tumors in animals, TCP is classified by the USEPA as "likely to be carcinogenic to humans." The main exposure route for the public is via the ingestion of contaminated drinking water, with minimal exposure contributed by inhalation and dermal contact. No federal MCL has been set for TCP in drinking water. Various states have established health-based drinking water guidance values. The state of Hawaii has established a state MCL of 0.6 μg/L. California in 2009 set a PHG of 0.0007 μg/L and in 2017 set a MCL of 0.005 μg/L for drinking water.

Various studies, both in North America and in Europe, have determined that TCP occurrence is widespread. Within the United States, TCP has been detected in many surface water and drinking water sources at levels ranging from 0.1 to 100 μg/L. The recent monitoring data indicate that only 0.4% of the samples have concentrations greater than the method detection level of 0.03 μg/L.

Low levels of TCP have been detected in soil at some hazardous waste sites. Its detection in groundwater near hazardous waste sites implies limited adsorption to soils (high mobility) and a low rate of biodegradation in soil and groundwater. TCP should readily volatilize from surface water because of its high Henry's Law constant. Its half-life was estimated to be 6.7 h in a river and 5.7 days in a lake. Due to its low adsorptivity, it should not be substantially adsorbed to sediment or particulates in the water column. TCP is resistant to biodegradation and hydrolysis, so breakdown is not expected to be an important fate process in the aquatic environment.

TCP is very recalcitrant in groundwater systems due to its long hydrolysis half-life and low biodegradability. Current conventional water treatment practices (e.g., coagulation, sedimentation, and filtration) have proved to be relatively ineffective at removing TCP from water. Despite its high Henry's constant, air stripping is not cost effective. Several studies showed some success in removing TCP using reverse osmosis, zero-valent zinc, and advanced oxidation. Ultraviolet light (UV), along with hydrogen peroxide, has shown some degradation of TCP, although the degradation rate constant was fairly low. Currently, granular activated carbon (GAC) is the only viable treatment option for TCP removal.

2.4.3 1,4-Dioxane

1,4-dioxane ($C_4H_8O_2$), also called *dioxane*, is a synthetic industrial solvent used mainly as a stabilizer in chlorinated solvents. 1,4-dioxane has been detected in groundwater, surface water and drinking water.

1,4-dioxane is used as a stabilizer for chlorinated solvents such as 1,1,1-trichloroethane (TCA) to reduce the degradation of important properties of TCA. Along with its association with solvents, 1,4-dioxane is used in many products, including paint strippers, dyes, greases, varnishes, and waxes. It is also found as an impurity in antifreeze and aircraft deicing fluids as well as in some consumer products (deodorants, shampoos, and cosmetics). It is a by-product in the manufacture of polyethylene terephthalate (PET) plastic and is used as a purifying agent in the manufacture of pharmaceuticals. Finally, 1,4-dioxane residues may be present in manufactured food additives, in 1,4-dioxane-containing food packaging materials, or on food crops treated with pesticides that contain 1,4-dioxane (such as vine-ripened tomatoes).

Based on evidence in humans and experimental animals, the IARC classified 1,4-dioxane as a Group B2 carcinogen (probable human carcinogen). Currently, there is no federal MCL for 1,4-dioxane, and several international regulatory guidelines for 1,4-dioxane exist. The World Health Organization suggested a 50 μg/L drinking water threshold value for 1,4-dioxane, whereas the EPA National Center for Environmental Assessment proposed a health-based advisory level of 3 μg/L in tap water. According to the EPA Integrated Risk Information System, cancer development could occur in 1 out of 1,000,000 people exposed to a concentration of 0.35 μg/L in drinking water over a lifetime. In the United States, several state governments have set varying advisory levels for 1,4-dioxane ranging from 0.3 to 70 ppb. California has set up a notification level of 1 ppb.

The detection of 1,4-dioxane in surface water and groundwater has been reported throughout the United States and in countries such as Japan, Germany, the Netherlands, the United Kingdom, and Canada. 1,4-dioxane readily dissolves into groundwater, and its movement is not retarded significantly by sorption to soil particles. It is highly mobile, is recalcitrant to microbial degradation, and has a low tendency to volatilize from water. Conventional water treatment practices, for example, coagulation, sedimentation, filtration, aeration, GAC adsorption, ozone, UV, and biofiltration have proved to be ineffective at removing 1,4-dioxane from water. Its low Henry's Law Constant (HLC) (4.8×10^{-6}) indicates that it will not readily volatilize out of water, making it difficult to remove using aeration. Additionally, the unfavorable octanol-water partition coefficient (log Kow = −0.27) and organic carbon partition coefficient (log Koc = 1.23) imply that 1,4-dioxane is hydrophilic and cannot be efficiently adsorbed onto activated carbon. Advanced oxidation processes (AOPs) have been shown to be effective for removing 1,4-dioxane. Combinations of hydrogen peroxide and ferrous iron, ozone and hydrogen peroxide, and UV and hydrogen peroxide were found to effectively oxidize 1,4-dioxane.

2.4.4 PFOA and PFOS

Perfluorooctanoic acid (PFOA—$C_7F_{15}COOH$ or C8) and perfluorooctane sulfonate (PFOS—$C_8F_{17}SO_3H$) are the most notable compounds of the perfluoroalkyl substances (PFASs), a group of anthropogenic chemicals with past and current uses in industrial processes and consumer products. They are extremely recalcitrant and persistent in the environment and occur ubiquitously in environments worldwide.

PFASs are used in firefighting foams, coating for food packaging, ScotchGard™, and Teflon™, among other products. PFASs help these products resist stains, grease, or water. In industrial applications, they act as emulsifiers or surfactants. Exposure to PFASs can occur through the use of products or the consumption of food or water containing PFASs. They are soluble in water and can enter source waters through industrial releases, discharges from wastewater treatment plants, stormwater runoff, release of firefighting foams, and land application of contaminated biosolids.

Chronic toxicology studies have been conducted on PFOA and PFOS, and both compounds caused tumors in rats. In 2006, the USEPA Science Advisory Board classified PFOA as a likely human carcinogen. There are not currently any federal regulations limiting PFASs in water, but the EPA is considering whether to establish MCLs for PFASs in drinking water. In 2016, the EPA established drinking water health advisory levels for PFOS and PFOA of 0.07 μg/L based on lifetime exposure concerns for sensitive subpopulations. Many states already have their own drinking water and groundwater guidelines to limit PFOA and PFOS, including Minnesota, New Jersey, and North Carolina.

PFASs have been detected in all types of waters, including surface, ground, tap, and bottled waters, wastewater influents and effluents, industrial waste influents and effluents, and rivers, lakes, and tributaries. PFOA and PFOS were the most frequently detected, and fewer than 1% of public water systems detected PFOS or PFOA above the drinking water health advisory level of 0.07 μg/L.

Conventional treatment at wastewater treatment plants and most drinking water treatment plants are ineffective at removing PFASs from water. Granular activated carbon (GAC) and anion exchange (AIX) can remove many PFASs but are less effective at removing shorter-chain PFASs. The most effective treatment technologies appear to be nanofiltration (NF) and reverse osmosis (RO). RO is a costly treatment method, and disposal or treatment of the membrane-concentrated stream is a consideration for both NF and RO.

However, any treatment technology will need to be evaluated for matrix effects and site-specific performance. Oxidation/reduction technologies also have the potential to degrade some PFASs, but further research is needed for these and all techniques for treating PFASs.

PRACTICE AND EXERCISES

Question 1: Using Table 2.2, describe the difference in chemical properties between benzene and MTBE and how they could impact drinking water sources.

Answer 1: Benzene is less volatile and soluble than MTBE. However, benzene has a higher Henry's Law partition coefficient, which means that benzene is easier to volatilize from the dissolved phase to the vapor phase. Therefore, benzene is easier to extract from contaminated water than MTBE. Knowing these chemical properties relative to each other, one can see that both benzene and MTBE can impact groundwater resources via the dissolved phase with infiltrating water or via the vapor phase downward by the concentration gradient. However, when both benzene and MTBE are dissolved in water, MTBE will move faster with groundwater, because it has higher solubility and a lower HLC, which make it harder to volatilize from the water.

Question 2: What are common treatment technologies for VOC removal from water? What parameters are critical for effective VOC removal using each type of treatment?

Answer 2: Common treatment technologies for VOC removal include PTA (Packard tower aeration) and GAC (granular activated carbon). For aeration, HLC is a key parameter: high HLC is more effective for aeration. For activated carbon, water solubility is a key parameter: low water solubility indicates that it can be readily removed by GAC.

Question 3: Why is the drinking water standard for lead the action level rather than the maximum contaminant level?

Answer 3: Because lead contamination of drinking water often results from corrosion of household plumbing materials belonging to water customers, the EPA established a treatment technique rather than an MCL for lead. A treatment technique is an enforceable procedure or level of technological performance which water systems must follow to ensure the control of a contaminant.

Question 4: What treatment processes can be used for the removal of both nitrate and perchlorate? Why?

Answer 4: Similar treatment processes that can remove both nitrate and perchlorate include anion exchange and reverse osmosis. Both nitrate and perchlorate are water-soluble salts and have negative ions, which can be replaced by the anion exchange process.

Question 5: What are two common arsenic compounds, and which one is more toxic? Please discuss the strategy of treatment for water containing both.

Answer 5: Two common arsenic compounds are arsenite (As(III)) and arsenate (As(V)); As(III) is more toxic. The common treatment is precipitation and filtration with ferric iron as coagulant. Since As(V) is relatively easy to remove, a typical strategy is to convert As(III) to As(V) by simple oxidation using chlorine or other oxidants and then to add ferric iron to aid precipitation/filtration to remove arsenic.

3 Human Health Risk Assessment in Site Remediation

Chawn Y. Jeng, Yue Rong, and Ravi Arulanantham

CONTENTS

3.1 INTRODUCTION

About 500 years ago, the European scientist Paracelsus (1493–1541) made a remark: "All substances are poisons; there is none which is not a poison. The right dose differentiates a poison and a remedy." This statement has become the basis of modern toxicology. Therefore, the toxicity of a

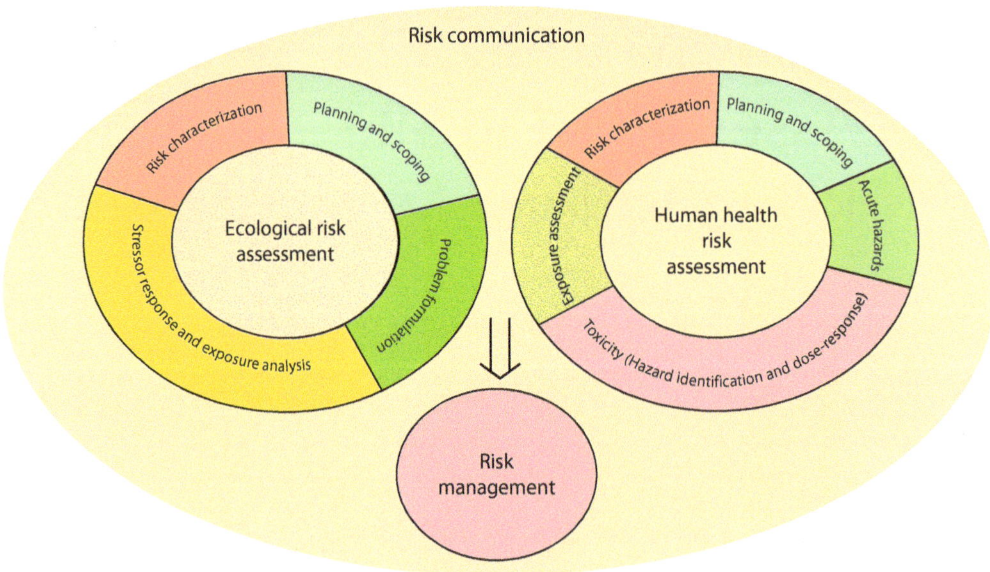

FIGURE 3.1 "Universe" of risk-based tools for environmental management.

contaminant to a human is related to the dose that the human takes into the body. Therefore, one pollutant's impact on human health depends on the human body's intake of pollutants in magnitude (dose) and the length of time exposed to this pollutant. This concept of assessing dose and exposure has become the anatomy of the risk assessment process. That is, human health risk assessment is about assessing human exposure and dose response to chemicals and other agents. This chapter will introduce the risk-based framework for site remediation, which is used to guide site assessment sampling, pollution investigation, remediation, and post-remediation monitoring.

The idea of applying the risk assessment process to the evaluation and regulation of environmental issues was introduced around the 1980s in the United States, first with the publication by the National Academy of Sciences (NAS, 1983), followed by the Risk Assessment Guidance for Superfund (RAGS) published by the United States Environmental Protection Agency (USEPA, 1989, 1991a,b). Subsequently, human health risk assessment and other risk-based tools have become important tools for regulators and environmental professionals alike to assess potential impacts of contaminants. Figure 3.1 is an illustration of the universe of common risk-based tools for environmental management in modern times, including human health and ecological risk assessments, risk management, and risk communication.

This chapter presents applications of human health risk assessment in the site remediation process. It is organized into the following sections: Section 3.2 provides an overview of human health risk assessment methodology and calculations; Section 3.3 describes how human health risk assessment can be used to support decision-making in site assessment and cleanup; Section 3.4 discusses several applications of risk-based approach to assess subsurface pollutants; and Section 3.5 presents two case studies to illustrate the important roles that health risk assessment plays in the site remediation process.

3.2 HUMAN HEALTH RISK ASSESSMENT OVERVIEW

A human health assessment is defined as the process of estimating the nature and probability of adverse health effects in humans who may be exposed to contaminants in environmental media now or in the future. It typically includes four basic steps:

- *Hazard identification*: Evaluate whether a chemical has the potential to cause harm to humans, and under what circumstances.

- *Dose-response assessment*: Determine the numerical relationship between chemical exposure and adverse (cancer and noncancer) health effects.
- *Exposure assessment*: Determine the frequency, duration, and levels of contact with chemicals of concern.
- *Risk characterization*: Quantify the health risk based on dose and exposure, and examine how well the data support conclusions.

These steps are described briefly in Sections 3.2.1 through 3.2.4 to provide background information for subsequent discussion. In addition, the use of risk assessment information to make risk management decisions is discussed in Section 3.2.5. Additional references are available for readers who are interested in learning more details about human health risk assessment (NAS, 1983; USEPA, 1989, 1991a,b).

3.2.1 Hazard Identification

Hazard identification is the process of determining whether exposure to a stressor (environmental contaminants in this case) can cause an increase in the incidence of specific adverse health effects (e.g., cancer, birth defects). The objective is to identify the types of adverse health effects that can be caused by exposure to the contaminants in question and to characterize the quality and weight of evidence supporting this identification. This process examines available scientific data and develops a weight of evidence to characterize the link between the negative effects and the chemical(s).

Sources of data for this evaluation typically include (1) epidemiological studies involving a statistical evaluation of human populations to examine the association between exposure to a contaminant and a health effect and (2) animal studies (rats, mice, rabbits, monkeys, etc.) that are designed to draw inferences about the potential hazard to humans. While human studies are generally preferable, they are often limited by not having accurate exposure data and the effects of other stressors that cannot be controlled in the studies. On the other hand, animal studies can be controlled to address specific data gaps, but they are subject to uncertainties due to extrapolating results from animals to humans.

A key component of hazard characterization involves evaluating the weight of evidence regarding a chemical's potential to cause adverse health effects. The weight of evidence narrative may include some standard "descriptors" signifying that certain qualitative threshold levels of evidence or confidence have been met, such as "Carcinogenic to humans" or "Suggestive evidence of carcinogenic potential." For assessing potential carcinogens, analysis of "mode of action" based on physical, chemical, and biological information may help to explain key events in a chemical's influence on tumor development.

3.2.2 Dose-Response Assessment

A dose-response relationship describes how the likelihood and severity of adverse health effects (the responses) are related to the amount and condition of exposure to a chemical (the dose). Figure 3.2 is a conceptual dose-response plot showing the relationship between chemical dose (*x*-axis) and tumor increase (*y*-axis). In general, the increase in chemical dose concentration is proportional to

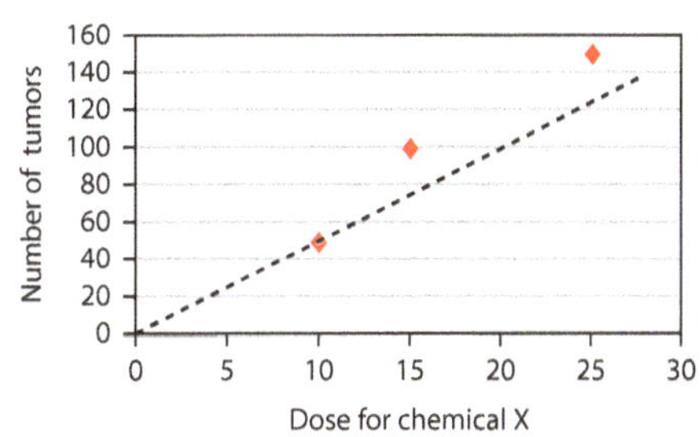

FIGURE 3.2 Hypothetical linear dose-response plot.

tumor growth. The problem, however, occurs in the low dose region. No tumor was observed in the low dose area. The lowest data points are now stretched to the origin by the extrapolation method.

As with hazard identification, dose-response data are often unavailable for humans or cover only a portion of the possible range of exposure. Thus, animal studies may be conducted to augment the available human data. However, dose-response relationships observed from animal studies are often at much higher doses than would be anticipated for humans, so extrapolations to lower doses and from animals to humans introduce uncertainty into the dose-response analysis.

For carcinogens, linear extrapolation is typically used as the default approach for dose-response assessment (referred to as a *linear* dose-response assessment). In these cases, the toxicity is assumed not to have a threshold, and a straight line is drawn from the point of departure for the observed data to the origin, as in Figure 3.2. The slope of this straight line, called the *cancer slope factor* (SF), is used to estimate the excess lifetime cancer risk (i.e., the probability that an individual will develop cancer over a lifetime) resulting from exposure following the linear dose-response relationship. A similar term, known as the *inhalation unit risk* (IUR), is used to assess inhalation risks, where the exposure-response relationship refers to concentrations in the air.

For non-carcinogens, their toxicities are typically assumed to have a threshold, defined as the dose below which no deleterious effect is expected to occur (usually referred to as a *non-linear* dose-response assessment). Thus, a no-observed-adverse-effect level (NOAEL) is the highest exposure level at which no statistically or biologically significant increases are seen in the frequency or severity of adverse effect between the exposed population and its appropriate control population. In cases in which a NOAEL has not been demonstrated experimentally, the term *lowest-observed-adverse-effect level* (LOAEL) is used, which is the lowest dose tested. However, the NOAEL approach fails to take into consideration the shape of the dose-response curve and other related information, so the benchmark dose (BMD) models, originally proposed as an alternative to the NOAEL methodology in the 1980s to address many of the limitations of the NOAEL method, and the computed statistical lower confidence limits on the benchmark dose (BMDL) have been adopted in the recent development of toxicity values. Thus, a reference dose (RfD), in units of milligrams per kilogram of bodyweight per day, is defined as an oral or dermal dose derived from the NOAEL, LOAEL, BMDL, or another suitable point of departure, with uncertainty factors (UFs) applied to reflect the limitations of the data used:

$$\text{RfD} = \text{NOAEL}\ \left(\text{or LOAEL or BMDL}\right)/\text{UFs} \tag{3.1}$$

These UFs take into account the variability and uncertainty that are reflected in possible differences between test animals and humans (generally 10-fold) and variability within the human population (generally another 10-fold); thus, the UFs are multiplied together: $10 \times 10 = 100$ (multiplied by another 10 if an LOAEL is used). A similar term, *reference concentration* (RfC, in units of milligrams per cubic meter of air), is used to assess inhalation risks, where concentration refers to levels in the air.

The use of these cancer and noncancer toxicity factors to estimate health risks is further discussed in Section 3.2.4, Risk Characterization.

3.2.3 Exposure Assessment

Exposure assessment is the process of determining the magnitude, frequency, duration, and route of human exposure to chemicals in the environment. Exposure can be measured directly, but more commonly, it is estimated through measured concentrations in the environment, models of chemical fate and transport within and across the environmental media, and estimates of human intake over time. The general procedure for exposure assessment includes three basic steps:

Step 1 – characterizing exposure setting: This step includes characterization of the physical environment and potentially exposed populations at and near the site based on the current and future land uses. Thus, it is possible that different populations may be evaluated if the land use changes in the future (e.g., from commercial to residential).

Step 2 – identifying exposure pathways: An exposure pathway generally consists of four components: a chemical source/release, fate and transport of contaminant(s), a point of human exposure, and an exposure route (ingestion, dermal contact, and inhalation). In this step, it is useful to prepare a conceptual site exposure model (CSEM) to list these components and potentially exposed populations identified in the first step for all exposure pathways relevant to the site. Figure 3.3 shows an example CSEM for a former industrial facility with releases of chemicals from several operation areas.

Step 3 – quantifying exposure: This step includes estimation of exposure concentrations and calculation of intakes for each exposure pathway. The exposure concentration term is typically estimated in two ways: central tendency (the arithmetic mean or 95% upper confidence limit on the mean) or reasonable maximum exposure (RME) by using the maximum or an upper percentile (e.g., 90th–98th) value within an exposure unit. Besides the exposure concentration, the intake also depends on the contact rate, contact time, body weight, and averaging time. Equations 3.2 and 3.3 illustrate how the intake or exposure is quantified for the ingestion and inhalation routes, respectively.

Ingestion Exposure: The USEPA RAGS Part A (1989) generic equation for calculating chemical dose is expressed as an average daily intake (milligrams of chemical per kilogram of body weight per day):

$$\text{Intake}(I) = (C \times CR \times EF \times ED)/(BW \times AT) \tag{3.2}$$

where:

I (mg/[kg × day])	is the average daily chemical intake
C (mg/kg or mg/liter water)	is the chemical concentration in soil or water
CR (kg/day or liters/day)	is the contact (i.e., ingestion) rate for soil or drinking water
EF (days/year)	is the exposure frequency
ED (years)	is the exposure duration
BW (kg)	is the body weight
AT (days)	is the average time

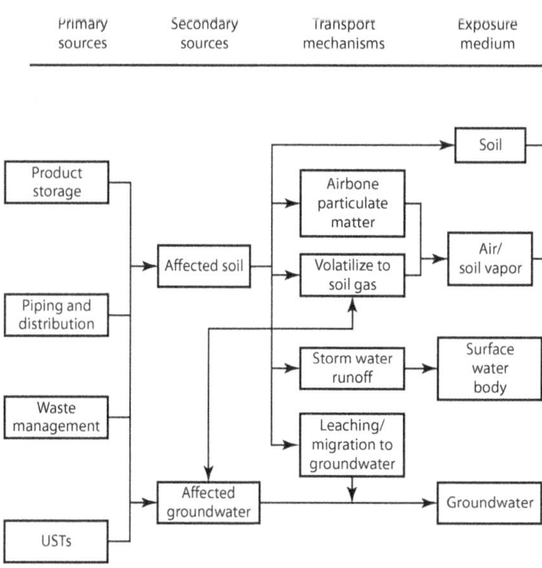

Exposure route	Receptor			
	Human		Biota	
	Residents	Site workers	Terrestrial	Aquatic
Soil				
Ingestion	X		X	X
Inhalation	X	X		
Dermal contact	X	X	X	
Air/soil vapor				
Ingestion				
Inhalation	X	X	X	
Dermal contact				
Surface water body				
Ingestion				
Inhalation				
Dermal contact	X	X		X
Groundwater				
Ingestion	X	X		
Inhalation	X	X		
Dermal contact	X	X		

FIGURE 3.3 Example of conceptual site exposure model. (From California DTSC, Preliminary Endangerment Assessment Guidance Manual, 2015).

For cancer, *AT* is 70 years × 365 days/year
For noncancer, *AT* is *ED* (years) × 365 days/year

Note that the average times for carcinogens and non-carcinogens are different, as the former is for calculating the average intake for an incremental lifetime cancer risk (ILCR) over an average lifespan of 70 years, while the latter is for calculating the average intake during the period of chemical exposure.

Inhalation Exposure: Current USEPA guidance recommends use of the chemical concentration in air (e.g., micrograms per cubic meter) rather than a contaminant intake (e.g., milligrams/[kilograms × days]) based on inhalation rate and body weight. The exposure concentration (EC) for inhalation is calculated using the following equation (USEPA, 2009):

$$\text{Exposure concentration } (EC) = (CA \times ET \times EF \times ED)/AT \tag{3.3}$$

where:

EC (μg/m³) is the exposure concentration
CA (μg/m³) is the contaminant concentration in air
ET (hours/day) is the exposure time
AT (hours) is the average time
For cancer, *AT* is 70 years × 365 days/year × 24 hours/day
For noncancer, *AT* is *ED* (years) × 365 days/year × 24 hours/day

Table 3.1 lists some commonly used values of the parameters in these equations for calculation of the intake for ingestion and the exposure concentration for inhalation. These "default" exposure parameters are based on reasonable high-end exposures for typical residents and commercial/industrial workers and thus, are considered protective of the general populations.

3.2.4 Risk Characterization

The final step of the human health risk assessment is to integrate toxicity and exposure assessments into quantitative or qualitative (if some information is missing) expressions of cancer risk and noncancer hazard.

The equations to quantify risks are also different for different exposure pathways. For ingestion and dermal contact, the cancer risk and noncancer hazard can be calculated with Equations 3.4a and 3.4b using the information discussed in Sections 3.2.2 and 3.2.3:

$$\text{Risk} = I_{(\text{life})} \times SF \tag{3.4a}$$

$$HQ = I_{(\text{life})} / RfD \tag{3.4b}$$

where:

Risk is the cancer risk (unitless; probability of an individual developing cancer over lifetime)
HQ is the noncancer hazard quotient (unitless)
$I_{(\text{life})}$ is the average lifetime intake (mg/kg-day; see Equation 3.2)
SF is the slope factor ($[\text{mg/kg-day}]^{-1}$)
RfD is the reference dose (mg/kg-day)

For the inhalation route, the cancer risk and noncancer hazard can be calculated with Equations 3.5a and 3.5b using the information discussed in Sections 3.2.2 and 3.2.3:

$$\text{Risk} = EC \times IUR \tag{3.5a}$$

$$HQ = EC/RfC \tag{3.5b}$$

TABLE 3.1
Example of Default Exposure Parameters

Exposure Parameters	Residential	Commercial/Industrial
CR—soil ingestion (mg/day)	100 (adult) 200 (child)	50 (indoor) 100 (outdoor)
CR—water ingestion (L/day)	2.5 (adult) 0.78 (child)	—
BW—body weight (kg)	80 (adult) 15 (child)	80
ET—exposure time (h/day)	24	8
EF—exposure frequency (days/year)	350	250 (indoor) 225 (outdoor)
ED—exposure duration (years)	20 (adult) 6 (child)	25

Source: USEPA, Update of Standard Default Exposure Factors. Office of Solid Waste and Emergency Response (OSWER) Directive 9200.1-120, February 6, 2014.

where:

EC is the exposure concentration (μg/m^3; see Equation 3.3)
IUR is the inhalation unit risk ([μg/m^3]$^{-1}$)
RfC is the reference concentration (μg/m^3)

Total cancer risk and noncancer hazard are calculated by summing the risks and hazards associated with individual pollutants for each exposure route (i.e., inhalation, ingestion, and dermal absorption) and then combining the risk across exposure pathways if necessary. Typically, the health risk assessment is calculated for the "reasonable maximum exposure" (RME) scenario by using high-end exposure concentrations and exposure parameters (Table 3.1). Sometimes, the health risk can also be calculated using "central tendency" exposure concentrations and site-specific exposure parameters if such information is available and determined to be reliable for use in the risk assessment.

Besides quantifying health risks, it is very important to assess the uncertainty associated with each step of the process on the results of risk assessment. For example, most of the toxicity factors and exposure parameters recommended by regulatory agencies are generally conservative and thus, likely to over-estimate actual health risks. Additionally, summation of the risks for individual chemicals ignores possible synergistic or antagonistic effects among these chemicals, and thus, the risks may be under- or over-estimated.

3.2.5 Risk Management Decision

As shown in Figure 3.1, the results of risk assessment are used to support risk management. Specifically, risk assessment provides information on potential health or ecological risks, while risk management determines how to manage those risks in a way best suited to protect human health and the environment. Besides the scientific results from the risk assessment process discussed earlier, risk management should also consider other factors, including:

- Laws and policies that define the basis for regulatory decisions to manage environmental problems; for example, the applicable or relevant and appropriate requirements (ARARs), such as drinking water maximum contaminant levels (MCLs) in the USEPA Superfund program.

- Technological factors such as the feasibility, implementability, and short-term and long-term effectiveness of risk management options.
- Economic factors such as the costs and benefits of reducing risks by various mitigation or remediation alternatives. Section 3.5.2 presents a case study to illustrate how to select a cost-effective remedy at a polychlorinated biphenyl (PCB)-contaminated site while still achieving the risk reduction goal.
- Social factors such as income level, ethnic background, and availability of health care that may affect the susceptibility of individuals to health risks from contaminants, as well as public values and perceptions on environmental issues.
- Political factors such as acceptance by other agencies (e.g., state and local governments), legislative sectors, special interest groups, and concerned citizens.

Due to the complexity of these factors in the decision-making process, risk managers typically allow for flexibility and may consider a range of risks to be acceptable depending on the response actions to be taken accordingly. Table 3.2 is an example of the decision matrix for managing the risks associated with potential vapor intrusion (see Section 3.4.3) by California regulators (DTSC, 2011).

3.3 USE OF RISK-BASED TOOLS AND RISK-INFORMED DECISION-MAKING IN SITE REMEDIATION PROCESS

This section discusses how the risk assessment and risk management tools described in Section 3.2 may be used to inform decision-making in the site remediation process. Section 3.3.1 introduces the concept and history of risk-informed decision making (RIDM) for addressing contaminated sites. Section 3.3.2 discusses the strengths and weakness of the RIDM process. Section 3.3.3 provides an overview of the various roles that risk-based tools play in the site assessment and cleanup process to guide decision-making. Examples and case studies of using these risk-based tools and RIDM for the cleanup of contaminated sites are presented in Sections 3.4 and 3.5.

3.3.1 Concept of Risk-Informed Decision Making (RIDM)

Across the United States, there are more than 500,000 properties contaminated with historical releases of chemicals. Clearly, there is a need to "clean" or manage these sites. In the 1980s and 1990s, the conventional approach was to treat all sites as "equal" and therefore, attempt to clean all

TABLE 3.2
Example Decision Matrix for Risk Management

Vapor Intrusion Risk	Risk Management Decision	Activities
Risk $<1\times10^{-6}$	No further action	None
$1\times10^{-6}<\text{risk}<1\times10^{-4}$	Evaluate need for action	Possible actions: • Additional data collection • Monitoring • Additional risk characterization • Mitigation • Source remediation
Risk $>1\times10^{-4}$	Response action needed	Vapor intrusion mitigation Source remediation

Source: California Department of Toxic Substances Control (DTSC), Guidance for the Evaluation and Mitigation of Subsurface Vapor Intrusion to Indoor Air, 2011.

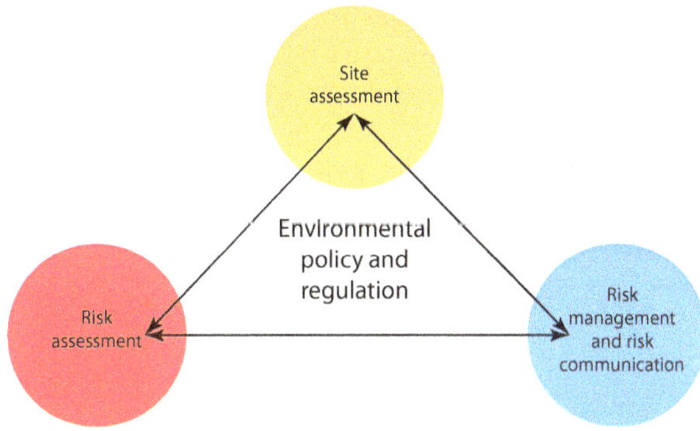

FIGURE 3.4 Key elements of risk-informed decision making.

sites to background levels, generic levels, or empirically derived levels, which was often technically and economically infeasible. According to the USEPA, the US spent more than $64 billion between 1981 and 1990 on cleanup at leaking underground petroleum sites. These costs are dwarfed when the costs of remediating hazardous waste sites are added. This historical approach often results in inconsistent decisions, delays in site closure, and a cleanup process that depends on the ability and efficiency of treatment technologies (engineering controls) without much regard to site-specific environmental conditions and the risks posed by the contaminants of concern. In response to this cleanup dilemma faced by all stakeholders at a national level, the development and publication of the Standard Guide for Risk-Based Corrective Action (RBCA) at Petroleum Release Sites by the American Society for Testing and Materials (ASTM, 1995) and Use of Risk-Based Decision Making in Underground Storage Tank (UST) Corrective Action Programs by the USEPA (1995) has resulted in many state programs embracing some application of risk-based decision making. In 1998, the Lawrence Livermore National Laboratory suggested "risk-informed decision making" as more appropriate terminology to describe the RBCA process.

Currently, most states have programs that use a risk-informed approach to implement "Brownfields" programs that bring impacted properties back into productive use. RIDM is a set of practical, technically defensible, and consistent activities (i.e., site investigation, exposure/risk assessment) and solutions (e.g., risk management options such as engineering controls and/or land use and access controls) that help to manage a site cleanup cost-effectively. It advocates the combination of "best management practices" with "best available technology" to protect public health, water quality, and the environment. Specifically, the RIDM approach emphasizes that cleanup decisions are made with input from stakeholders representing both local and state concerns and that the inputs to the decision-making process include: (1) potential health impacts; (2) resource damage and effect on water supply; (3) long-term fate of contaminants; (4) cost of remediation options (if necessary); (5) aesthetics; and (6) other relevant factors. Inherent in these factors are interrelated considerations of contaminant longevity and migration, projected land/resource use and possible restrictions, and the reliability and uncertainty of existing data. Corrective actions taken using a RIDM process should (1) improve the situation; (2) not foreclose future options; and (3) minimize societal strain.

As depicted in Figure 3.4, RIDM has four main elements: site assessment, risk assessment, risk management, and environmental policy and regulation. These elements are briefly discussed here:

- Site assessment includes innovative, cost-effective techniques to collect a variety of background information on the actual or perceived problem (i.e., *is there a public health/ water quality/environmental problem?*). In the RIDM process, site assessment is governed by the pathways and receptors that are identified in a CSEM, as discussed in Section 3.2.3. A good CSEM becomes the cornerstone of RIDM at any site, as it identifies environmental

issues that need to be addressed (and those issues that need not be addressed). In addition, a CSEM provides a framework for the entire project and an excellent communication tool for the regulators, potential responsible parties (PRPs), and other stakeholders.
- Risk assessment is a way of integrating and analyzing the variety of information to quantify the problem (i.e., *what are the nature and the magnitude of health and environmental impacts caused by contaminants?).* The scope of a risk assessment is determined by developing a CSEM, which describes what is currently known about the chemical sources, likely migration pathways, exposure routes, and possible exposure scenarios both current and future, so that the data necessary to complete a risk assessment are identified during site assessment.
- Risk management consists of practical ways of responding to the problem, which may include engineering and/or institutional solutions (i.e., *how do we manage the problems?).* Historically, institutional controls were typically applied after the risk assessment had been completed. However, they can also be used upfront during the development of the CSEM to define the scope of the risk assessment and therefore, determine the extent of site assessment. This avoids investigating and evaluating redundant pathways.
- Environmental policy and regulation would use current science and relevant information, address knowledge gaps and uncertainty, balance technologic/economic limitations and societal issues, and establish appropriate standards.

Therefore, to develop and implement cost-effective and viable risk management options, one must accurately define the nature and magnitude of the problem in the first place. In this context, risk assessment is a powerful tool to identify which chemicals and pathways need control and the level of control that is required to reduce health and environmental risks to acceptable levels. The basic risk assessment guidelines and the criteria for selecting a risk management option can be found in the Risk Assessment Guidance for Superfund (USEPA, 1989) and the National Contingency Plan (Code of Federal Regulations [CFR], 1990), and are summarized in Section 3.2.

In summary, an RIDM process, as envisioned in the ASTM standard E 1739–1795, can provide a predictable road map to best manage environmental problems where site investigation and risk management options are based on current and reasonably foreseeable risks to public health and environmental resources, including beneficial uses of groundwater.

3.3.2 Strengths and Weaknesses of RIDM Process

To implement the four elements of the RIDM process, one soon realizes that a multitude of technical, non-technical, and policy issues must be sorted out and resolved with all the stakeholders before seeing "the light at the end of the tunnel." This subsection lists the key advantages (strengths) and some disadvantages (weaknesses) of the RIDM process as it applies to soil and groundwater cleanup.

Strengths

- The RIDM process identifies critical technical and policy objectives and decisions *upfront,* and therefore, the stakeholders, PRPs, regulators, property owners and users, lending institutions, citizens' groups, the public, and any others are given an opportunity to find appropriate answers or solutions to many of their concerns at early stages of the project. This process of identifying and making clear and predictable objectives upfront is one of the biggest advantages of RIDM. A good place to start this process, in many cases, is the development of the CSEM. Sites vary widely in terms of extent of contamination, hydrogeologic characteristics, and potential risks posed to public health and the environment. Therefore, based on site-specific situations, the CSEM should outline most or all of the following technical and policy issues that need to be resolved for decisions to be made on how, where, and when to address them: definition of contamination; choice of chemical(s) of concern; site representative concentrations; acceptable risk levels for carcinogens and non-carcinogens; site-specific and/or default exposure and transport factors; current and future land use and groundwater use; exposed populations, including ecological receptors; point of current and future exposure; point of compliance; active cleanup options; application of natural attenuation; the use and scope of institutional controls; and issues of public participation and risk communication.

- The RIDM process allows a tiered approach to problem solving. One of the strengths of RIDM is that one can start simply and move on to more complex investigations and cleanups as needed. Since many sites may not warrant a site-specific risk assessment, financial and regulatory resources can be optimized by moving first from simple screening techniques to more complex models based on the seriousness of the problem. This approach is invaluable in managing many petroleum-impacted sites and can lead to an early "No Further Action Required" letter.
- The RIDM process avoids unnecessary investigations and associated costs. Historically, site investigations often did not have clear upfront objectives—it was a paradigm of "collect data and more data," but many of the data collected had no relevance to human health or ecological risks posed by site contaminants and thus, did not inform risk management decisions. However, using the CSEM to guide site investigation allows one to focus on complete exposure pathways and potentially exposed receptors, to collect the right data, and to avoid repeated site investigations.
- The use of RIDM allows the development of defensible cleanup objectives other than maximum contaminant levels (MCLs) or other generic standards at sites where (1) contaminated groundwater has no current beneficial use; (2) surface and near-surface soil concentrations have to be protective of on-site and off-site exposures (residential, commercial/industrial, recreational, construction); (3) subsurface soil concentrations have to be protective of groundwater beneath the site; (4) multiple chemicals (both carcinogens and non-carcinogens) are present in soil and groundwater; (5) multiple pathways, multiple media of exposure, and multiple routes of exposure are present at the same time; and/or (6) the potentially exposed receptors are separated from the source by both time and distance.
- In the RIDM process, risk is the primary driver. Thus, institutional controls can be as effective as active remediation or engineering controls, and in many cases, more practical and cost effective in preventing both short-term and long-term exposure to potential receptors. Historically, in attempting to clean every site to drinking water standards (with no consideration of risk), the first and sometimes only solution was mass removal such as pump and treat. One of the lessons learned in the past decades is that institutional controls can be an effective alternative solution to many cleanup dilemmas. In addition, many institutional controls can also play an important role when used upfront during the development of the scope of the risk assessment.
- Once cleanup goals are established (site-specific cleanup levels, compliance point, and compliance time), the responsible parties and the regulatory agencies can evaluate different approaches and time frames for achieving these goals. This process allows flexibility in reaching the objectives, while being consistent in the results and standards that the responsible parties are held to.

Weaknesses

- Many people (both regulators and consultants) who have been involved in site cleanups in the past decades are used to conducting them in a conventional way. Therefore, training in the RIDM process is needed. When we are faced with the challenge of making tough

policy choices, we often choose to make conservative decisions to be on the safe side. Cumulatively, these conservative assumptions lead to unpredictable and costly decisions, negating the benefits of an RIDM process. The irony of the RIDM process is that the same features that make it effective to manage risks can also, if not resolved in a timely manner, cause the decision-making process to come to a screeching halt. These features are the same technical and policy issues that were identified earlier. For example, one has to resolve early in the RIDM process (1) acceptable risk levels (for carcinogens and non-carcinogens) to be considered; (2) whether to use maximum concentration, or average concentration, or area-weighted concentration, or time-weighted concentration, or a statistical formula as the site representative concentration; (3) human health/ecological protection standards to be compared against site background concentrations; (4) where the ecological standards are to be applied; and (5) how the institutional control (e.g., if soil contaminated with petroleum and lead is allowed to remain under parking lots and road ways) is to be maintained to prevent future exposures. Otherwise, many decisions regarding risk assessment and risk management cannot be made or enforced, and this will unnecessarily delay the project or force one to pursue aggressive cleanup to reach generic default levels even when the site may pose no risk.
- At some large sites, community involvement in the overall decision-making process is vital for the successful completion of the project. If the community is not consulted upfront, many of the assumptions made during the site investigation and risk assessment stage may have to be explained and possibly reconfigured. Many times, the issues separating actual risks, perceived risks, and the community perception that risk-based decisions are a way to leave large amounts of contamination in neighborhoods are difficult subjects to communicate and can lead to frustration among all stakeholders. Therefore, if the community is not consulted and their concerns addressed upfront, a predictable road map is not likely.
- Some issues, such as anti-degradation and resource protection, off-site migration of plumes, property devaluation, future land use changes, and the implementation and long-term maintenance of institutional controls, are inherently difficult subjects to resolve and are sometimes beyond the authority of regulatory agencies.
- RIDM is not a universal panacea for all concerns. While numerical, valid risk assessments can show insignificant risks, risk managers may still need to comply with special regulatory requirements in some circumstances (drinking water supplies, sensitive wetlands, etc.)

Current experience indicates that sites with the following features are good RIDM candidates: (1) primarily commercial/industrial areas; (2) the probability of using the shallow groundwater is very low; and (3) high real estate demand. These are sites where it is easy to overcome the technical and policy issues and implement successful risk-based decisions. However, sites with the following features might be less successful at implementing RIDM: (1) having a history of frequent land use changes; (2) containing persistent chemicals, especially with off-site contamination; (3) groundwater is widely used as domestic/municipal supply; (4) including sensitive habitats; and (5) low real estate demand. In the last 30 years, much effort has been devoted to making RIDM a legitimate regulatory policy in many of the states in the United States. To make RIDM an effective process, these types of sites will need special planning and a team effort by regulators, consultants, PRPs, and environmental groups. Several practical applications of the RIDM process for the assessment and cleanup of subsurface pollutants are presented in Section 3.4.

3.3.3 Roles of Risk Assessment and Risk Management in Site Remediation Process

An important feature of the RIDM process, which is absent from the traditional regulatory approach, is to incorporate site-specific risk assessment and risk management (based on current and probable future risks posed by contaminants) into all phases of the cleanup process on the discovery of contamination at any site. Figure 3.5 shows a simplified cleanup process in the Superfund program (similar processes are used for other cleanup programs) and the four key questions where risk-based tools can inform decision-making in this process, as discussed in this subsection.

Does the site have enough data? This is the key question that is often faced during the site investigation phase. For example, the responsible party feels that enough data have been collected to support a decision on whether remedial action is warranted, while regulators may want additional sampling, because some areas were not fully characterized. Thus, many regulatory agencies have developed risk-based screening levels using the risk assessment methods discussed in Section 3.2, such as the regional screening levels (RSLs) by USEPA, to evaluate environmental data. These default screening levels are based on the exposure assumptions that are protective to general populations (e.g., Table 3.1) and a target cancer risk of 1×10^{-6} for carcinogens and a target noncancer hazard of 1 (or 0.1) for non- carcinogens. In addition, on-line calculation tools are also available to allow the development of "site-specific" screening levels using more realistic assumptions based on specific site use and activities (e.g., construction and recreation).

It should be noted that the primary objective of data collection is to support a decision on whether remedial action is needed, as shown in Figure 3.5, so additional sampling to below the screening levels (or non-detects in some cases) for every chemical at all locations is not achievable or needed, as long as the locations with key chemicals posing significant risk levels are adequately delineated, and their spatial distribution shows a decreasing trend with step-out samples, so that exposure point concentrations can be estimated reliably or conservatively in the baseline risk assessment discussed in the next paragraph.

Does the site need remedial action? These days, most regulatory agencies require the preparation of a baseline health risk assessment using investigation data to determine whether environmental cleanup is needed, especially for complex sites with releases of multiple contaminants across environmental media (e.g., soil, soil gas, and groundwater). The main difference between the risk-based screening levels discussed earlier and a baseline risk assessment is that the screening levels are chemical specific and medium specific and thus, do not account for cumulative effects associated with exposures to multiple contaminants and across environmental media. On the other hand, a baseline health risk assessment will calculate the total cancer risk and noncancer hazard for all contaminants of concern in each environmental medium identified in the conceptual site model. It should also address both current and possible future site uses by evaluating different exposure scenarios (e.g., current vacant lot to be developed for either commercial or residential use) to help decision-making.

How to choose a remedy? As discussed in Chapter 1, most environmental laws and regulations require the selection of a remedy to achieve overall protection of human health and the environment and to comply with applicable or relevant and appropriate standards (ARARs) while considering other factors such as performance, effectiveness, implementability, and cost. For example, some agencies have policies or guidance to recommend specific remedial technologies depending on the

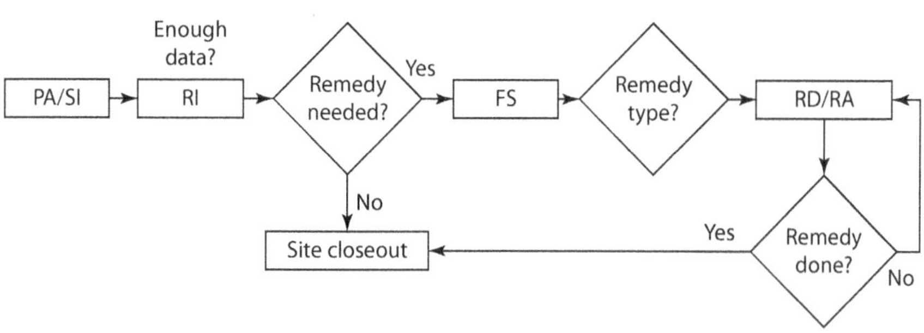

FIGURE 3.5 Simplified superfund cleanup process. (FS: feasibility study; PA/SI: preliminary assessment/ site inspection; RD/RA: remedy design/remedial action; RI: remedial investigation.) (From Song, S., et al., Strategies to Enhance Risk Managements. Remediation Innovative Technology Seminar (RITS) sponsored by the Naval Facilities Engineering Command (NAVFAC), 2006.)

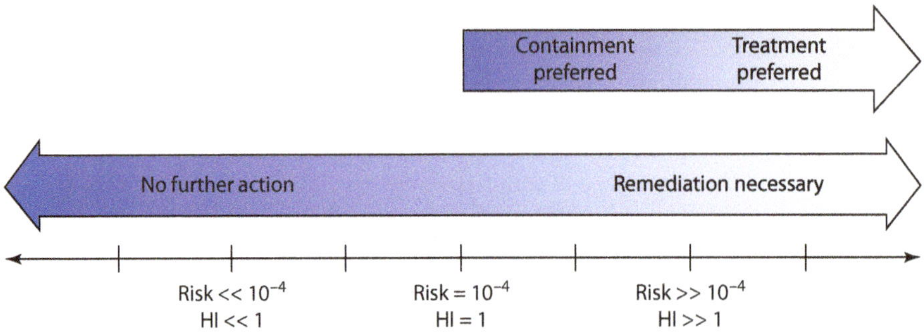

FIGURE 3.6 Recommended remedial actions for "principal threat" wastes. (From Song, S., et al., Strategies to Enhance Risk Managements. Remediation Innovative Technology Seminar (RITS) sponsored by the Naval Facilities Engineering Command (NAVFAC), 2006.)

health risk levels posed by a contaminated site. Figure 3.6 shows an example of such application in the USEPA's National Oil and Hazardous Substances Pollution Contingency Plan (NCP), which states that a combination of methods may be used, as appropriate, to achieve protection of human health and the environment. Specifically, USEPA recommends treatment to address the "principal threats" posed by a site and engineering controls (such as containment) for waste that poses a relatively low long-term threat or where treatment is impracticable, and defines "principal threat" wastes as source materials considered to be highly toxic or highly mobile, that is, posing a potential risk several orders of magnitude greater than the acceptable risk level (the cancer risk of 1×10^{-4} and the noncancer hazard index of 1) (USEPA, 1991c, 1997).

In addition, health risk assessment can be used as a tool to evaluate various remedial alternatives along with cost analyses to support the selection of a cost-effective remedy that is also protective to public health and the environment (see an example case study in Section 3.5.2).

What is the remediation goal? This is the last key question in the cleanup process where health risk assessment plays an important role, so that all parties involved know when the remediation is complete. For most projects, risk-based concentrations are developed for key contaminants in each environmental medium to guide the cleanup. While often referred to as *cleanup goals* or *remedial action goals* to facilitate real-time decisions in the field, these risk-based concentrations are subject to the same shortcoming as the screening levels discussed earlier, as they do not usually account for cumulative effects from exposures to multiple contaminants and across environmental media. Thus, a post-remediation health risk assessment is often required to demonstrate that the total cancer risk and noncancer hazard for all contaminants of concern remaining in all environmental media after the cleanup meet the overall cleanup objective set by the project team.

The remaining sections present several practical applications of using these risk-based tools to support site investigation and cleanup decisions (Section 3.4) and two case studies to illustrate how the risk assessment may be used to determine whether a site needs remedial action and to support the selection of a cost-effective remedy (Section 3.5).

3.4 CONCEPTUAL SITE MODEL AND APPLICATION OF RISK-BASED TOOLS TO IMPACT SITES

To apply the risk-based framework to remediate an impact site, a conceptual site model (CSM) needs to be developed. A CSM is a summary of current understanding about the site conditions and pollution characteristics, including site history, contaminant sources, site physical condition, groundwater hydrology, contaminants' fate and transport at the site, environmental sampling results, and potential effects on human health and the environment. The CSM should be updated whenever new data and information become available.

This section first provides an overview of contaminant fate and transport processes in Section 3.4.1, followed by three practical applications of using the RIDM process and risk-based tools for the assessment and cleanup of subsurface pollutants. Pollutants discharged to the subsurface typically impact humans and the environment in two ways: downward to the underlying groundwater to impact water resources and eventually human health, and upward to the surface to affect human health. Sections 3.4.2 and 3.4.3 discuss how the risk-based approaches are applied to evaluate these pathways. Section 3.4.4 presents an application of using health risk assessment and risk management tools to address multiple environmental media and pathways for releases from leaking underground storage tanks.

3.4.1 Contaminant Fate and Transport Processes

This section will discuss several major fate and transport processes, including advection, dispersion and diffusion, sorption, volatilization, and bio-transformation.

3.4.1.1 Advection

Advection is a process whereby dissolved contaminants are transported physically with moving groundwater. The transport speed depends on the groundwater velocity. Figure 3.7 demonstrates how the water moves. Therefore, if a contaminant is dissolved in the water, it will be transported with the water.

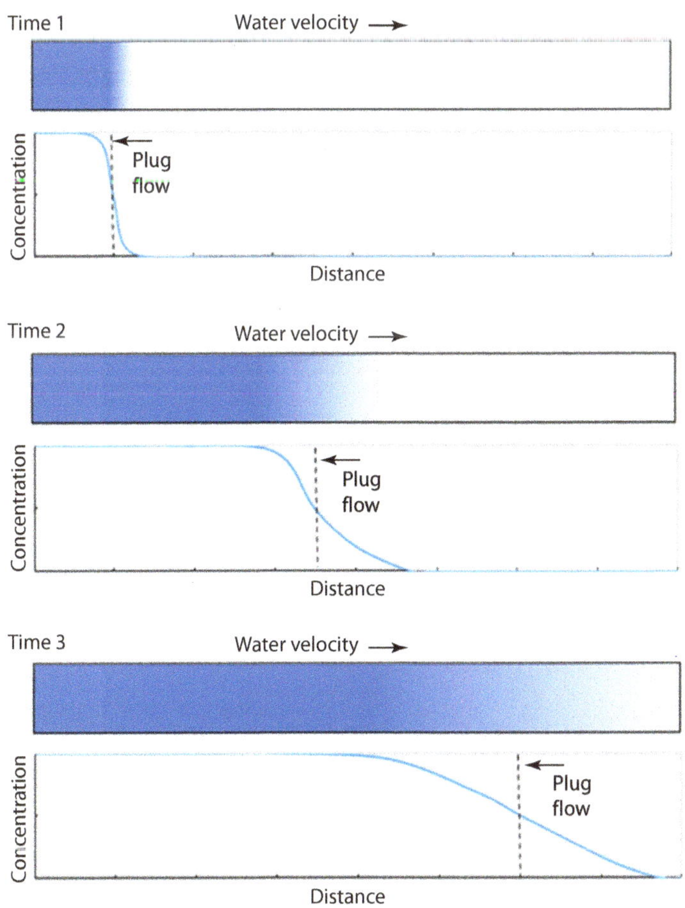

FIGURE 3.7 Advection water movement.

3.4.1.2 Dispersion and Diffusion

Dispersion is a process whereby dissolved contaminants disperse due to mechanical disturbance or agitation by an uneven velocity of groundwater. The uneven velocity may be caused by pore size, path length, or friction in pores. Figure 3.8 demonstrates these dispersion mechanisms.

Diffusion is a process whereby dissolved contaminants diffuse due to molecular movement along a concentration gradient in groundwater. The diffusion rate can be estimated using a diffusion coefficient, which is a chemical- and media-specific number. In most chemical fate and transport studies, particularly in mathematical modeling studies, the processes of dispersion and diffusion are expressed under one coefficient, called the *dispersion coefficient*. This is because it is very difficult to measure or estimate both separately in a real site situation.

Modeling the fate and transport processes usually includes advection and dispersion/diffusion processes, because in real groundwater transport cases, both advection and dispersion happen concurrently. Figure 3.9 shows the difference in a contaminant's fate and transport in an advection process alone and in advection plus dispersion.

3.4.1.3 Sorption

Sorption is a process whereby contaminants become attached physically or chemically to solid soil particles in the vadose zone or groundwater. Figure 3.10 shows sorption types when contaminants are in different forms and situations. The equation governing sorption is commonly expressed as follows:

$$C_s = K_d \cdot C_l = K_{oc} \cdot f_{oc} \cdot C_l \tag{3.6}$$

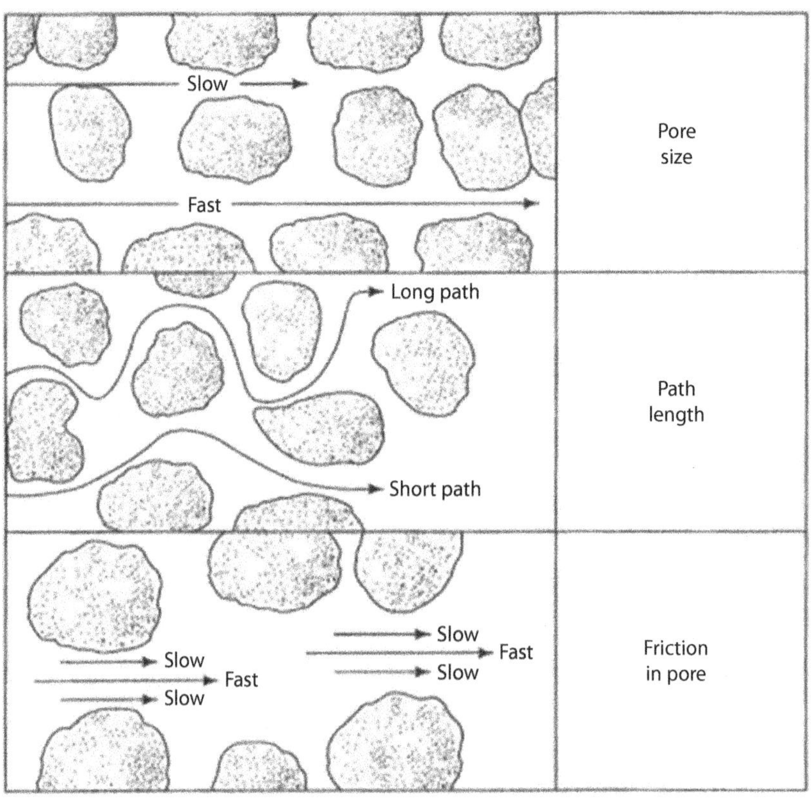

FIGURE 3.8 Three dispersion mechanisms.

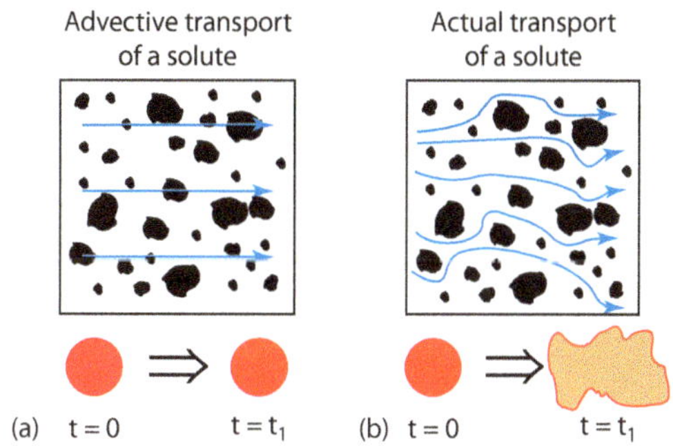

FIGURE 3.9 Contaminant fate and transport with advection only and advection plus dispersion.

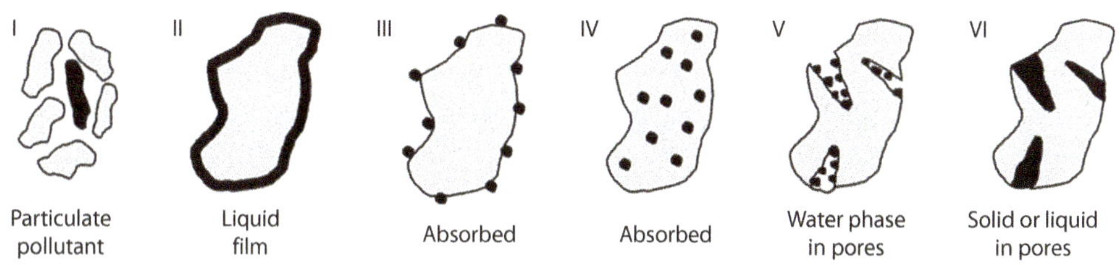

FIGURE 3.10 Sorption in different forms and situations.

where:

C_s	is chemical concentration in sorbed phase
C_l	is chemical concentration in liquid phase
$K_d\ (K_{oc} \cdot f_{oc})$	is contaminant partition coefficient between solid phase and liquid phase (L/kg)
K_{oc}	is chemical octanol-carbon partition coefficient (L/kg)
f_{oc}	is solid organic carbon content (unitless)

Therefore, most material sorption studies are focused on the quantification of soil organic carbon content, since K_{oc} is readily available from the literature.

3.4.1.4 Volatilization

Volatilization is a process whereby contaminants volatilize physically from the dissolved phase or the solid phase in the vadose zone or groundwater. The governing equation is commonly expressed as follows:

$$H = C_{\text{gas}} / C_l \left(20°\text{C}\right) \tag{3.7}$$

where:

H	is the Henry's Law constant
C_{gas}	is chemical concentration in gaseous phase
C_l	is chemical concentration in liquid phase

Therefore, the vapor phase concentration can be calculated from the chemical dissolved concentration, given the Henry's Law constant in the literature.

3.4.1.5 Bio-Transformation

Bio-transformation is a process whereby contaminants change their chemical formation by biological processes in the environment. This process is very important in bio-degradation and bio-remediation, which will be discussed in Chapters 8 and 10 of this book. When a first order decay rate for biodegradation is assumed, the following equation is commonly used to estimate the concentration decrease rate (exponentially):

$$C_t = C_0 exp(-\lambda t) \tag{3.8}$$

where:

C_t is contaminant concentration at time t
C_0 is initial contaminant concentration at time 0
λ is the decay constant (1/year)
t is the time (years)

Based on Equation 3.8, the concept of degradation "half-life" is defined as the time needed to reduce the initial concentration to the level of 50% of the initial concentration. It then can be calculated as

$$C_t = 0.5 \cdot C_0 \tag{3.9}$$

Thus, substituting Equation 3.8 into Equation 3.9,

$$\begin{aligned} 0.5 &= exp(-\lambda t) \\ Ln\ 0.5 &= -\lambda t \\ t_{1/2} &= 0.693/\lambda \end{aligned} \tag{3.10}$$

where $t_{1/2}$ is the contaminant degradation half-life (years).

The $t_{1/2}$ value can be readily calculated if the decay constant λ is available in the literature and can then be used to estimate the cleanup time for a remedial system design.

3.4.2 Impact of Contaminant Leaching to Groundwater

As discussed earlier, a pollutant discharged to the subsurface can impact water resources via downward movement (leaching) to the underlying groundwater. Soil cleanup goals are developed for groundwater protection purposes taking into account the contaminant fate and transport discussed earlier. For example, the California Los Angeles Regional Water Quality Control Board (LARWQCB) calculated the soil cleanup levels by using mathematical models (LARWQCB, 1996; Rong and Sakaida, 1996). Table 3.3 shows these calculated soil cleanup levels in the units of parts per million or milligrams per kilogram. (1) The top portion, for total petroleum hydrocarbons (TPH), was based on literature and professional judgment; (2) the bottom portion shows calculated soil cleanup levels of gasoline pollutants (aromatics) based on an attenuation method, depending on site soil type and depth to groundwater as well as the drinking water standard for a particular chemical to determine the acceptable soil concentration.

Refer to the following link for the derivation and details of the leaching model calculation: www.waterboards.ca.gov/losangeles/water_issues/programs/remediation/brownfields/RBs%201996%20Guide%20Book1_1.pdf

TABLE 3.3
Maximum Soil Screening Levels (mg/kg) for TPH and BTEX above Drinking Water Aquifers

	Distance above Groundwater	Carbon Range		
T		C4–C12	C13–C22	C23–C32
P	150 f	1,000	10,000	50,000
H	20–150 ft	500	1,000	10,000
	<20 ft	100	100	1,000

	Distance above Groundwater	Lithology			
B		Gravel	Sand	Silt	Clay
T	150 ft	B=0.044	B=0.077	B=0.165	B=0.8
E		T=2	T=4	T=9	T=43
X		E=8	E=17	E=34	E=170
		X=23	X=48	X=93	X=465
		MTBE=0.039	MTBE=0.078	MTBE=0.156	MTBE=0.78
	80 ft	B=0.022	B=0.033	B=0.066	B=0.34
		T=1	T=2	T=4	T=18
		E=4	E=7	E=15	E=73
		X=11	X=20	X=40	X=200
		MTBE=0.013	MTBE=0.039	MTBE=0.065	MTBE=0.338
	20 ft	B=0.011	B=0.011	B=0.011	B=0.044
		T=0.15	T=0.3	T=0.45	T=2.3
		E=0.7	E=0.7	E=2	E=9
		X=1.75	X=1.75	X=5.3	X=24.5
		MTBE=0.013	MTBE=0.013	MTBE=0.013	MTBE=0.065

TPH, total petroleum hydrocarbons; BTEX, benzene, toluene, ethylbenzene, and xylenes, respectively; MTBE, methyl tertiary butyl ether.
Respective MCLs (ppm): B=0.001, T=0.15, E=0.7, X=1.75, MTBE=0.013.
BTEX screening concentrations determined per the attenuation factor method as described in RWQCB Guidance for VOC Impacted Sites (March 1996), with a natural degradation factor of 11 for BTEX and of 3 for MTBE. Table values can be linearly interpolated between distances above groundwater and are proportional to fraction of each lithological thickness.
Values in Table 3.3 are for soils above drinking water aquifers. All groundwaters are considered as drinking water resources unless exempted by one of the criteria as defined under SWRCB Resolution 88 to 63 total dissolved solids (TDS > 3000 mg/L, or deliverability <200 gal/day, or existing contamination that cannot be reasonably treated). Regional Board staff will make a determination of potential water use at a particular site considering water quality objectives and beneficial uses. For non-drinking water aquifers, regardless of depth, TPH for ">150 ft" category in the table should be used; BTEX screening levels are set at 100 times respective MCLs and MTBE 30 times MCL as preliminary levels determined to be protective of human health and the environment.
Distance above groundwater must be measured from the highest anticipated water level. Lithology is based on the Unified Soil Classification System (USCS) scale definitions.

The soil cleanup concentrations in Table 3.3 are assumed to be linear. Therefore, mixed soil types can be calculated by the internal difference method. For example, at 80 ft depth, the concentration of soil cleanup for toluene (T) for 50% sand and 50% silt is 3 mg/kg (the average of 2 and 4 mg/kg). The same principle of interpolation also applies to different groundwater depths. For example, under the Gravel column, the cleanup concentration for benzene (B) at the 115 ft depth should be 0.033 mg/kg (i.e., the average of 0.044 and 0.022 mg/kg at the 150 and 80 ft depth, respectively), and so on. We can see the highest concentration in the table at the depth of 150 ft and the grid of the clay, while the lowest concentration is at 20 ft depth and the grid of the gravel. Any depths greater than 150 ft or smaller than 20 ft can also be extrapolated from the table values.

3.4.3 Risk-Based Cleanup Levels for Vapor Intrusion to Indoor Air

As discussed earlier, a pollutant discharged to the subsurface can impact human health via upward movement to the surface where humans live and work. In the United States, subsurface vapor intrusion into indoor air has received attention lately, as it is a potential concern for any building, existing or planned, located near soil or groundwater contaminated with volatile chemicals. Pathways of volatile contaminants that may lead to indoor air concern via vapor intrusion include (1) volatilization upward from the underlying contaminated soil, (2) migration (diffusion and advection) of contaminants in soil gas, and (3) volatilization upward from the underlying groundwater (see Chapter 6 for additional details).

The health risk of indoor air inhalation from vapor intrusion can be estimated by Equation 3.5. For existing or future buildings at and near a site contaminated with volatile chemicals in the subsurface, the goal is to remediate the site to a level where the indoor air risk due to subsurface volatile contaminants is acceptable. Therefore, the risk assessment for vapor intrusion is not only to assess the current indoor risk from contaminant vapors but also to guide the cleanup effort; that is, the contaminant concentration corresponding to the acceptable risk becomes the cleanup goal for site remediation.

There are two ways to assess vapor intrusion risk: mathematical modeling of subsurface (soil gas and groundwater) data and direct indoor air sampling.

3.4.3.1 Mathematical Modeling

The most widely used analytical model for predicting vapor intrusion into indoor air is the so-called J&E model (Johnson and Ettinger, 1991). The primary equations of the J&E model describe upward diffusion of soil gas through the soil column, advection and diffusion into buildings through foundation cracks, and mixing within the indoor space. The groundwater model also calculates the soil gas concentrations at the water table corresponding to the underlying groundwater source using Henry's Law. These equations combine the estimates of the indoor air concentrations and corresponding health risks from vapor intrusion of a soil gas or groundwater source.

The USEPA has developed the model into a spreadsheet for screening purposes (USEPA, 2017). Besides the forward calculation of indoor air risks using measured soil gas and groundwater concentrations, this model also uses a "back-calculation" algorithm to determine the cleanup criteria for contaminated sites based on a risk-based target indoor air concentration. Specifically, a cancer risk of 10^{-6} (one in a million) to 10^{-4} (one in ten thousand) and a noncancer hazard quotient of 1 may be selected as the target risk levels. The model calculates the concentrations of contaminants in indoor air that correspond to these target risk levels and then their concentrations in soil gas and groundwater. The calculated concentrations in soil gas and groundwater can then be used to develop the cleanup goals for these contaminants.

3.4.3.2 Indoor Air Sampling

As discussed in Chapter 6, indoor air concentrations directly measured by canisters or passive samplers are often preferred for existing buildings, as they do not require assumptions that are necessary for mathematical modeling on the fate and transport of subsurface contaminants entering buildings. On the other hand, indoor air sampling can be intrusive to occupants and is also subject to interference from indoor and outdoor sources with the same contaminants. Detailed protocols on indoor air sampling can be found in various regulatory guidance documents (see Chapter 6) and thus, are not discussed here.

Measured indoor air concentrations can be compared directly with the risk-based target indoor air concentration to determine the health risk and proper response actions (e.g., those shown in Table 3.2). Besides remediation of contamination sources, various mitigation options are also available and are discussed further in Chapter 6.

3.4.4 California Underground Storage Tank Low Threat Closure Policy

The concepts and methods of human health risk assessment and risk management have been discussed in Section 3.2, followed by an overview on the use of human health risk assessment in the site cleanup process for decision-making in Section 3.3. Examples of applying risk assessment to set up site remediation levels for groundwater protection and vapor intrusion are discussed in Sections 3.4.2 and 3.4.3, respectively. This section presents an application of using risk assessment and risk management tools to develop policies on regulating multi-media releases from underground storage tanks (USTs). In 2012, the California State Water Resources Control Board issued a California UST Low Threat Closure Policy (CSWRCB, 2012) to guide petroleum-impacted site remediation and case closure at leaking UST sites. In the Policy, there are quantitative numerical criteria for the case closure evaluation, which is considered a first of this kind, as cleanup criteria usually are developed case by case on a site-specific basis.

The Policy includes a preamble and pre-conditions (called *general criteria*), such as the removal of pollution sources; "the secondary source of pollution" (i.e., residual high concentrations of contaminants that remain in soil and water) is cleaned up to the maximum extent practicable; gasoline free product is removed to the maximum extent practicable; the contaminant plume is stable, and the concentration within the plume is decreasing; the pollutants are only limited to petroleum compounds; the required site conceptual model is developed, and so on. There are a total of eight general criteria. When all the pre-conditions are satisfied, three media-specific criteria are developed for the following environmental media and exposure pathways: (1) groundwater, (2) vapor intrusion to indoor air, and (3) direct contact and outdoor air exposure. Based on the above-mentioned human health risk assessment calculations, the numerical criteria for these three environmental media are provided in the Policy. If an impact site meets all pre-condition requirements and the numerical criteria for the three environmental media, this site can be considered low risk and closed. A detailed discussion about the eight general criteria and three media-specific criteria can be found in Chapter 1.

3.4.4.1 Groundwater

Figure 3.11 shows the numerical criteria for five classes of groundwater plume in the Policy, based on chemical concentration and plume length, presence of free product, and distance to the closest drinking water well. For example, the first class requires no free product, the plume length less than 100 ft (about 30 m), and the distance to a drinking water well more than 250 ft (about 75 m); as long as the plume is controlled within this range of numerical criteria, chemical concentrations in groundwater may not be considered as a criterion. However, keep in mind that there is a provision in the pre-condition of the Policy that "requires the contaminated plume is stable and the concentration is decreasing."

3.4.4.2 Vapor Intrusion to Indoor Air

Figure 3.12 is one example of the numerical criteria for the vapor intrusion pathway in the Policy. In this case, the oxygen content in soil vapor is less than 4% or not measured, so natural

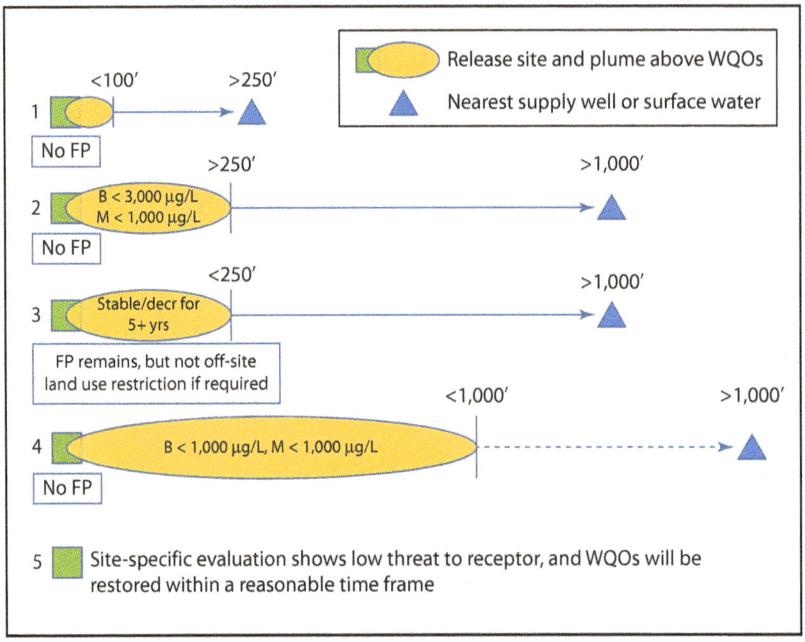

FIGURE 3.11 Numerical criteria for groundwater media in the California UST Low Threat Closure Policy. (B: benzene; FP: free product; M: methyl tert-butyl ether; Stable/decr: stable or decreasing in areal extent.)

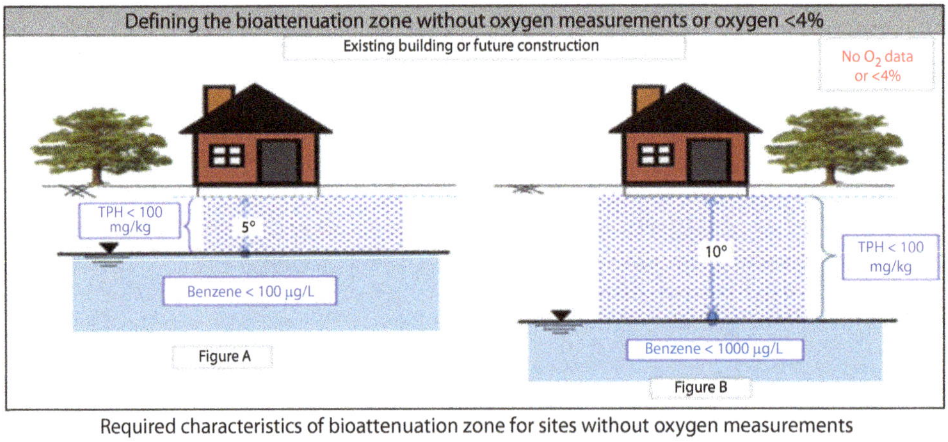

FIGURE 3.12 Numerical criteria for vapor intrusion in the California UST Low Threat Closure Policy.

biodegradation of petroleum hydrocarbons may take place slowly. Figure A and B within Figure 3.12 show the numerical criteria of a minimum bioattenuation zone thickness (i.e., groundwater depth) and the corresponding TPH and benzene concentrations in soil and groundwater, respectively, to be considered low risk. For example, if groundwater is at or more than 10 ft depth (Figure B), the TPH concentration in soil and the benzene concentration in groundwater should be less than 100 mg/kg and 1000 µg/L, respectively, to meet the low risk criteria and be closed under the Policy.

TABLE 3.4
Criteria for Direct Contact and Outdoor Air Exposure in California UST Low Threat Closure Policy

Concentrations of Petroleum Constituents in Soil That Will Have No Significant Risk of Adversely Affecting Human Health

	Residential		Commercial/Industrial		Utility Worker
Chemical	0–5 ft below Ground Surface (bgs) (mg/kg)	Volatilization to Outdoor Air (5–10 ft bgs) mg/kg	0–5 ft bgs (mg/kg)	Volatilization to Outdoor Air (5–10 ft bgs) (mg/kg)	0–10 ft bgs (mg/kg)
Benzene	1.9	2.8	8.2	12	14
Ethylbenzene	21	32	89	134	314
Naphthalene	9.7	9.7	45	45	219
PAH[a]	0.063	NA	0.68	NA	4.5

[a] Based on the seven carcinogenic poly-aromatic hydrocarbons (PAHs) as benzo(a)pyrene toxicity equivalent (BaPe). Sampling and analysis for PAH are only necessary where soil is affected by either waste oil or Bunker C fuel.

The area of impacted soil where a particular exposure occurs is 25 by 25 m (approximately 82 by 82 ft) or less.

NA = not applicable.

3.4.4.3 Direct Contact/Outdoor Air

Table 3.4 shows numerical criteria in soil for direct contact and outdoor air exposure in the Policy. These criteria are developed for the key petroleum constituents that contribute most significantly to health risks for these exposure pathways. For example, the concentration of benzene in soil should be less than 1.9 mg/kg at 5 ft (1.5 m) and less than 2.8 mg/kg at 10 ft (about 3 m) below the ground surface to be considered low risk for a residential scenario. If the soil concentrations for all four key constituents are below the table values for the current and planned land use, the site is considered low risk and can be closed under the Policy.

3.5 CASE STUDIES

This section presents two actual case studies to illustrate how the risk assessment and risk management tools discussed earlier have been used to guide the assessment and cleanup of contaminated sites.

3.5.1 Assessment of Potential Health Impact from Releases at a Gasoline Station

The first case study involves a gasoline station located in Los Angeles, California, with releases from former underground storage tanks (USTs) that may have impacted nearby residential properties to the south. Figure 3.13 shows a map of the gas station and surrounding areas. In the past, the USTs that leaked were installed in the south of the gas station close to the residential homes. The depth to groundwater at the site is about 3 m, and groundwater monitoring indicated gasoline slick (free product) in some of the samples. One of the residential properties has a swimming pool (see Figure 3.13) with a bottom depth of about 4 m. This means that the bottom of the pool is below the groundwater level, so the pool cuts into the groundwater table and may potentially lead the gasoline slick into the pool.

To address this pollution problem, the lead government regulatory agency required the responsible party to install groundwater monitoring wells in the resident's backyard to determine the

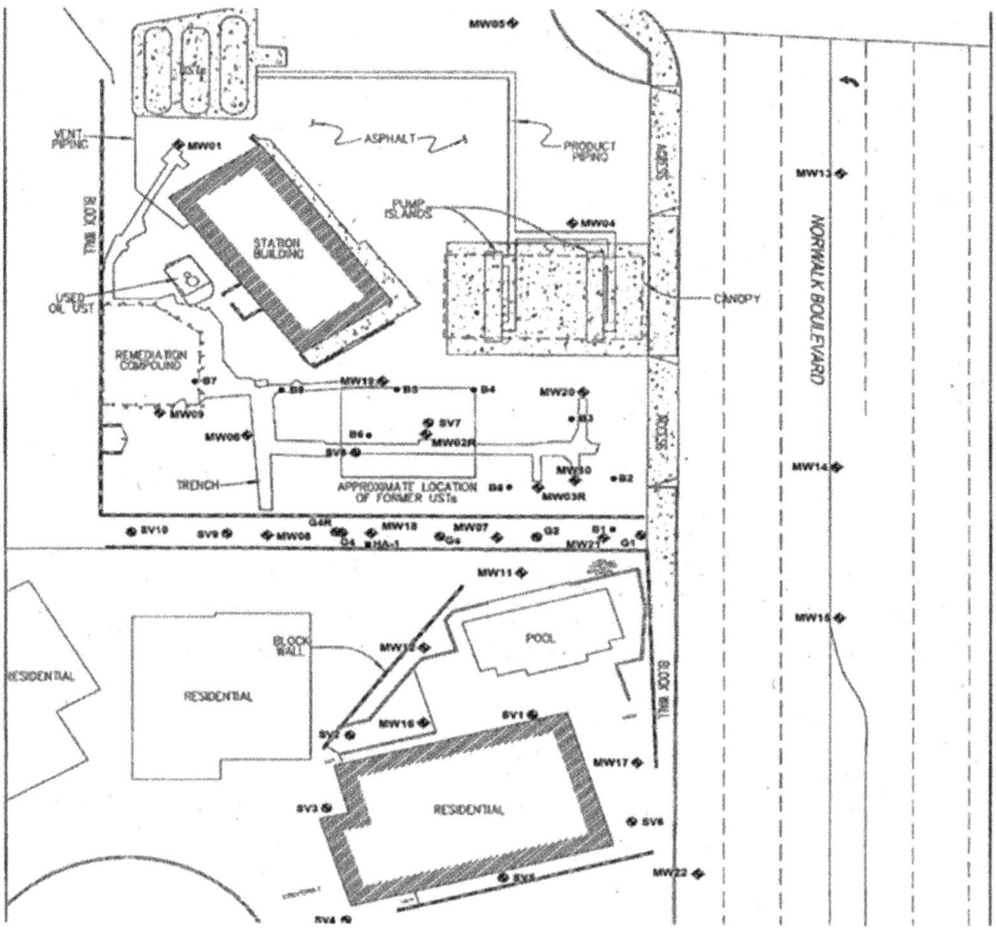

FIGURE 3.13 Site map and sampling locations (Case Study 1).

nature and extent of effects on groundwater, especially the gasoline free product pollution plume. In addition, the agency requested the responsible party to increase the "dual-phase" (gas and liquid) subsurface extraction method to clean soil and groundwater on-site at the same time. The cleanup effect was significant, as benzene concentrations in groundwater beneath the residential backyard decreased from 1670 (MW11) and 12 μg/L (MW12) to 0.65 and 0.48 μg/L, respectively, in 3 years. In addition, it was no longer possible to see the gasoline free product in groundwater. Overall, it was a very successful cleanup effort for the removal of the contamination source.

During the cleanup process, the responsible party was also required to evaluate potential vapor intrusion health risks to the residents. First, the J&E model (see Section 3.4.3) was used to calculate the acceptable concentrations in groundwater using a target cancer risk of 10^{-6} for indoor air exposure. The model calculated a target concentration of 16 μg/L for benzene in shallow groundwater (~3 m). Compared with the benzene concentration of 0.48–0.65 μg/L measured at the two wells (MW 11 and 12) in the residential backyard, the potential health risk of indoor exposure to benzene pollutants appears insignificant under the model assumptions. The next step was to conduct soil gas sampling in the resident's backyard as an additional line of evidence for evaluating vapor intrusion. The analysis of soil gas samples showed that benzene concentrations in soil gas are very low, almost undetectable. Thus, these results confirmed that the health risk to residents via vapor intrusion is very low.

Finally, the responsible party was also required to perform a comprehensive health risk assessment for all relevant (i.e., potentially complete) exposure pathways at the resident's property, namely

(1) incidental soil ingestion and dermal contact with soil (assuming no hand washing after work), (2) consumption of fruits and vegetables grown in the resident's backyard, and (3) indoor air inhalation via vapor intrusion. Using the methods and equations described in Section 3.2, the total cancer risk for all pathways combined is calculated as 1×10^{-8}, which is less than the target risk threshold of 10^{-6}. In summary, the multi-faceted approach, including mathematical modeling of groundwater data, sampling of soil gas in the resident's backyard, and performing a multi-pathway risk assessment, confirmed that the releases from former USTs at the gas station did not pose a significant health risk to the neighboring residents.

3.5.2 Selection of a Cost-Effective Remedy at a PCB-Contaminated Site

The second case study is an active industrial manufacturing facility in a Midwest state in the United States (Song et al., 2006). The soils at a portion of the facility were impacted by historical use of PCBs as hydraulic fluids. The surrounding areas are currently zoned industrial, and the future land use is expected to remain the same. Multiple stakeholders, including the USEPA as the lead agency, the state agency, and a community advisory group, are involved in the cleanup process. Following a thorough site investigation, a baseline health risk assessment was performed based on hypothetical exposures under reasonably expected future site uses (Table 3.5).

Due to potential elevated health risks for routine workers and large-scale construction workers (the values in bold face in Table 3.5), the USEPA requested an evaluation of the following remedial alternatives:

- *Alternative 1*: Soil cover (capping with 3 ft clean soil) only
- *Alternative 2*: Soil cover+removal of >1500 ppm PCB-impacted soils
- *Alternative 3*: Soil cover+removal of >500 ppm PCB-impacted soils
- *Alternative 4*: Soil cover+removal of >50 ppm PCB-impacted soils

To make an informed decision for selecting a cost-effective remedy, an analysis of risk reduction for each remedial alternative along with their costs was performed. Figure 3.14 shows the noncancer hazard index (HI) and the projected cost for the "no action" (i.e., baseline risk assessment with institutional controls) and each of the remedial options. Note that the costs for the "no action" include the implementation of institutional controls such as deed restriction, preparation of a soil management plan, and operation and maintenance (O&M) requirements.

The risk reduction and cost analysis showed that the soil cover and removal to 1500 ppm PCB-impacted soils (Alternative 2) would provide full protection even if institutional controls fail, while soil removal to 500 or 50 ppm (Alternative 3 or 4) would incur significant additional costs but would not provide beneficial risk reduction. Based on this analysis, Alternative 2 was selected as the final remedy for implementation at the site.

TABLE 3.5
Human Health Risk Assessment Results

Receptor	Total Cancer Risk	Noncancer HI
Routine worker	**2×10^{-4}**	**10**
Construction worker (small-scale)	9×10^{-6}	1
Construction worker (large-scale)	2×10^{-5}	**10**

Note: Bold values exceed the acceptable risk thresholds.

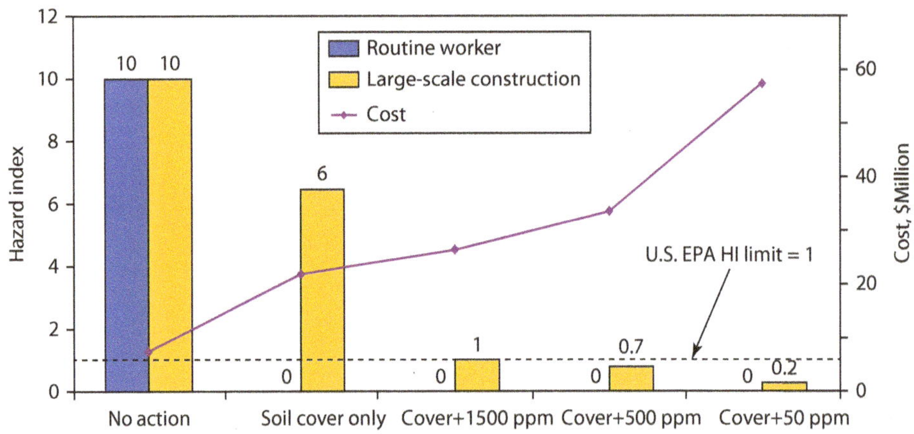

FIGURE 3.14 Noncancer HI and costs for remedial alternatives (Case Study 2). (From Song, S., et al., Strategies to Enhance Risk Managements. Remediation Innovative Technology Seminar (RITS) sponsored by the Naval Facilities Engineering Command (NAVFAC), 2006.)

3.6 CONCLUSIONS

"How clean is clean" has always been one of the key questions for site remediation. To answer this question, it is important to not only focus on science and technology, but also consider potential health risks and risk management dimensions for decision-making. This chapter demonstrates that human health risk assessment and risk management play important roles throughout the site assessment and cleanup process. While the discussion in this chapter is focused on the standard risk assessment methodology, most regulatory agencies do consider site-specific conditions when developing remediation goals for contaminated sites. The site-specific factors typically evaluated in a remedial action plan include pollutant use histories, pollutant properties (e.g., water solubility, volatility, etc.), soil and hydrogeologic environments, fate and transformation process in the subsurface, and so on. Therefore, adequate investigation data and a good conceptual site model that takes into account site-specific conditions are critical to support a reliable health risk assessment and for developing remedial goals. Key considerations for conducting site investigations and developing a conceptual site model are discussed in Chapters 4 and 5 of this book.

The risk assessment and risk management tools discussed in this chapter provide a systematic approach to determining "how clean is clean" by quantifying the risk associated with contaminants to ensure that they no longer pose a negative impact to human health and the environment. While one-size-fits-all remediation criteria may not be feasible or practical in many cases, this chapter also presents some practical applications where the risk-based tools may be used to streamline or enhance the cleanup process.

PRACTICE AND EXERCISES

Question 1: For toluene, a 95% lower confidence limit on the benchmark dose (BMDL) of 238 mg/kg-day was used as the point of departure, based on increased kidney weight in a 13-week gavage study in rats (NTP, 1990). USEPA has applied the following uncertainty factors (UF) to this effect level: 10 for extrapolation for interspecies differences (animal to human), 10 for consideration of interspecies variation (human variability), 10 for use of a subchronic study to estimate chronic effects, and 3 for database insufficiencies. Calculate the oral reference dose (RfD) using the information provided.

Solution 1:

$$\text{Total UF} = 10 \times 10 \times 10 \times 3 = 3000$$

Using Equation 3.1,

$$\text{RfD} = \text{BMDL/Total UF} = 238/3000$$
$$= 0.08 \text{ mg/kg-day}$$

Question 2: An average no-observed-adverse-effect level (NOAEL) of 128 mg/m^3 was calculated from the available studies that indicated neurological effects in occupationally exposed workers exposed to toluene in air. (a) Calculate an adjusted NOAEL value from the occupational exposure scenario (5 days per week and 10 m^3 breathing rate during the 8-h workday) to continuous exposure conditions (7 days per week and 20 m^3 breathing rate during the entire day). (b) Derive the inhalation RfC using the adjusted NOAEL and an uncertainty factor (UF) of 10 for intraspecies variation to account for potentially susceptible human subpopulations and life stages.

Solution 2:

(a) NOAEL (adjusted) = 128 mg/m^3 × (5 days/7 days) × (10m^3/20m^3) = 46 mg/m^3
(b) RfC = NOAEL (adjusted)/UF = 46/10 = 5 mg/m^3

Question 3: Use the USEPA Integrated Risk Information System (IRIS) database (www.epa.gov/iris) to identify the inhalation unit risk (IUR) and the inhalation reference concentration (RfC) for the following compounds: tetrachloroethylene, trichloroethylene, 1,1,1-trichlorothane, 1,1-dichloroethylene, and vinyl chloride. Assuming they are present at the same concentration in air, rank these compounds from the highest to the lowest potential to pose (a) lifetime carcinogenic risk and (b) chronic non-carcinogenic hazard.

Solution 3:

1. Lifetime carcinogenic risk:

   Vinyl chloride (8.8×10^{-6} [μg/m^3]$^{-1}$), trichloroethylene (4.1×10^{-6} [μg/m^3]$^{-1}$), tetrachloroethylene (2.6×10^{-7} [μg/m^3]$^{-1}$), 1,1,1-trichlorothane and 1,1-dichloroethylene (neither is a carcinogen).

2. Chronic non-carcinogenic hazard:

   Trichloroethylene (2×10^{-3} mg/m^3), tetrachloroethylene (4×10^{-2} mg/m^3), vinyl chloride (1×10^{-1} mg/m^3), 1,1-dichloroethylene (2×10^{-1} mg/m^3), 1,1,1-trichlorothane (5 mg/m^3).

Question 4: Using the exposure parameters for *residential adult* in Table 3.1, (a) calculate the intake (I) of benzene detected at 50 mg/kg in soil; and (b) calculate the exposure concentration (EC) of benzene detected at 1 μg/m^3 in indoor air. (Note: the answers are different for cancer and noncancer endpoints.)

Solution 4:

1. Using Equation 3.2,

$$\text{Intake}(\text{cancer}) = \frac{50\ \text{mg/kg} \times 100\ \text{mg/day} \times 350\ \text{days/year} \times 20\ \text{years} \times \left(1\ \text{kg}/10^6\ \text{mg}\right)}{80\ \text{kg} \times 70\ \text{years} \times 365\ \text{days/year}}$$

$$= 1.7 \times 10^{-5}\ \text{mg}/(\text{kg} \times \text{day})$$

$$\text{Intake}(\text{noncancer}) = \frac{50\ \text{mg/kg} \times 100\ \text{mg/day} \times 350\ \text{days/year} \times 20\ \text{years} \times \left(1\ \text{kg}/10^6\ \text{mg}\right)}{80\ \text{kg} \times 20\ \text{years} \times 365\ \text{days/year}}$$

$$= 6.0 \times 10^{-5}\ \text{mg}/(\text{kg} \times \text{day})$$

2. Using Equation 3.3,

$$\text{EC}(\text{cancer}) = 1\ \mu\text{g/m}^3 \times 24\ \text{h/day} \times 350\ \text{days/year}$$
$$\times 20\ \text{years}/(70\ \text{years} \times 365\ \text{days/year} \times 24\ \text{h/day})$$
$$= 0.27\ \mu\text{g/m}^3$$

$$\text{EC}(\text{noncancer}) = 1\ \mu\text{g/m}^3 \times 24\ \text{h/day} \times 350\ \text{days/year}$$
$$\times 20\ \text{years}/(20\ \text{years} \times 365\ \text{days/year} \times 24\ \text{h/day})$$
$$= 0.96\ \mu\text{g/m}^3$$

Question 5: Repeat the calculation of the intake and the exposure concentration in Question 4 for *residential child.*

Solution 5:

1. Using Equation 3.2,

$$\text{Intake}(\text{cancer}) = \frac{50\ \text{mg/kg} \times 200\ \text{mg/day} \times 350\ \text{days/year} \times 6\ \text{years} \times \left(1\ \text{kg}/10^6\ \text{mg}\right)}{15\ \text{kg} \times 70\ \text{years} \times 365\ \text{days/year}}$$

$$= 5.5 \times 10^{-5}\ \text{mg}/(\text{kg} \times \text{day})$$

$$\text{Intake}(\text{noncancer}) = \frac{50\ \text{mg/kg} \times 200\ \text{mg/day} \times 350\ \text{days/year} \times 6\ \text{years} \times \left(1\ \text{kg}/10^6\ \text{mg}\right)}{15\ \text{kg} \times 6\ \text{years} \times 365\ \text{days/year}}$$

$$= 6.4 \times 10^{-4}\ \text{mg}/(\text{kg} \times \text{day})$$

2. Using Equation 3.3,

$$\text{EC}(\text{cancer}) = 1\ \mu\text{g/m}^3 \times 24\ \text{h/day} \times 350\ \text{days/year} \times 6\ \text{years}/(70\ \text{years} \times 365\ \text{days/year} \times 24\ \text{h/day})$$

$$= 0.082\ \mu\text{g/m}^3$$

$$\text{EC}(\text{noncancer}) = 1\ \mu\text{g/m}^3 \times 24\ \text{h/day} \times 350\ \text{days/year} \times 6\ \text{years}/(6\ \text{years} \times 365\ \text{days/year} \times 24\ \text{h/day})$$

$$= 0.96\ \mu\text{g/m}^3$$

Question 6: Use the USEPA Integrated Risk Information System (IRIS) (www.epa.gov/iris) to find the oral slope factor (SF) and the reference dose (RfD) for benzene. Based on the intakes calculated in Questions 4 and 5, estimate (a) the lifetime incremental cancer risk and (b) the noncancer hazard for exposure to 50 mg/kg benzene in soil by adult and child residents.

Solution 6:

For benzene, oral SF = 0.055 $(\text{mg/kg-day})^{-1}$; oral RfD = 4.0×10^{-3} mg/(kg-day)

1. Using Equation 3.4a,

$$\text{Cancer risk}(\text{adult}) = \text{I}_{\text{adult}} \times \text{SF} = 1.7 \times 10^{-5}\ \text{mg}/(\text{kg} \times \text{day}) \times 0.055(\text{mg/kg-day})^{-1}$$

$$= 9.7 \times 0^{-7}$$

$$\text{Cancer risk}(\text{child}) = \text{I}_{\text{child}} \times \text{SF} = 5.5 \times 10^{-5}\ \text{mg}/(\text{kg} \times \text{day}) \times 0.055(\text{mg/kg-day})^{-1}$$

$$= 3.0 \times 10^{-6}$$

$$\text{Lifetime cancer risk} = 9.7 \times 10^{-7} + 3.0 \times 10^{-6} = 4 \times 10^{-6}$$

2. Using Equation 3.4b,

$$\text{HQ (adult)} = I_{\text{adult}} / \text{RfD} = 6.0 \times 10^{-5}\ \text{mg}/(\text{kg} \times \text{day}) / 4.0 \times 10^{-3}\ \text{mg}/(\text{kg-day}) = 0.015$$

$$\text{HQ(child)} = I_{\text{child}} / \text{RfD} = 6.4 \times 10^{-4}\ \text{mg}/(\text{kg} \times \text{day}) / 4.0 \times 10^{-3}\ \text{mg}/(\text{kg-day}) = 0.16$$

Question 7: Use the USEPA Integrated Risk Information System (IRIS) (www.epa.gov/iris) to find the inhalation unit risk (IUR) and the reference concentration (RfC) for benzene and the answers in Questions 4 and 5 to calculate (a) the incremental lifetime risk and (b) the noncancer hazard for exposure to 1 $\mu g/m^3$ benzene in air.

Solution 7:

For benzene, IUR = 7.8×10^{-6} $(\mu g/m^3)^{-1}$; RfC = 30 $\mu g/m^3$

1. Using Equation 3.5a,

$$\text{Cancer risk}(\text{adult}) = \text{EC}_{\text{adult}} \times \text{IUR} = 0.27\ \mu\text{g/m}^3 \times 7.8\times10^{-6}\left(\mu\text{g/m}^3\right)^{-1} = 2.1\times10^{-6}$$

$$\text{Cancer risk}\ (\text{child}) = \text{EC}_{\text{child}} \times \text{IUR} = 0.082\ \mu\text{g/m}^3 \times 7.8\times10^{-6}\left(\mu\text{g/m}^3\right)^{-1} = 6.4\times10^{-7}$$

$$\text{Lifetime cancer risk} = \text{EC}\times\text{IUR} = 2.1\times10^{-6} + 6.4\times10^{-7} = 3\times10^{-6}$$

2. Using Equation 3.5b,

$$\text{HQ}(\text{adult}) = \text{EC}_{\text{adult}}/\text{RfC} = 0.96\ \mu\text{g/m}^3/30\ \mu\text{g/m}^3 = 0.03$$

$$\text{HQ}(\text{child}) = \text{EC}_{\text{child}}/\ \text{RfC} = 0.96\ \mu\text{g/m}^3/\ 30\ \mu\text{g/m}^3 = 0.03$$

Question 8: For the same condition as in Question 4, calculate (a) the intakes and (b) the corresponding cancer risk and noncancer hazard for a construction worker using the following exposure parameters: the soil ingestion rate of 330 mg/day, the exposure frequency of 225 days per year, and the exposure duration of 1 year.

Solution 8:

1. Using Equation 3.2,

$$\begin{aligned}\text{Intake}(\text{cancer}) &= 50\ \text{mg/kg}\times330\ \text{mg/day}\times225\ \text{days/year}\times1\ \text{year}\times\left(1\ \text{kg}/10^6\,\text{mg}\right)/\\&\quad\left(80\ \text{kg}\times70\ \text{years}\times365\ \text{days/year}\right)\\&= 1.8\times10^{-6}\ \text{mg}/(\text{kg}\times\text{day})\end{aligned}$$

$$\begin{aligned}\text{Intake}(\text{noncancer}) &= 50\ \text{mg/kg}\times330\ \text{mg/day}\times225\ \text{days/year}\times1\ \text{year}\times\left(1\ \text{kg}/10^6\ \text{mg}\right)/\\&\quad\left(80\ \text{kg}\times1\ \text{year}\times365\ \text{days/year}\right)\\&= 1.3\times10^{-4}\ \text{mg}/(\text{kg}\times\text{day})\end{aligned}$$

2. Using Equations 3.4a and 3.4b,

$$\text{Cancer risk} = I\times\text{SF} = 1.8\times10^{-6}\ \text{mg}/(\text{kg}\times\text{day})\times0.055\,(\text{mg/kg-day})^{-1} = 1\times10^{-7}$$

$$\text{HQ}(\text{adult}) = I_{\text{adult}}/\text{RfD} = 1.3\times10^{-4}\ \text{mg}/(\text{kg}\times\text{day})/4.0\times10^{-3}\ \text{mg}/(\text{kg-day}) = 0.032$$

Question 9: Comparing the answers from Questions 6 and 8, rank the receptors from the highest to the lowest (a) lifetime carcinogenic risk and (b) non-carcinogenic hazard associated with the soil ingestion pathway. Discuss possible reasons for different orders in the cancer and noncancer rankings.

Solution 9:

1. *Cancer*: Lifetime resident (child + adult) > resident (child) > resident (adult) > construction worker
2. *Noncancer*: Resident (child) > construction worker > resident (adult)

Child resident has the highest risk in both cases due to highest intakes (high soil ingestion rate and low body weight).

Construction worker has lower cancer risk than adult resident due to much shorter exposure duration (1 vs. 20 years), but has higher noncancer hazard due to higher soil ingestion rate (the exposure duration term is not a factor in the noncancer intake calculation).

Question 10: (a) Using the decision matrix shown in Table 3.2 and the cancer risks calculated in Question 7, what actions, if any, would you recommend as a risk manager? (b) If background concentrations in outdoor air in the study area range from 0.5 to 1.5 $\mu g/m^3$, would this information affect your recommendation?

Solution 10:

1. Because the lifetime cancer risk for a resident is greater than 1×10^{-6}, one or more of the following actions should be considered:
   - Collect additional data for evaluation of potential vapor intrusion (see Chapter 6)
   - Monitor indoor air periodically for any changes
   - Identify any indoor, outdoor, or subsurface sources for benzene and other petroleum products before deciding whether mitigation or source remediation should be considered
2. Because the benzene concentration detected in indoor air is within the background levels measured outdoors, it is possible to attribute it to outdoor sources (note: benzene is a common air pollutant in urban environments). However, it may be prudent to monitor indoor air (and outdoor air) periodically for any significant changes before closing the case.

Question 11: Given the soil cleanup values in Table 3.3, calculate the soil cleanup level for ethylbenzene (E) at 60 ft in soil containing 40% sand and 60% silt.

Solution 11:

Based on values for ethylbenzene in Table 3.3 (note: the table values are in a linear relationship):
Sand at 80 ft = 7 ppm and silt at 80 ft = 15 ppm,
Therefore, 40% sand + 60% silt = $0.4 \times 7 + 0.6 \times 15 = 11.8$ ppm.
Sand at 20 ft = 0.7 ppm and silt at 20 ft = 2 ppm,
Therefore, 40% sand + 60% silt = $0.4 \times 0.7 + 0.6 \times 20 = 1.48$ ppm.
Then, soil cleanup level at 60 ft = Z
(Z − 1.48 ppm)/(11.8 ppm − 1.48 ppm) = (60 ft − 20 ft)/(80 ft − 20 ft)
Z = 8.36 ppm

Question 12: Due to sorption effect, a contaminant "retardation factor" in groundwater is defined as

$$R = 1 + \rho_b \left(\mathrm{Kd}/\theta\right)$$

where:

- R is the retardation factor (unitless)
- ρ_b is the bulk density of the soil (g/mL)
- Kd is the contaminant chemical partition coefficient between solid phase and liquid phase (L/kg)
- Θ is the effective porosity (unitless)

Given that bulk density is 1.68 g/mL, effective porosity is 0.21, soil organic carbon content (f_{oc}) is 0.0036, and the chemical octanol-carbon partition coefficient (K_{oc}) for tetrachloroethylene (PCE) is 155 L/kg, calculate the retardation factor for PCE in this environment.

Solution 12:

$$R(PCE) = 1 + \rho_b (Kd/\theta) = 1 + 1.68 \times (K_{oc} \times f_{oc}/0.21) = 1 + 1.68 \times (155 \times 0.0036)/0.21 = 5.46$$

Question 13: Given that the first order decay rate constant for benzene is 0.043 (1/day) in soil, calculate the half-life of benzene in the environment.

Solution 13:

Using Equation 3.10,

$$t_{1/2} = 0.693/\lambda$$

where:

- $t_{1/2}$ is the contaminant degradation half-life (days)
- λ is the decay constant (1/day)

$t_{1/2} = 0.693/\lambda = 0.693/0.043 = 16$ (days)

Question 14: What are the key points of the California Underground Storage Tank Low Threat Closure Policy?

Solution 14:

(1) Petroleum-based contaminants can be naturally degraded when oxygen contents are sufficient. (2) Human health risk assessment can be a tool to determine the residual concentrations left behind at an impacted site.

REFERENCES

American Society for Testing and Materials (ASTM) 1995 (Revision 2015). Standard E1739-95, Guide for Risk-Based Corrective Action Applied at Petroleum Release Sites. ASTM International, West Conshohocken, Pennsylvania. (www.astm.org)

California Department of Toxic Substances Control (DTSC) 2011. Guidance for the Evaluation and Mitigation of Subsurface Vapor Intrusion to Indoor Air. (www.dtsc.ca.gov/assessingrisk/upload/final_vig_oct_2011.pdf)

California DTSC 2015. Preliminary Endangerment Assessment Guidance Manual.

California Los Angeles Regional Water Quality Control Board (LARWQCB) 1996. Interim Site Assessment & Cleanup Guidebook - Appendix A. (www.waterboards.ca.gov/losangeles/water_issues/programs/remediation/brownfields/RBs%201996%20Guide%20Book1_1.pdf)

California State Water Resources Control Board (CSWRCB) 2012. Low-Threat Underground Storage Tank Case Closure Policy. (www.waterboards.ca.gov/board_decisions/adopted_orders/resolutions/2012/rs2012_0016atta.pdf)

Code of Federal Regulations (CFR), 1990. 40 CFR 300, National Contingency Plan. March 8.

Johnson, P.C. and Ettinger, R.A., 1991. Heuristic Model for Predicting the Intrusion Rate of Contaminant Vapors into Buildings. *Environmental Science and Technology*, vol. 25, no. 8: 1445–1452.

National Academy of Sciences (NAS), 1983. Risk Assessment in the Federal Government. National Toxicology Program (NTP), 1990. Toxicology and carcinogenesis studies of toluene (CAS No. 108-88-3) in F344/N rats and B5C3F1 mice (inhalation studies). Public Health Service, United States Department of Health and Human Services; NTP TR 371.

Rong, Y. and Sakaida, R.R., 1996. Screening for VOCs. *Soil and Groundwater Cleanup*, January-February: 6–8.

Song, S., Jeng, C.Y, and Cizerle, K. 2006. Strategies to Enhance Risk Managements. Remediation Innovative Technology Seminar (RITS) sponsored by the Naval Facilities Engineering Command (NAVFAC).

United States Environmental Protection Agency (USEPA), 1989. Risk Assessment Guidance for Superfund (RAGS), Volume I: Human Health Evaluation Manual, Part A. Interim Final. EPA/540/1-89/002. December.

USEPA, 1991a. Risk Assessment Guidance for Superfund (RAGS), Volume I: Human Health Evaluation Manual, Part B: Development of Risk-based Preliminary Remediation Goals. Interim. EPA/540/R-92/003. December.

USEPA, 1991b. Risk Assessment Guidance for Superfund (RAGS), Volume I: Human Health Evaluation Manual, Part C: Risk Evaluation of Remedial Alternatives. Interim. October.

USEPA, 1991c. A Guide to Principal Threat and Low Level Threat Wastes. (www.epa.gov/superfund/resources/gwdocs/threat.pdf)

USEPA, 1995. Use of Risk-Based Decision-Making in UST Corrective Action Programs. Office of Solid Waste and Emergency Response (OSWER) Directive 9610.17. March 1.

USEPA, 1997. Rules of Thumb for Superfund Remedy Selection. (www.epa.gov/superfund/resources/rules/rulesthm.pdf)

USEPA, 2009. Risk Assessment Guidance for Superfund (RAGS), Volume I: Human Health Evaluation Manual, Part F: Supplemental Guidance for Inhalation Risk Assessment. Final. EPA-540-R-070-002. January.

USEPA, 2014. Update of Standard Default Exposure Factors. Office of Solid Waste and Emergency Response (OSWER) Directive 9200.1-120. February 6.

USEPA, 2017. EPA Spreadsheet for Modeling Subsurface Vapor Intrusion. (www.epa.gov/vaporintrusion/epa-spreadsheet-modeling-subsurface-vapor-intrusion)

4 Groundwater Hydrology, Soil and Groundwater Contamination Assessment and Monitoring

Weixing Tong

CONTENTS

The earth is made of three basic types of rocks: igneous, sedimentary, and metamorphic rock. Through various geological processes (weathering, erosion, etc.), the rocks become soil. The earth's land surface is covered by either rock or soil. Most of the hard rocks are solid and contain very little void space, although some rocks contain void space due to their mineral structures or chemical or physical weathering processes; for example, fractured rock and Karst landform. Soil, on the other hand, is a very loose material and has more void space. When water reaches the earth's surface through precipitation or irrigation, some of the excess water will infiltrate downward below the ground surface due to gravity. When the infiltrated water encounters an impermeable layer, the water will fill up all the void space and accumulate above that layer, forming a saturated zone, which is called *groundwater*. In the uppermost portion of the earth's crust, large volumes of porous and fractural materials store groundwater. Although only a small fraction of the precipitation from any single storm may reach the saturated zone, the process has continued through geologic time, resulting in a large subsurface reservoir. Hydrogeology studies the interrelationship between geological materials and processes with water. This chapter will cover the basic principles of soil characteristics, groundwater occurrence, and soil and groundwater contamination assessment and monitoring.

4.1 GROUNDWATER HYDROLOGY

The existence of water in the subsurface is controlled by its solid metrics. Due to consolidation or crystallization, minerals in rocks are closely bound together with very little void space. Therefore, most solid rocks, except fractured or weathered rocks, can hold very little or no water. Groundwater is mostly stored in an unconsolidated soil layer. The amount of water a soil layer can hold is dependent on the soil grain size, sorting, and packing.

4.1.1 Soil Formation and Classification

The uppermost layer of the earth may contain a three-phase system of solid, liquid, and gaseous materials. The solid phase contains mineral grains and organic matter. The organic matter represents the remains of plants and animals that are undergoing decay. The liquid phase is water containing dissolved solutes. The vapor phase includes water vapor and other gases, which may not be present in the same proportions as in the atmosphere (Jury et al., 1991). The earth layer that contains the three-phase system is called the *zone of aeration*, *unsaturated zone*, or *vadose zone*.

The solid phase of the vadose zone may consist of soil, which is formed by *in situ* weathering; or sediment, which has been transported from the place of weathering; or un-weathered bedrock. Mineral grains may be disaggregated; for example, sand grains in a dune.

4.1.1.1 Unified Soil Classification System (Casagrande, 1948; ASTM, 1989)

In 1948, Casagrande developed the Unified Soil Classification System (USCS) for classifying soils. This classification scheme has become widely accepted and is almost universally used in the engineering and engineering geology professions. The USCS groups soils according to their grain sizes. Soils are either coarse-grained, fine-grained, or highly organic (Table 4.1). These groups are further subdivided based on soil properties that can be measured in the laboratory, such as Atterberg limits and gradation. The consistency of a solid is the degree of adhesion between the particles that can resist deformation or rupture. By increasing or decreasing the water content, it is possible to change

TABLE 4.1
Unified Soil Classification System

		Identification Procedures			Symbol	Typical Names
Coarse Grain Soils	Gravels >50% of coarse fraction is larger than No. 4 sieve	Clean Gravels	Wide range in grain size and substantial amounts of all intermediate particle sizes		GW	Well-graded gravels, gravel-sand mixture, little or no fines
			Predominately one size or range of sizes with some intermediate sizes missing		GP	Poorly graded gravels, gravel-sand mixture, little or no fines
		Gravels with Fines	Non-plastic fines (see ML)		GM	Silty gravels, poorly graded gravel-sand-silt mixtures
			Plastic fines (see CL)		GC	Clayey gravels, poorly graded gravel-sand-clay mixtures
	Sands >50% of coarse fraction is smaller than No. 4 sieve	Clean Sands	Wide range in grain size and substantial amounts of all intermediate particle sizes		SW	Well-graded sands, gravelly sands, little or no fines
			Predominately one size or range of sizes with some intermediate sizes missing		SP	Poorly graded sands, gravelly sands, little or no fines
		Sands with Fines	Non-plastic fines (see ML)		SM	Silty sands, poorly graded sand-silt mixtures
			Plastic fines (see CL)		SC	Clayey sands, poorly graded sand-clay mixtures
Fine Grain Soils	Silts and Clays LL < 50	Dry Strength	Dilatancy	Toughness	For Fraction Smaller than No. 40 Sieve	
		None–light	Quick–slow	None	ML	Inorganic silts and very fine sands, silty or clayey fine sands with slight plasticity, rock flour
		Medium–high	None–very slow	Medium	CL	Inorganic clays of low to medium plasticity, gravelly clays, sandy clays, silty clays, lean clays
		Slight–medium	Slow	Slight	OL	Organic silts and organic silt-clays of low plasticity
	Silts and Clays LL > 50	Slight–medium	Slow–none	Slight–medium	MH	Inorganic silts, micaceous or diatomaceous fine sandy or silty soils, elastic silts
		High–very high	None	High	CH	Inorganic clays of high plasticity. Fat clays
		Medium–high	None–very slow	Slight–medium	OH	Organic clays of medium to high plasticity
Highly Organic Soils		Readily identified by color, odor, spongy feel, and frequently by fibrous textures			PT	Peat and other highly organic soils

the consistency. The soil passes from a solid state, when dry, through the semisolid, plastic, and liquid states as water is added. The water contents at the boundaries between these states of consistency are called *Atterberg limits*. Although the soil classification is made on the basis of laboratory tests, the classification system can be applied successfully in the field.

Soils are represented by letter symbols. G (Gravels) and S (Sands) are the coarse-grained soil symbols; M (Silts), C (Clays), and O (Organic silts or clays) are the symbols for the fine-grained soils; and PT represents the highly organic soils. The second letter modifies the major group divisions. For the gravelly and sandy soils, the modifiers are W (well-graded), P (poorly graded), M (silty), and C (clayed). Silts and clays have the modifiers L (for liquid limit index < 50) and H (for liquid limit index > 50).

The coarse-grained and fine-grained soil boundary is placed at the No. 200 sieve (0.003 inch). The size is the smallest particle size visible to the unaided eye. If more than 50% of the soil is retained on the No. 200 sieve, it is coarse-grained. Coarse-grained soils are either gravels or sands, depending on whether more or less than 50% of the grains are retained on a No. 4 sieve. Gravel-size particles range from 3 inches to 3/16 inch (retained on a No. 4 sieve). Sand-size particles fall between 3/16 inch (pass the No. 4 sieve) and 3/1000 inch (retained on the No. 200 sieve). Table 4.2 shows the size limits for each type of coarse-grained particle. Particles over 3 inches in diameter (cobbles and boulders) are not classified in the USCS, but their presence in the sample should be noted in the sample description.

Particles passing the No. 200 sieve are silts (M) and clays (C). Tests to differentiate between them can be run in the field or laboratory. Field tests to characterize the fine-grained fraction of the samples are dilatancy, dry strength, and toughness. Silts have little or no strength when dry, while clays have considerable strength. Dilatancy, dry strength, and toughness are also used to identify the fine-grained fraction of coarse-grained soils.

The degree of plasticity is the other factor that modifies the classification of fine-grained soils. The plasticity index can be estimated in the field but can only be accurately determined in the laboratory by performing the Atterberg limits test. Fine-grained soils with a liquid limit >50 are modified by the symbol H (MH or CH), and those with a liquid limit <50 have a symbol of L (ML or CL). Fine-grained soils containing 30% or more coarse-grained fraction should be modified by descriptive term, such as "gravelly" or "sandy." If the coarse fraction is between 10 and 30%, the words "with sand and/or gravel" should be added to the group name.

TABLE 4.2
USCS Coarse-Grained Particle Size Classification

Particle	Subdivision	U.S. Standard Sieve	Size (mm)[a]
Boulders	—	—	Over 305
Cobbles	—	—	<305 to 75
Gravel	Coarse	<3″ to ¾″	<75 to 19
	Fine	<3/4″ to No. 4	<19 to 4.76
Sand	Coarse	<No. 4 to No. 10	<4.76 to 1.68
	Medium	<No. 10 to No. 40	<1.68 to 0.42
	Fine	<No. 40 to No. 200	<0.42 to 0.074

Source: McCarthy, D. F., *Essentials of Soil Mechanics and Foundations,* Regents/Prentice Hall, Englewood Cliffs, NJ, 1993, and American Society for Testing and Materials (A.S.T.M.), *Standard Test Method for Classification of Soils for Engineering Purposes*, 288–297, 1989.

[a] The actual size for each grain size range may vary with different authors (Morris, D. A. and Johnson, A. I., *Summary of Hydraulic and Physical Properties of Rock and Soil Material,* 1967.)

4.1.2 Groundwater Occurrence

As the water percolates through the soil and rock, it can accumulate above the impermeable layer. At the depth where the fluid pressure in the pores of the rock or soil is exactly atmospheric is a surface called the *water table*. Below the water table is groundwater. Just above the water table is a zone called the *capillary fringe*, in which water is held in capillary tension. The zone above the capillary tension is called the *vadose zone*.

When a large amount of water is stored in the porous geological material, this is called an *aquifer*. An aquifer is a geological unit that can store and transmit economically significant quantities of water to wells or springs. The aquifer is continuously being affected by water moving in and out of the system. A recharge area is an area where water infiltrates downward into the saturated zone. Sources of recharge typically are rainfall, runoff in stream channels, and irrigation water. A discharge area is where subsurface water is discharged to the land surface, other bodies of water (such as springs, seeps, or stream baseflow), or to the atmosphere (Figure 4.1).

Many types of geologic units serve as aquifers. Geologic materials known to be aquifers include alluvial deposits, limestone, volcanic rocks, sandstone, plutonic and metamorphic rocks, and clay.

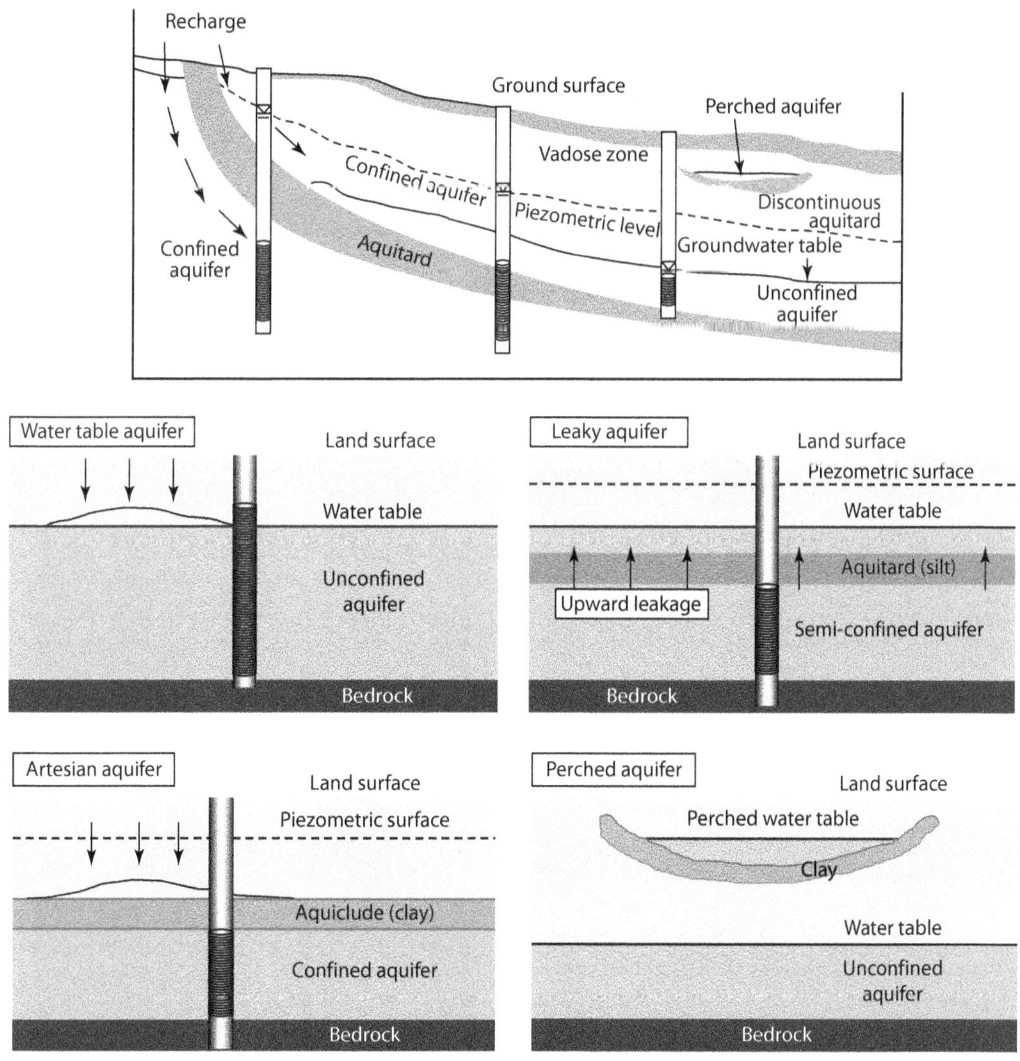

FIGURE 4.1 Different types of aquifers.

A key requirement is the ability of a formation to store water in the pore spaces between the sediments or rock grains. The porosity of earth materials is the percentage of the rock or soil that is void of material. It is calculated mathematically by the equation

$$n = \frac{100V_v}{V} \tag{4.1}$$

where:

n is the porosity (percentage)

V_v is the volume of void space in the a unit volume of earth material (L^3; cm^3 or m^3)

V is the unit volume of earth material, including both voids and solids (L^3; cm^3 or m^3)

The degree of porosity a rock or soil has depends on the grain packing, uniformity of grain size, degree of sorting, grain shape, environment of reposition, and depth of burial. Porosity may be derived from intergranular spaces (primary porosity), fractures (secondary porosity), or both. The porosity can be determined by heating the sample in the laboratory and measuring the difference between the dry sample and the saturated sample. The measured porosity in the laboratory is called *effective porosity*, because it excludes pores that are not large enough to contain water molecules and those that are not interconnected. Effective porosity (n_e) is the porosity available for fluid flow.

Another major aquifer property is transmissivity or hydraulic conductivity. The transmissivity is a parameter describing the capacity of an aquifer to transmit water through pore spaces. The hydraulic conductivity represents the rate at which fluid can move through a permeable medium. The calculation of transmissivity and hydraulic conductivity will be discussed in the following section.

Aquifer layers are often separated by impermeable materials called *confined layers*. A confined layer is a geological unit that has little or no intrinsic permeability and restricts the movement of groundwater. A leaky confining layer, called *aquitard*, is a layer of low permeability that can store groundwater and also transmit it slowly from one aquifer to another under vertical hydraulic gradients. *Aquiclude* consists of impermeable materials containing water but not transmitting water. Clays are typical aquiclude materials. *Aquifuge* represents impermeable materials that do not contain water, such as solid granite.

There are two types of aquifer: unconfined and confined. An aquifer that has a water table is called an *unconfined aquifer* (sometimes called a *water table aquifer*). The elevation of the water table may rise and fall with recharge or discharge to the aquifer. An aquifer overlain by a confining, impermeable layer is called a *confined* or *artesian aquifer*. Water in a confined aquifer is generally under pressure that is greater than atmospheric pressure. The water level in a well stands at some height above the top of the aquifer (often called *piezometric head*). If the water level is above the land surface, the well is a flowing artesian well. The level to which water will rise in tightly cased wells in an artesian aquifer is called the *potentiometric surface*.

4.1.3 Groundwater Movement

Laws and principles governing groundwater movement can be developed from the fundamental laws of physics. The direction and movement of groundwater in porous media are determined by areas of recharge and discharge and the system of interconnected passages comprising the microstructure of the porous matrix. The path followed by a sequence of water particles in steady flow is called a *streamline* and consists of a procession of similar particles, each overcoming frictional resistance to movement under a gentle but persistent hydraulic gradient.

Each streamline takes a slightly different path, diverging around individual grains and merging through narrow passages in the interstices. The rate of movement on each streamline also varies, speeding up when converging through constrictions and slowing down on entering larger pore spaces. However, although the microscopic movement may appear highly varied, rotational, and tortuous, on a

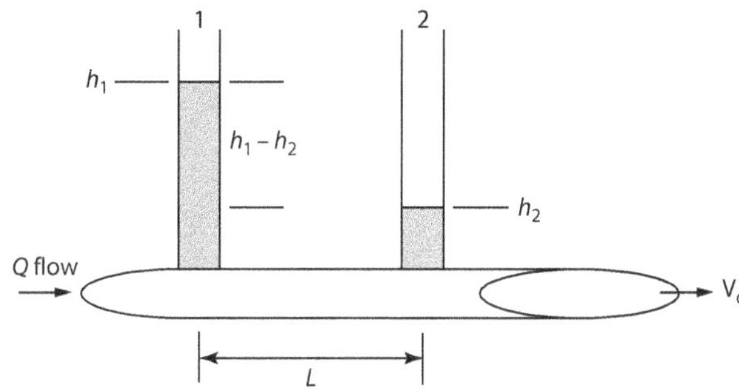

FIGURE 4.2 Darcy's Law for flow through a pipe.

macroscopic scale, the entire process when averaged over sets of adjacent streamlines seems perfectly orderly, with the sets of streamlines moving in essentially parallel lines. Such flow is called *streamline* or *laminar flow* and is of fundamental importance to the understanding of groundwater movement.

In turbulence flow, water particles move in irregular and rotational paths. Whereas laminar flow occurs at relatively low velocities, turbulent flow is characterized by high velocities. Flow in rivers or streams is generally turbulent, whereas groundwater flow is generally laminar. In the mid-1800s, a French engineer, Henri Darcy, conducted the first systematic study of the movement of water through a porous medium (Darcy, 1856). He studied the movement of water through beds of sand used for water filtration. Darcy found that the rate of water flow through a bed of a "given nature" is proportional to the difference in the height of the water between the two ends of the filter beds and inversely proportional to the length of the flow path. He determined also that the quantity of flow is proportional to a coefficient (K), which is dependent on the nature of the porous medium. The flow is also obviously proportional to the cross-sectional area of the pipe (A). His experimental setup is shown in Figure 4.2 and illustrates what is now known as Darcy's Law, the basic equation describing groundwater flow:

$$Q = -KA\frac{(h_1 - h_2)}{L} \tag{4.2}$$

where:

Q	is the volume of water crossing a unit cross-sectional area per unit time (gallons per minute, gpm)
K	is the hydraulic conductivity (gallons per day per square foot, gpd/ft^2)
A	is the cross-sectional area through which flow occurs (square feet, ft^2)
$(h_1 - h_2)/L = i$	is the hydraulic gradient
L	is the length of the flow path (feet)

The minus sign indicates that the flow is in the direction of decreasing head. It is often omitted from the numerical calculation of Darcy's Law.

Darcy's Law can be expressed in more general terms as

$$Q = -KA\left(\frac{dh}{dl}\right) = q\,A;\ q = -K\frac{dh}{dl} \tag{4.3}$$

where:

dh	is the change in head between two points that are very close together
dl	is the small distance between these two points
q	is called *specific discharge* or *discharge velocity* (ft/day or gpd/ft^2). It is sometimes called the *Darcian velocity*.

The velocity expressed here assumes that flow occurs through the entire cross section of aquifer material without regard to pore space. This velocity, therefore, is a hypothetical velocity and represents the average velocity in the direction of decreasing uniform head. The use of Darcian velocity in flow equations is justified if one assumes that on a macroscopic scale, flow is orderly, and the bulk of particles on average move locally essentially in parallel paths. On a microscopic scale, however, groundwater moves only through effective pore areas at a rate equal to "effective" or "seepage" velocity. Actual groundwater flow, therefore, is faster than the average flow as represented by the Darcian velocity. The relation between Darcian and seepage velocity is stated as

$$v_e = \frac{q}{\theta} \tag{4.4}$$

where:

v_e is the seepage velocity (feet/day)
θ is the effective porosity of material

Groundwater seepage velocities vary widely depending on aquifer materials and hydraulic gradients. Rates typically range between a few feet per year and several hundred feet per day (Meinzer, 1942).

Another parameter, transmissivity (T), is often used to describe the capacity of an aquifer to transmit water of the prevailing kinematic viscosity. It equals the hydraulic conductivity of the aquifer times the saturated thickness of the aquifer (b):

$$T = K \cdot b \tag{4.5}$$

where:

T is the transmissivity (gpd/ft or ft^2/day)
K is the hydraulic conductivity (gpd/ft^2)
b is the saturated thickness of aquifer (ft)

4.1.3.1 Aquifer Tests

Aquifer tests are performed to determine the hydraulic properties and boundary conditions of aquifers. One of the most important types of aquifer test is run by pumping a well at a constant discharge rate while concurrently measuring the change in water levels in observation wells. The change in water levels from the time the test began is called *drawdown*. At first, drawdown is very rapid, but as pumping continues, the rate of drawdown decreases. Analysis of the drawdown characteristics allows the calculation of aquifer parameters such as transmissivity, storativity, and hydraulic conductivity.

The mathematical model most often used for analysis of aquifer tests is the Theis equation. This model is an exact solution for the radial flow of groundwater in a confined aquifer to a well. The Theis equations for transmissivity (T) and storativity (S) are

$$T = \frac{15.3QW(u)}{s} \tag{4.6}$$

$$S = \frac{Ttu}{360r^2} \tag{4.7}$$

where:

T is the transmissivity (ft^2/day)
S is the storativity (dimensionless number)

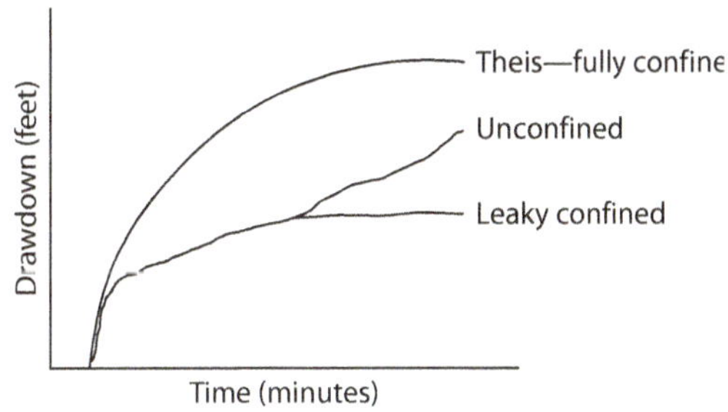

FIGURE 4.3 Theis, unconfined and leaky confined curves for a single well.

Q is the pumping rate (gal/min)
s is the drawdown (feet)
t is the time (minutes)
r is the distance from pumping well to observation well (feet)
$W(u)$ is the well function of u

$$u = \frac{360 r^2 S}{T t}$$

To obtain the values of T and S using the Theis equations, the drawdown versus time since pumping began (for one observation well) is plotted on log-log paper to obtain a curve similar to one of the curves shown in Figure 4.3.

To simplify solutions for Theis equations, approaches have been developed for a single observation well and multiple observation wells. The time-drawdown method (also known as the *Jacobs straight-line method*) was designed to use drawdown measurement in one observation well. In this method, the drawdown values are plotted on semi-log paper as a function of time since pumping started. A straight line is drawn through the data points and extended backward to the zero-drawdown axis. The straight line is fitted through the points farthest from the pumping well because these points are more accurate. In fact, data taken during the first 10 min of the test are not valid. The aquifer parameters are then calculated using the slope of the straight line and the intercept of the line with the time axis when drawdown is zero. This method is illustrated in Figure 4.4, and the equations for transmissivity and storativity are

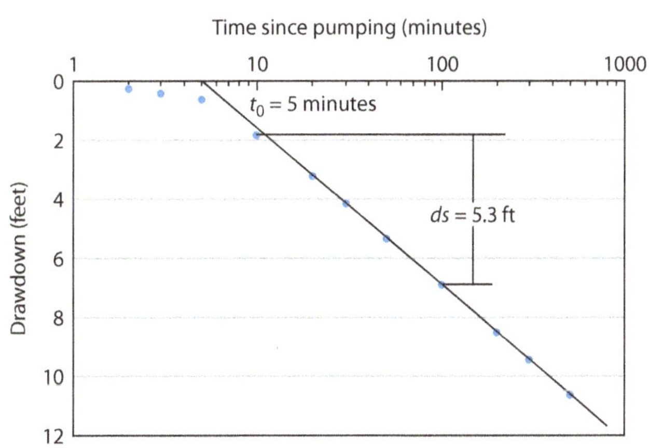

FIGURE 4.4 Plot of drawdown versus time in time-drawdown method.

$$T = \frac{35Q}{ds} \tag{4.8}$$

$$S = \frac{T\,t_0}{640\,r^2} \tag{4.9}$$

Units are the same as those in the Theis method. These equations are for transmissivity measured in square feet per day. The values read from the graph are ds and t_0. The intercept of the line with the time axis (t_0) is measured in minutes. d_s is the change in drawdown per one log cycle, a constant value on the graph, measured in feet.

The distance-drawdown method, a variation of the Jacobs method, may be used if the drawdowns in three or more observation wells are measured simultaneously. In this situation, the drawdown varies with the distance from the pumping well in accordance with the Theis equation. Drawdown is plotted on the arithmetic scale, as was done in the time-drawdown method, and distance is plotted on the logarithmic scale (x-axis). The data will form a straight line. Fit the line through the points closest to the pumping well. The values read from the graph are ds and r_0. Beyond r_0, the pumping well has no effect. The equations to use to solve for transmissivity and storativity, illustrated in Figure 4.5, are

$$T = \frac{70Q}{ds} \tag{4.10}$$

$$S = \frac{T\,t_e}{640\,r_e^2} \tag{4.11}$$

Units are the same as those in the Theis method.

Aquifer pumping tests are expensive and time consuming. One test may take several days to run and may involve installation of pumping and observation wells. Moreover, if the water in the aquifer is contaminated, it may be necessary to treat it prior to disposal, which sometimes may not be feasible. As an alternative to a pumping test, a slug or bail-down test can be performed in a small-diameter monitoring well. This type of test can be used to determine the hydraulic conductivity of the formation in the immediate vicinity of a monitoring well. A known volume of water is

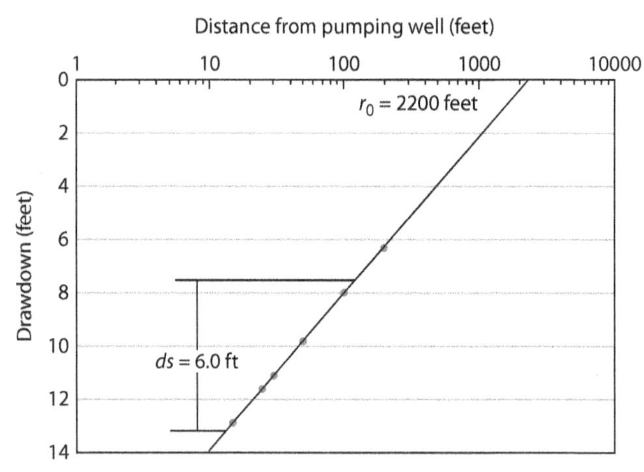

FIGURE 4.5 Plot of drawdown versus distance in distance-drawdown method.

quickly drawn from or added to the monitoring well. The rate at which the water level falls or rises is measured. These data are then analyzed by an appropriate method. The slug test only involves a small volume of water near the well borehole; therefore, the parameter values obtained from the test only represent a small area, whereas the parameter values obtained from aquifer pumping test often represent the properties of the entire aquifer screened by the well.

4.2 CONTAMINATION SITE ASSESSMENT

Due to human activities, contaminants can be released into the environment by spill, leak, or discharge from industrial facilities. Typical sources for soil and groundwater contamination in urban areas are underground storage tanks (USTs), dry cleaners, degreaser sinks, refinery facilities, tank farms, oil pipelines, metal plating facilities, surface spills, and so on. In the countryside, excessive use of fertilizers and irrigation with contaminated water (from mining or oil and gas production) can be the sources for contamination. The chemicals of concern (COCs) in these sources are hazardous to human health and the environment.

To conduct a site assessment, soil, vapor, and groundwater samples must be collected from the site and sent to a certified laboratory for analysis. Based on the analysis results from all media, a site conceptual model (SCM) will be developed to evaluate the potential risk to human health and groundwater resources. The site assessment results will help to select an appropriate remediation technology (if necessary), establish a monitoring network, and provide data that can be used for risk assessment and case closure review.

4.2.1 Soil Investigation

A soil investigation at a contamination site is conducted to characterize the source and the extent of a release. It also serves to establish the severity of the release and the impacts on receptors by providing concentrations of COCs. Therefore, collecting representative soil samples is a very important step in the site investigation. The soil sample can provide information on lithology and contaminant concentration. Soil samples can be collected through continuous-core or multiple discrete-depth samples. Although discrete-depth sampling is mostly used in the site investigation project, continuous logs of the subsurface are particularly valuable, because they can provide more accurate hydrogeological information, which can be used to produce cross sections and help to design a remediation system.

4.2.1.1 Soil Sampling Methods

There are several methods to collect soil samples, including hand auger, open (test) pit, and drilling. Hand auger is mostly used for shallow soil investigation (<10 ft). Soil samples can be collected using either a specialized bit or a hand-drive sampler loaded with brass tubes. Hand augers are also routinely used to clear drilling locations of underground obstructions.

Test pits are frequently used to investigate shallow soil conditions and stratigraphy. The pit is usually dug using a backhoe or excavator; the size of the equipment used is dependent on the depth of the test pit. Test pits are particularly useful for obtaining a "bigger picture" cross-sectional view of stratigraphic contacts and the location of soil staining or the occurrence of light non-aqueous phase liquid (LNAPL) as compared with borings. The sidewalls of test pits are usually logged by drawing a cross-sectional view and identifying (1) soil types and the depth of contacts, (2) observations of fill, (3) observation of staining, (4) important natural or manmade features, and (5) the water table. Soil samples can be collected from test pits either by driving brass tubes into the sidewalls or bottom of the pit or (in the case of deep pits) by driving a tube into the soil from the excavator bucket. Test pits are backfilled either with the excavated soil or with imported fill in accordance with regulatory requirements. Compaction of the backfilled pit is also accomplished per regulatory requirements.

The most commonly used method is drilling, which involves drill rig and field crew to operate it. There are many drilling methods developed for environmental investigation and widely used in the industry. The most common drilling methods include direct push technology (DPT) (e.g., cone penetration test [CPT], Geoprobe®, hydropunch, membrane interface probe [MIP], laser-induced fluorescence [LIF]), hollow stem auger (HSA), rotary drilling methods (air rotary and mud rotary), and sonic drilling.

Before any field work for soil investigation, a soil sampling strategy must be developed, and the purpose of the sampling must be clearly defined. The sampling locations should be planned to identify the potential source area, delineate the impact, or help site characterization. In the state of California, it is required that prior to any field work, the responsible party (RP) and its consultant must conduct a subsurface clearance survey which can be provided by service companies such as Underground Service Alert (USA) or DigAlert.

Typical practice has been that soil samples for chemical analysis are collected in sleeves or liners that line the sampling barrel (e.g., acetate sleeves for DPT and brass sleeves for split-barrel samplers). At the desired sampling depth, the DPT sleeve is cut and capped, or the brass sleeve is selected and capped. It is also possible to collect a soil sample for chemical analysis from an unlined sampling barrel by quickly driving a brass sleeve into the soil core and then capping the sleeve.

4.2.1.2 Soil Sample Preparation and Borehole Maintenance

For the analysis of volatiles (e.g., total petroleum hydrocarbons as gasoline/gasoline range organics [TPHg/GRO] or volatile organic compounds [VOCs]), the Environmental Protection Agency (EPA) Method 5035 for preservation of soils collected is required. Method 5035 is a preparation method for soil samples that minimizes the loss of volatiles and is the preferred preparation method for California. More details on the method can be found at www.epa.gov/osw/hazard/testmethods/sw846/pdfs/5035.pdf. The sample to be analyzed for volatiles is collected by removing the top 1 inch of soil from the sleeve, collecting the sample into the EPA Method 5035 preservation apparatus (e.g., En-core sampler, Figure 4.6), and chilling as required. Samples must be labeled and placed on ice in a cooler under chain-of-custody for transport to a state-certified laboratory for analysis. Samples should be delivered to the laboratory within 24–48 hours of collection, if possible, to limit the potential for analysis outside of method holding times.

Soil borings or DPT holes must be grouted in accordance with regulatory requirements. These requirements may vary with hole depth (e.g., above or below the water table), and different agencies allow the use of different grouting materials (e.g., high-solids bentonite slurry, Portland cement, or cement-bentonite mixtures). Grout is typically placed into the hole from the bottom to the top using a tremie or pump to ensure that the borehole is completely filled and that bridging in the hole does not occur.

For collecting multiple-depth samples, soil samples are normally collected at 5 ft intervals (and/or at change of lithology or if obvious contamination is found) for geological logging using the USCS and field screening using a photo ionization detector (PID). Soil samples must be logged by the field geologist.

Prior to each boring, all down-hole equipment should be decontaminated using a brush and detergent solution followed by a tap water rinse and a final deionized water rinse to prevent cross-contamination. On collection, soil samples should be sealed, labeled, and placed in an insulated ice chest, pending submission (under chain-of-custody record) to a state-certified laboratory for analysis.

4.2.2 Soil Vapor Investigation

When VOC contaminants are released to soil or dissolved into groundwater, they tend to generate vapor due to their volatilization potential from the impacted soil and groundwater. To assess the

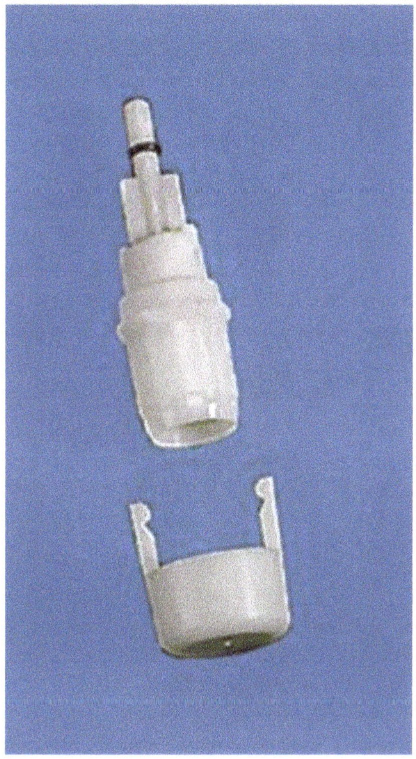

FIGURE 4.6 En-core sampler.

risk, soil-vapor samples (soil-vapor survey) are often collected from the contamination site. Soil-vapor surveys can be used for a number of purposes, including the following:

- Initial site screening, where the objective is to assess whether VOCs are present
- Site assessment/characterization, where the objective is to assess the source, extent, and magnitude of impacted soil, groundwater, and/or vapor
- Risk assessments, where the objective is to assess the risk to public health, safety, and the environment
- Remediation and post-remediation monitoring, where the objective is to assess remediation progress or completion
- Ongoing monitoring for risk assessment, remediation monitoring, landfill gas monitoring, and background methane monitoring

4.2.2.1 Overview of Soil-Vapor Survey Methods

There are three principal methods for collecting soil-vapor data: (1) active, (2) passive, and (3) flux chamber. Each method offers advantages and disadvantages, which are briefly described in this subsection. The design and protocols of a soil-vapor survey program are dependent on the objectives of the program, the types of contaminants anticipated to be present, and the site conditions. There are a variety of sampling methods and equipment designs for collecting soil-vapor samples that can potentially yield different values. If shallow low-flow or non-flow (fine-grained soil or clay) conditions exist at the site, soil-vapor analysis may not be a useful tool, and soil samples may be more applicable.

Active: The active approach consists of the withdrawal of an aliquot of soil vapor from the subsurface, typically with a sampling probe, followed by analysis of the withdrawn vapor. Analysis is often performed on-site using a variety of analytical instruments. Alternatively, soil-vapor samples can be stored in gas-tight containers and analyzed at an off-site laboratory. The active method is quantitative, and values are reported in concentration units (e.g., parts per million by volume [ppmv], micrograms per liter [μg/L]). This approach is the most common soil-vapor collection method for a number of reasons, including ease of sample collection, the opportunity for real-time data to direct further sampling, and the ability to acquire quantitative measurements.

Passive: The passive approach consists of the emplacement of an adsorbent into the subsurface and subsequent removal and analysis of the adsorbent. The absorbent is typically placed in the upper end of an inverted container that has an open bottom. Measured values cannot be reported as concentrations, only as total adsorbed mass (e.g., micrograms) or in some form of relative units, because the amount of vapor that comes into contact with the adsorbent is unknown. Due to this limitation, passive surveys are useful for qualitative purposes only. Because one effect of the adsorbent is to concentrate the soil vapor, this approach offers advantages over the active approach in locations of low–vadose zone permeability and at sites with lower contaminant concentrations (less than 0.1 μg/L vapor). However, contaminants must still have an appreciable vapor pressure to be detected by this method. The technique requires two visits to the field, one to deploy the adsorbents and a second trip to retrieve them, and does not allow the acquisition of real-time data.

Flux Chambers: A flux chamber consists of an enclosed chamber that is placed on the surface for a specific period of time. Vapor concentrations are measured in the chamber after a period of time. This method is also quantitative and yields both concentration data in the chamber and flux data (mass/area-time). Flux chambers are the least common soil-vapor survey method and are typically used only for risk-based applications, when direct vapor fluxes out of the subsurface are desired.

The protocols for soil-vapor sampling most commonly followed in California are described in Regional Board's Advisory for Active Soil Gas Investigations (July 2015), available at www.waterboards.ca.gov/losangeles/water_issues/programs/ust/docs/VI_ActiveSoilGasAdvisory_FINAL.pdf.

The active soil-vapor method is most typically employed in site assessment. The passive soil-vapor method can also be used, especially in locations with limited access and at sites where relatively low concentrations of VOCs are expected. The sampling locations and grid spacing should be sufficiently small to encounter areas of former USTs, piping, dispensers, and so on, and any areas of gross contamination. When historical data regarding the layout of a source configuration are unavailable, a useful strategy is to collect samples in a grid pattern. For a typical service station, a grid spacing of approximately 50 feet is reasonable. For more detailed site assessment/characterization, a sample spacing of 10–20 ft is reasonable in the source area. Soil-vapor samples are typically collected from 5 to 15 ft below ground surface (bgs) to assess surface and UST releases. The chosen depth will be dependent on the suspected source and what is being assessed: soil and/or groundwater. To assess the vertical extent of contamination, samples should be collected every 5–10 ft, depending on the depth to groundwater at the location of highest concentration.

4.2.3 Groundwater Investigation

To conduct a groundwater investigation, a groundwater sample must be collected underneath the site. There are two types of groundwater sampling: grab groundwater sampling and monitoring well sampling. A grab groundwater sample is collected through a temporary setting, for example, direct-push method or open borehole method, and it cannot be collected repeatedly. It only provides semi-quantitative results. The majority of groundwater samples are collected through monitoring wells. Groundwater samples can be collected repeatedly through a monitoring well, and they represent the water condition for the sampled zone.

4.2.3.1 Groundwater Monitoring Well Design and Construction

Groundwater monitoring well design and construction involve several steps, including selection of target zone; selection of screen length, screen slot size, and filter pack; well installation; and well development. At petroleum release sites, monitoring wells have typically been screened across "the water table" irrespective of stratigraphy.

Site stratigraphy should be understood prior to selecting target zones for monitoring well screens. This is best achieved by collecting continuous-core samples via several exploratory borings prior to installing wells. Although wells screened across "the water table" may be desirable in the source area to evaluate the presence of mobile product, most of the dissolved-phase mass will migrate away from the source area through soil units with higher relative hydraulic conductivity, typically coarse-grained sediments such as sand or gravel. For sites with heterogeneous stratigraphy, therefore, these coarse-grained units should be selected as target zones for monitoring wells both within and beyond the source area. If a well is being installed to monitor a "deeper" zone (which is presumably less impacted than the "shallow" zone), it is important to seal off the shallow zone during drilling and well installation to avoid or minimize cross-contamination or drag-down. This is particularly important in an LNAPL source area.

The selection of screen depth and length depends on the purpose of the monitoring well and the fluctuations in the depth to groundwater at the site. Contaminants in the dissolved phase can form a "smear zone" in soil between high and high and low water-level elevations, and mobile product (if present) will only be measurable in a monitoring well if the well is screened across the smear zone and the water table occurs within the smear zone at the time of the measurement. If monitoring product thickness is the purpose of the monitoring well, then it is important to consider fluctuations in groundwater elevations when designing the well. For sites where groundwater elevations are stable, well screens that extend 5 ft above and below the current water table are acceptable in most cases. However, if the smear zone at the site or data at nearby properties indicate that the water level is likely to rise or fall more than 5 ft above or below the current water table over the year or over several years, the screened interval may need to be expanded.

The proper selection of screen size and filter pack for the target zone allows adequate well efficiency and the production of low-turbidity samples. Well design is based on the lithology of the target zone and the purpose of the well. At a minimum, the filter pack material (e.g., silica sand) and well casing material (e.g., polyvinyl chloride [PVC]) must be non-reactive and appropriate for the subsurface chemical environment. In addition, the permeability of the well-screen slot size and filter pack should be slightly less than the permeability of the surrounding formation to limit sediment from entering, yet close enough to the permeability of the surrounding formation so as not to impede the natural flow of groundwater through the subsurface.

The selection of well-screen slot size and filter pack can involve a "rigorous" procedure (based on sieve analysis) or an "intermediate" procedure (based on field experience), depending on project requirements.

The selection of screen size and filter pack for groundwater monitoring wells is mostly based on field experience. It relies on visual estimates of grain-size distribution in the target zone to determine the proper filter pack and screen slot size. Based on field experience, a target zone that is comprised of silt might call for a #0/30 sand filter pack (or grain-size equivalent) and a 0.010-inch slot size screen. Target zones with silty sand or fine sand might be completed using a #2/16 filter pack (or equivalent) and a 0.020-inch slot size screen. To maximize well efficiency, medium sand or coarser target zones might be completed using a #3 or #2/12 filter pack (or equivalent) and a 0.030-inch slot screen.

The typical drilling methods used for the installation of monitoring wells include hollow stem auger (HAS), rotary, and sonic. The key criteria for selection of drilling methods are related to the well construction: target zone depth, inside diameter of the well casing needed for passing of pumps and sampling tools, and a borehole diameter that is at least 4 inches larger than the outside diameter of the well casing. Monitoring wells are typically constructed with threaded Schedule 40 PVC blank casing of 2- or 4-inch inside diameter. In cases of deep wells (e.g., >100 ft), Schedule 80 PVC may be used because of its greater strength to withstand the increased pressure. The inside diameter of the casing must be large enough to pass probes and purge/pump/sampling equipment; the use of 2-inch inside diameter casing has become very common with the advent of narrow pumps and sampling equipment. The well screen will usually consist of threaded machine-slotted PVC or wire-wrapped screen. The slot size for the well screen will be compatible with the filter pack and the target zone (see discussion earlier). The bottom of the well screen will be fitted with a threaded end cap. For deep wells, stainless-steel centralizers should be installed above and below the well screen and every 50 ft above the top of screen to ensure that there is proper annular space. After placement of the casing and screen assembly into the borehole, the annular filter pack and annular seal is installed (see Figure 4.7 for an example of a monitoring well construction diagram).

The filter-pack sand will be poured or tremied into the annular space around the well screen. The filter-pack sand is usually placed above the uppermost perforation a minimum distance of 10% of the screened interval to account for potential settlement. For example, if a well has a screened interval of 20 ft, then the filter pack should extend a minimum of 2 ft above the uppermost perforation. The well should be pre-developed before placement of the seal. Pre-development consists of surging or agitating the water column within the well casing to promote settlement of the filter pack and the removal of gaps or bridges in the filter pack. Pre-development should be performed for at least 5 min and continue until the filter pack stops settling. Additional sand may need to be added to the annular space on completion of the pre-development activities. A seal is installed above the filter pack to prevent contaminants at the surface from migrating down into the filter pack and into groundwater. A transition seal comprised of fine sand or bentonite chips/pellets is then poured or tremied into the annular space above the filter sand. The bentonite should be hydrated and allowed to cure in accordance with the manufacturer's specifications. The thickness of the transition seal is typically 2 ft long, but it may be larger depending on the overlying grout column thickness.

A neat cement grout, cement/sand grout, cement/bentonite grout, or high-solids bentonite grout is then placed from the top of the transition seal to the ground surface. The grout seal is usually pumped

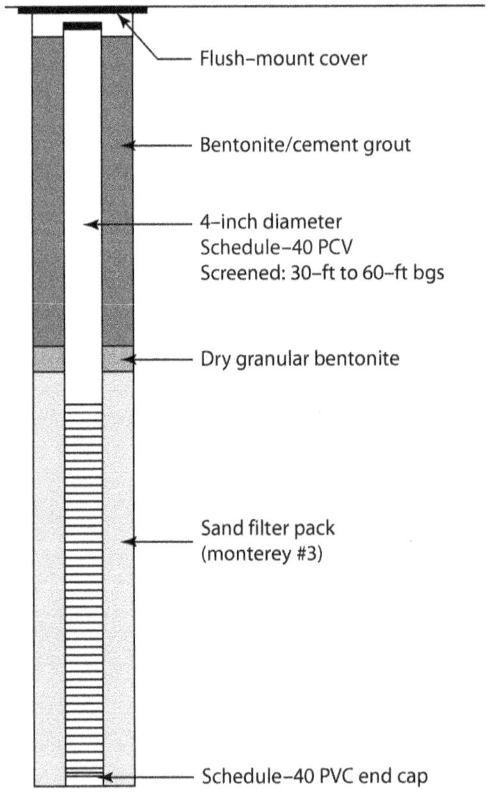

FIGURE 4.7 Monitoring well construction diagram.

or tremied into the annular space. The seal thicknesses and grout/additive/water mixtures are determined based on Department of Water Resources (DWR) Well Standards Bulletin 74-81 and 74-90, site-specific conditions, and local regulatory requirements. Depending on local regulatory requirements, an inspector may observe the mixing and placement of the grout seal. The time period that the grout must set may vary between 48 and 72 hours before the well can be developed. It is recommended that curing times be based on the manufacturer's specifications and that the lead regulatory agency be contacted in advance to determine curing requirements prior to conducting well development activities.

Surface completions can be at or below grade (e.g., traffic-rated vault or Christy box) or above grade (e.g., steel protective stovepipe). The type of surface completion for the well is based on site-specific conditions. The objective for the surface completion design is to reduce the potential for entry of surface water runoff or foreign matter into the well and to secure the well from unauthorized entry. Surface completions often have to make up grout volume at the top of the borehole due to grout shrinkage in the annular space.

All soil cuttings, decontamination water, well development water, and purge water during the soil boring and well installation are typical sources of investigation waste and must be properly containerized and stored, sampled, analyzed for COCs, and disposed of in accordance with regulatory requirements, based on the concentration of COCs in the waste and the receiving facility. Most agencies require documentation of waste disposal (e.g., manifest). Additional information is available online at the U.S. EPA website (USEPA, 2011): www.epa.gov/superfund/policy/remedy/sfremedy/waste.htm

4.2.3.2 Wells Designed to Monitor Multiple Zones

Often, it is desirable to monitor multiple water-bearing zones at one location to profile the vertical distribution of COCs or hydraulic head. The most common way to achieve this is to use a well cluster, where wells of different depths are installed in separate, but closely spaced, boreholes. A second way to achieve this is to install "nested" wells, where several wells of different depths are placed within one large borehole, and the annular fill is emplaced in stages of filter pack and bentonite or grout seal, according to the depths of the well screens. A third way to monitor multiple zones is where one "casing" is placed in one borehole, and the "casing" has multiple sampling ports at different depths. Examples of this type are the continuous multi-channel tubing (CMT) system and the Westbay system. In these cases, the annular fill is installed in stages of filter sand and grout seal according to the depths of the sampling ports.

The use of nested wells or multi-port systems may be controversial and should be approved by the regulatory agency. The pros of these methods are that they can be less expensive for monitoring many target zones than installing separate wells. The cons are that the correct placement of the annular fill takes great care, the grout between the filter sand intervals may shrink and leave voids in the annular space, and the destruction of nested wells is usually more difficult, because multiple casings are in one borehole. Another consideration regarding multi-port systems is that sampling equipment may be limited for these types of monitoring wells.

4.2.3.3 Groundwater Well Development

All the monitoring wells must be developed after installation. The selection of the appropriate well development method depends on site-specific conditions and local regulatory requirements.

The purposes of well development are to

- Remove sediment accumulated in the well during construction
- Remove/mitigate borehole wall damage due to drilling
- Allow water to flow more freely toward the well
- Consolidate the filter pack around the well screen and remove the fines from the filter pack and from the target zone adjacent to the filter pack
- Provide hydraulic connection between target zone and well screen

The development of wells is usually accomplished by an iterative process that involves bailing fines from the well, surging the screen area with a surge block or other device, and over-pumping at a rate exceeding the anticipated purge rate for sampling. Various pumps can be used during well development, including submersible and air-lift. Fines should be bailed out of the well several times during development. The bailing of fines is necessary to remove them from the well and to protect pumps. This iterative process continues until the return water is observably clear. Field parameters such as temperature, specific conductance, and pH provide valuable information, but their stabilization does not indicate the completion of well development. Likewise, there is no set number of casing volumes removed that indicates the completion of well development. The bailing, surging, and pumping process continues until the produced water is visually clear. The development of wells screened across fine-grained soils can be very time consuming because of the low percentage of open area of the required slotted well screen and the low transmissivity of the formation. Field staff need to spend the time required to complete the well development. If the target zone recharges too slowly, potable water can be added to the casing to assist with surging. The amount of added water must be documented, so that the same volume, at a minimum, is pumped out of the well before development is completed.

4.2.3.4 Well Survey

In the state of California, well locations should be surveyed for horizontal and vertical control by a California Registered Civil Engineer or licensed professional surveyor. These coordinates are typically referenced to a United States Geological Survey (USGS) datum. The *x*, *y*, and *z* coordinates for each well at a site are also required to be uploaded to GeoTracker, a California statewide database. The coordinates should also be reported in the site assessment report. It is customary practice for the surveyor (or the well-installation geologist) to place a permanent mark on the top of the casing that indicates the survey point and the point where groundwater elevation measurements are routinely collected.

4.2.4 Groundwater Monitoring

Groundwater monitoring is a critical component of site assessment as well as site remediation for a contaminated site. Effective groundwater monitoring requires consistent methods of gauging, sampling, and sample handling and analysis. Information obtained during groundwater monitoring includes depth to groundwater, thickness of separate-phase hydrocarbons (SPH) ("free product") if present, concentrations of COCs, and measurements of general chemistry parameters.

4.2.4.1 Gauging

Gauging a well is completed prior to sampling to determine the depth to the bottom of the casing, the depth to the water surface, and the thickness of free product (if present). Gauging is performed using electronic devices or a graduated tape treated with water- or hydrocarbon-sensitive paste. Prior to taking a measurement, adequate time should be allowed for the open well to equilibrate with atmospheric pressure. Depths are recorded to the nearest 0.01 foot relative to a surveyed reference point. If free product is present, the measured depth to water is to be corrected using the following formula:

$$\text{DTWc} = \text{DTWm} - (\text{TSPH} \times \text{SGSPH}) \tag{4.12}$$

where:

DTWc is the corrected depth to water
DTWm is the measured depth to water
TSPH is the thickness of separate-phase hydrocarbons (free product)
SGSPH is the specific gravity of separate-phase hydrocarbons

The specific gravities of some fuels are

- *Gasoline*: 0.75 g/cm^3
- *Diesel*: 0.81 g/cm^3
- *Jet Fuel (JP-4)*: 0.79 g/cm^3

4.2.4.2 Hydraulic Gradient

The rate of groundwater movement depends on the hydraulic gradient. The hydraulic gradient is the slope of the water table or potentiometric surface measured relative to two wells along the flow path. It is defined as the change in head per distance along the flow path between two points where heads are measured. The gradient is a dimensionless quantity, measured in feet per foot.

$$i = \frac{dh}{l} = \frac{h_1 - h_2}{l} \tag{4.13}$$

where:
- i is the hydraulic gradient
- dh is the change in head = head difference between points 1 and 2
- l is the distance between two points where heads are measured

The gradient is an important factor in groundwater studies, because by using it, one can determine the rate and direction of groundwater flow. The direction of the flow can be determined by using water elevation data (elevation heads) from a minimum of three wells. A contour map of the water elevation can be prepared, using the same principles as one would use to contour the ground elevations. The contours marking the heights of the water table are called *equipotentials* and are analogous to strike. Everywhere along an equipotential, there exists equal head. Perpendicular to the contours are streamlines, which identify the direction of groundwater movement (Figure 4.8). An exact solution to the location of the water-level contours and the direction of groundwater movement is obtained by treating the water elevations as ground elevations and solving as a three-point problem, as shown in Figure 4.8.

4.2.4.3 Groundwater Sampling Procedures

As with soil sampling, there are many groundwater sampling guidance documents available. "Practical Handbook of Environmental Site Characterization and Ground-Water Monitoring" (Nielsen, 2006) and "Ground-Water Sampling Guidelines for Superfund and RCRA Project Managers" (Yeskis and Zavala, 2002) are two comprehensive references. If measurable free product or sheen is present in a well, it should be recorded on the field forms and in the data summary tables. The well should not be purged and sampled under most circumstances, because the water sample concentrations would

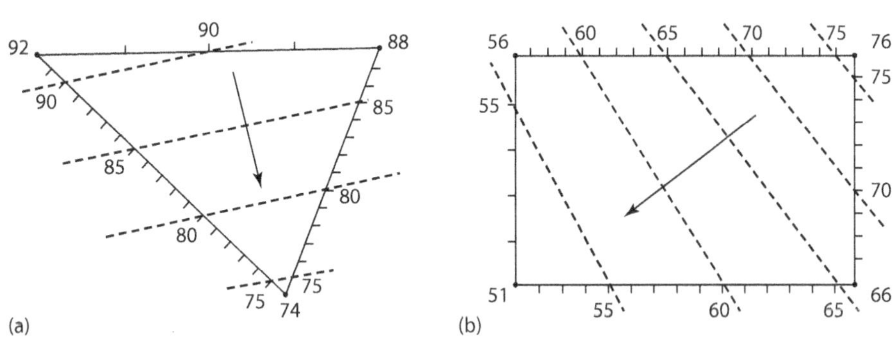

FIGURE 4.8 Graphic method for determining the slope of a potentiometric surface from (a) three wells and (b) four wells.

most likely not be representative of the dissolved phase. The act of purging and sampling typically mixes at least some of the product with the water, thus yielding an unrepresentative water sample. In some cases, it may be specifically desired to sample the water beneath the product, but to do so involves special sampling techniques, and often, there is still a minute amount of product that gets included in the "water" sample. In these cases, special laboratory handling can be used to isolate the water phase of the sample so that the analytical results are as representative of the dissolved phase as possible. In some cases, a sample of product is desired for other site assessment purposes; such a sample would be collected without purging. To minimize the possibility of cross-contamination, purging and sampling proceed from the least contaminated to the most contaminated wells, and non-dedicated purging and sampling equipment is completely decontaminated between monitoring wells. Decontamination requires thorough washing of the equipment with an appropriate cleaning solution, rinsing it twice with clean tap water, and rinsing it a third time with deionized or distilled water.

4.2.4.4 Purging

The overall objective of groundwater sampling is to collect samples that are representative of *in situ* groundwater conditions. The sampling of groundwater monitoring wells has traditionally involved purging as the first step to remove stagnant water in the well casing. Purging involves the removal of sufficient water from a well to ensure that the samples are representative of the groundwater in the impacted zone rather than simply standing water from within the well. Three methods are discussed in the following subsections.

4.2.4.4.1 Conventional Method of Purging

Three or more casing volumes of water is the traditional volume required for purging a monitoring well with a bailer or pump to remove stagnant water above the screened interval prior to collecting samples. Water-quality indicators, including pH, temperature, visual turbidity, and specific conductivity of the extracted water, are monitored throughout the purging process. Purging continues until conditions are stable (i.e., the variance between sequential conductivity measurements is 10% or less). The purged volumes at which measurements are collected, the measured values, the total volume of water removed, any anomalies noted (odor, color, high sediment content, etc.), and the time purging began and ended are reported for each well. It is important that a consistent purging protocol be followed during each sampling event. Changes in sample pumps (flow), depths where purge pumps are placed, volumes removed, and so on can alter the sample results gathered during that particular sampling event.

The casing or boring volume (i.e., water-column volume) in gallons can be calculated using the following industry standard calculations:

$$\text{Height of water column}\left(\text{in feet}\right)=\left(\text{well depth}\right)-\left(\text{depth to water}\right)$$

$$\text{Water-column volume}\left(\text{in gallons}\right)=\left(\text{height of water column}\right)\times\left(\text{volume/foot value of a well diameter}\right)$$

The following are approximate volumes of water in a casing or boring (based on well diameter) used in this water-column volume calculation:

- *1-inch well*: 0.041 gallon/foot
- *2-inch well*: 0.163 gallon/foot
- *3-inch well*: 0.367 gallon/foot
- *4-inch well*: 0.653 gallon/foot
- *Other*: Radius squared × 0.163 to calculate gallons per foot

The casing volume is generally required to be recorded on a groundwater monitoring well purge form or field documentation.

- After purging, a well is generally allowed to recover to at least 80% of the static water level before samples are collected. This should be verified and noted on field documentation.
- Care should be taken not to purge slow-recharge wells dry. If that does happen, water may cascade into the well from unknown intervals, and the samples obtained may not be representative of groundwater conditions. Therefore, the rate of purging and/or the volume of water removed from wells that are historically slow to recharge are adjusted accordingly.

4.2.4.4.2 Low-Flow Purging

Low-flow purging (or micropurging) does not require the removal of large volumes of water. In micropurging, groundwater is pumped at a low flow rate (less than 1 liter per minute or 0.25 gallon per minute) from within the well screen. This technique minimizes the mixing of overlying stagnant casing water and water within the screened interval. In addition, sample turbidity is reduced. Water-quality indicators, including pH, temperature, dissolved oxygen (DO), oxidation/reduction potential (ORP), turbidity, and specific conductivity of the extracted water are monitored throughout the purging process within a flow-through cell (e.g., YSI® Sonde). Purging continues until conditions are stable (i.e., the variance between sequential conductivity measurements is 10% or less). The purged volumes at which measurements are collected, the measured values, the total volume of water removed, any anomalies noted (odor, color, high sediment content, etc.), and the time purging began and ended are reported for each well. It is important that a consistent purging protocol be followed during each sampling event. Changes in flow rates, depths where purge pumps are placed, volumes removed, and so on can alter the sample results gathered during that particular sampling event (Puls and Barcelona, 1996; Kaminski, 2003). This low-flow technique has several advantages over traditional purging:

- Less turbid samples are produced.
- Less volatilization occurs.
- Sampling accuracy and precision are improved.
- Low pumping rates preserve the integrity of the filter pack and well seal and reduce the movement of fine sediments into the well.
- Lower volumes of purge water requiring storage, treatment, or disposal are generated.

4.2.4.4.3 No-Purge Sampling

No-purge sampling is conducted by carefully lowering a bailer to the water table and allowing it to fill with minimal disturbance of the water column, without prior purging of the well (American Petroleum Institute [API], 2000). The method is applicable at sites with unconsolidated, unconfined water-bearing units and only at wells screened across the water table. No-purge sampling should not be used at wells where free product is present. It should be considered for sites where high-precision sampling is not needed, for example, for routine monitoring. It should be supplemented with conventional or low-flow techniques. This method is quick and inexpensive, and may eliminate the variability introduced by purging. In addition, purge-water management and disposal costs are eliminated.

4.2.4.5 Passive Samplers

A type of no-purge groundwater sampling can be performed with passive samplers. These passive samplers (passive diffusion bag samplers [PDBs] are the most commonly used) remain submerged in the monitoring well for a specified time period. The passive sampler type must be matched to the chemicals being monitored.

4.2.4.6 Sample Collection

Water samples are collected using equipment that minimizes the chance of volatile constituents escaping from the sample. Submersible pumps, bladder pumps, peristaltic pumps, low-flow pumps, and bailers (preferably single use) are acceptable. Air-lift pumps may not be used. Any equipment that is not dedicated to a single well must be thoroughly decontaminated before it is used at another well. Samples are transferred from the sampling equipment to the appropriate containers carefully to minimize mixing with ambient air. For samples that will be analyzed for VOCs, the container is filled completely at a rate less than 100 milliliters per minute so that no headspace is present. Fill the sample container until there is a meniscus (the sample liquid rises above the lip of the sample container). Then carefully place the cap on the container and screw it down tightly. The presence of headspace can be checked for by inverting the sealed container, tapping it lightly, and looking for bubbles. If bubbles are present, that sample must be discarded and another sample collected to ensure accurate data. However, if the water or the sediment in the water contains calcium carbonate, bubbles will be generated, even in the absence of headspace. In this latter case, the usual hydrochloric acid (HCl) preservative is not added to the sample containers on subsequent samplings.

Samples are collected in containers and in the quantities appropriate for the requested analyses. All samples are labeled, properly sealed, and preserved according to stated laboratory requirements. Proper chain-of-custody procedures need to be followed, including no time gaps between field sampler(s), courier, laboratory, or other handler(s) of the samples. The appropriate holding times for the sample medium, the analytical method, and the preservative used must be strictly observed.

4.2.5 Conceptual Site Model

The conceptual site model (CSM) is a fundamental element of a comprehensive site investigation. The CSM establishes the source and attributes of the unauthorized release, describes all affected media (including soil, groundwater, and soil vapor as appropriate), describes local geology, hydrogeology, and other physical site characteristics that affect contaminant environmental fate and transport, and identifies all confirmed and potential receptors (including water supply wells, surface-water bodies, structures, and their inhabitants).

A CSM is a representation of the characteristics of the site in diagrammatic or written form that shows the possible and confirmed relationships between the source(s) of contamination, pathways, and receptors. The supporting data and analyses used to develop the CSM can be contained in multiple reports submitted to the regulatory agency over a period of time.

The objectives of a CSM are

- To convey an understanding of the origin, nature, and lateral and vertical extent of contamination.
- To identify potential contaminant fate and transport processes and pathways.
- To identify potential human and environmental receptors that may be impacted by contamination associated with the site.
- To guide site investigation activities and identify additional data needed (if any) to draw reasonable conclusions regarding the source(s), pathways, and receptors.
- To frame the evaluation of risk to human health, safety, and the environment posed by releases at a contamination site.

These objectives emphasize the need for an approach whereby a CSM is developed early and is iteratively refined through the project life cycle. Each piece of data that is collected should serve to refine the CSM. The most recent ITRC Vapor Intrusion Pathway Guideline document (ITRC, 2007b) provides additional information on developing a CSM.

4.2.5.1 Overview of the CSM Process

An initial CSM should be developed prior to any investigative work at a contamination site. The initial CSM begins with a general understanding of the environment in which the site is found and the sources of potential contaminants. From there, the CSM describes the potential ways in which the contaminants may migrate in the environment and the receptors that may ultimately be affected. The intent is to provide a general idea of conditions at the site and indicate what type of additional data, if any, may be needed to determine the degree of risk associated with the site.

To develop an initial CSM, readily available information about the site, the nature and quantity of the release, and preliminary information regarding the migration pathways and potential receptors are gathered. Data may also be collected from available sources concerning surrounding populations, such as ambient water quality, use and approximate locations of nearby wells, subsurface soil conditions, locations of subsurface utilities, climatological conditions, and land use. As additional data are obtained through site investigation, the understanding of contaminant transport at a site is refined, and sometimes revised.

The initial CSM should be used to guide the site-assessment activities. Any part of the CSM with significant uncertainty should be investigated to gain a better understanding and to refine the CSM. For example, the potential migration pathways may need to be investigated to determine whether or not they are complete.

4.2.5.2 Role of CSM in Risk Evaluation and Corrective Action

Once environmental samples have been collected and analytical data are available during site assessment (or other phases of work, such as verification sampling during remediation or confirmation sampling after a remedial action), the data are usually compared with the closure criteria established by the lead agency.

If the information gained during the site assessment and risk evaluation indicates that the site may pose a current or future threat to human health, safety, and/or the environment, remediation or other corrective actions may be performed to mitigate the adverse effects of the release. The CSM is used to support proposed corrective action and guide remediation decisions and to ensure that potential receptors' exposures are appropriately eliminated or mitigated.

4.2.5.3 Updating the CSM

As additional information becomes available during site investigation, or the site conditions change because of remediation, implementation of engineering controls, or other physical changes, the CSM should be re-evaluated and updated to incorporate the new data and any new understanding of the site conditions.

4.2.5.4 Components of the CSM

The following subsections describe the following components of a CSM:

- Hydrogeologic setting
- Source
- Contaminant transport and exposure pathways
- Receptors

4.2.5.5 Hydrogeologic Setting

The hydrogeology (geologic factors that affect groundwater flow) of a site generally controls contaminant migration. Gaining an understanding of the geologic setting will also help to determine the pathways of migration. Much of the geologic information for a contamination site can be gathered from historical reports, state and federal environmental databases (including boring logs obtained from other cases in the near proximity), and electronic and paper files covering the site and adjacent properties from various federal, state, and local agencies. Geologic aspects to consider when conceptualizing the geology at a contamination site include

- Site topography
- Regional and local geologic conditions, including key aquifer and aquitard units
- Site-specific soil texture/lithology (e.g., identify the predominant types of soil at the site, such as clay, sand, gravel, fractured bedrock, sediments, etc.), stratigraphy, and structures (dipping strata, faults, etc.) that may affect contaminant transport

An understanding of the regional hydrogeology is also important in developing the CSM, especially if groundwater could potentially become impacted or is already impacted. Hydrogeologic features to be considered when developing the CSM include

- Depth to the water table and its seasonal and known historical fluctuation
- Groundwater flow within the shallowest aquifer (gradient direction, hydraulic conductivity, flow velocity), vertical gradient, and degree of interconnection between unconfined, semi-confined, and confined groundwater
- Whether or not the source is beneath a low-permeability surface (such as asphalt or concrete)
- Designated beneficial uses of groundwater beneath the site
- Location of proximal supply wells that may influence groundwater flow or be potential receptors
- Location of nearby surface-water bodies (if any) and potential transport pathways to surface-water bodies

4.2.5.6 Source

A *source* is the environmental medium/media containing elevated contaminant concentrations associated with a release. Some risk-based corrective action (RBCA) programs define the source to be the original cause of the contamination; however, it is possible that by the time a site becomes a contamination site, the original source has been eliminated, and the current source of contamination is soil and/or groundwater. Items to consider when determining the source are included in the following list. Some of the specifics may be determined based on historical information; others will need to be determined during site assessment.

- The origin(s) of the release (e.g., a leaking UST, dispenser, product piping, and/or surface spill)
- The number of USTs, the capacity of the tanks (e.g., 12,000 gallons), the products stored, the date of installation, and the removal date(s) (if applicable)
- The location of historical and active USTs, dispensers, and product piping
- Details about the specific release location(s) (e.g., spill locations, and time frame/dates if known)
- The type of fuel released and the COCs associated with the fuel
- The historical use of fuel additives (e.g., methyl tertiary butyl ether [MTBE] or other fuel oxygenates, lead, lead scavengers)
- The media that are impacted (e.g., soil, groundwater)
- Other potential sources, such as surface spills, aboveground storage tank (AST) leakage, or pipeline leakage

The information needed to define the source—to be obtained during the site assessment—includes the following:

- Lateral and vertical extent of
    - LNAPL

    - COCs in unsaturated-zone soil
    - COCs in saturated-zone soil and the smear zone
    - COCs in groundwater
- The distribution of the COCs in the impacted media

After evaluation of the information obtained during site characterization, the extent and magnitude of the contamination can be defined. This is not an exact science; usually, some assumptions will need to be made. In these cases, it is important, from a risk-evaluation perspective, to be conservative.

4.2.5.7 Contaminant Transport and Exposure Pathways

Pathways are the mechanisms by which a receptor may contact the COCs at a site. Exposure pathways consist of (1) a source of contaminants (as described previously), (2) contaminant transport or the physical migration of the contaminants, (3) a point of exposure where the receptor may come into contact with contaminants, and (4) an exposure route (such as ingestion or inhalation).

4.2.5.8 Contaminant Transport

Different COCs have different behaviors, and the same COC behaves differently with various phases in the subsurface. Therefore, an understanding of the fate and transport of each COC is critical for evaluating migration pathways or indirect exposure pathways. Typical migration pathways for contamination sites include

- LNAPL or dense non-aqueous phase liquid (DNAPL) migration from the source area through soil
- Dissolved-phase migration of COCs in the groundwater zone
- Vapor migration of COCs from soil, groundwater, or LNAPL
- Migration of COCs with groundwater and discharging of COCs to surface water

In the surface-water example, the receptors may include ecological receptors as well as human receptors.

4.2.5.9 Points of Exposure

A *point of exposure* is where a receptor comes into contact with contamination. The exposure point may, or may not, be at the same location as the source. Exposure points should include potential future uses of the land, including adjacent land if there is a potential for exposure to off-site receptors (e.g., groundwater containing LNAPL moving downgradient, or volatilization into a future residence). Some examples of points of exposure include

- Surface soil
- Water faucet used for drinking water
- Air inside a residence or commercial/industrial building
- Outdoor (ambient) air (from volatilization from surface soil to air)

For ecological receptors, the exposure point may be surface water or sediment that has been impacted (or could become impacted) from the source.

4.2.5.10 Exposure Route

Exposure routes are the mechanisms by which receptors may come into contact with contamination. Exposure routes at leaking underground fuel tank sites include

- Dermal contact with contaminated soil
- Ingestion of contaminated soil
- Inhalation of outdoor air impacted by volatile emissions
- Ingestion of contaminated groundwater
- Inhalation of vapors (in indoor air at a residence or commercial building) from contaminated soil, groundwater, or LNAPL
- Dermal contact with impacted surface water and/or sediments

During the development of the CSM, each of the elements of a pathway should be considered and investigated as necessary. For example, if groundwater at the site is not potable, and the COCs in groundwater are not expected to migrate and impact a current or future potable water source above established limits, then the groundwater migration pathway may be eliminated.

4.2.5.11 Receptors

A receptor is a human or other living organism with the potential to be exposed to and adversely affected by contaminants as a result of contact with contaminated media either at the source or along a contaminant migration pathway. Potential receptors at leaking underground fuel tank (LUFT) sites may include

- Adults and children in a residential scenario
- Adults in an occupational scenario
- Adults in a construction/utility worker scenario
- Adults and children using groundwater that has been contaminated by a release at the site as a potable water supply
- Aquatic receptors such as fish and benthic invertebrates

"Sensitive" human receptors are not evaluated separately, because the California Environmental Protection Agency (Cal/EPA) and the United States Environmental Protection Agency (EPA) toxicity values used in risk evaluations already consider sensitive subgroups.

Terrestrial ecological receptors may not be a very common type of receptor, considering that LUFT sites are typically small, paved, and located in largely urban and/or otherwise disturbed environments. Significant impacts on ecological receptors are unlikely to occur in most cases. However, if the potential to impact sensitive habitats or nearby surface water exists, these receptors should be included in the CSM. Situations in which potential impacts on ecological receptors may warrant evaluation include cases in which impacted groundwater may migrate and discharge to nearby surface-water bodies and cases in which the LUFT site is located in areas where special-status ecological receptors may reside.

It is important to consider the current and reasonably likely future uses of the site and adjacent properties when identifying receptors. Local zoning and planning agencies can generally assist in these determinations. Determining conditional uses at the contamination site and adjacent properties is important, because changes in use may require consideration of different receptors. For example, a light-industrial park being re-developed for residential living needs to be evaluated for both adults and children who may live on the property.

4.2.5.12 Receptor Identification

The types of potential receptors located on adjacent properties should be identified if they could come onto the site or be exposed to the chemicals at the site. The extent of the area where receptors should be identified will vary based on the exposure pathways, as well as the extent and type of contamination.

To identify whether receptors may be drinking potentially impacted groundwater, a survey of water supply wells near the site may be conducted. This survey is generally based on reviewing Department of Water Resources (DWR) well records and asking local water district and applicable

City and/or County staff if they are aware of any wells within the search radius. Areas with known multiple private wells nearby may require door-to-door contact of local residents to determine their source of water.

Information about water-supply wells can often be obtained from the well owner. Desired information includes

- Current status of the well (operational or idle) and pumping rate
- Purpose of the well, such as drinking water, irrigation, industrial, livestock, and so on
- Well construction details (i.e., the depth and length of the well screen and sand pack interval)

PRACTICE AND EXERCISES

Question 1: A soil sample is collected and taken to the lab. The volume of the sample is 75 cm^3. At the natural water content, the sample weighs 158.88 g. It is then saturated with water and reweighed. The saturated weight is 164.34 g. The sample is gravity drained, and its weight is found to be 147.30 g. Finally, it is oven-dried; the dry weight is 142.89 g. Assume the density of water is 1.00 g/cm^3. Find each of the following:

1. Volumetric water content
2. Gravimetric water content
3. Saturation ratio
4. Porosity
5. Specific yield
6. Dry bulk density
7. Particle density
8. Porosity computed from density

Answer 1:

1. Determine the volumetric water content.
   The mass of water at saturation is the saturated mass minus the dry mass.
   The volume of the water is the mass of the water divided by the density of water.

$$W_w = 158.88 \text{ g} - 142.89 \text{ g} = 15.99 \text{ g}$$

$$V_w = (15.99 \text{ g})/(1 \text{ g/cm}^3) = 15.99 \text{ cm}^3$$

$$\theta_v = V_w/V_v = 15.99 \text{ cm}^3/75 \text{ cm}^3 = 0.2132$$

2. Determine the gravimetric water content under natural conditions.
   The mass of the water is the moist mass minus the dry mass. The gravimetric water content is the ratio of the mass of the water to the dry mass of the soil.

$$\Theta g = 100(W_w/W_s)\% = 100\left[(15.99 \text{ g})/(142.89 \text{ g})\right]\% = 11.19\%$$

3. Determine the saturation ratio.

$$R_s = V_w/V_v$$

Since V_v is equal to $V_{w(\text{saturated})}$

$$R_s = (15.99 \text{ cm}^3)/(21.45 \text{ cm}^3) = 0.75$$

4. Determine the soil porosity.

$$W_{w(\text{saturated})} = 164.34\text{ g} - 142.89\text{ g} = 21.45\text{ g}$$

$$V_{w(\text{saturated})} = (21.45\text{ g})/(1\text{ g/cm}^3) = 21.45\text{ cm}^3$$

Porosity is $100(V_v/V)$. Since $V_v = V_{w(\text{saturated})} = 21.45\text{ cm}^3$, and $V = 75\text{ cm}^3$,

$$n = 100(21.45/75)\% = 28.6\%$$

5. Determine the specific yield.

$$W_g = 164.34\text{ g} - 147.30\text{ g} = 17.04\text{ g}$$

$$V_{gw} = 17.04\text{ g}/(1\text{ g/cm}^3) = 17.04\text{ cm}^3$$

$$S_y = 100\ (V_{gw}/V) = 100\ (17.25\text{ cm}^3)/(75\text{ cm}^3) = 23\%$$

6. Determine the dry bulk density.
   The mass of the soil particles is 142.89 g, which is the oven-dried weight. Therefore,

$$\rho_b = W_s/V = (142.89\text{ g}/(75\text{ cm}^3)) = 1.91\text{ g/cm}^3$$

7. Determine the particle density.
   The volume of the solids is the total volume minus the volume of the voids.

$$V_s = 75\text{ cm}^3 - 21.71\text{ cm}^3 = 53.29\text{ cm}^3$$

$$\rho_m = W_s/V_s = (142.89\text{ g})/(53.29\text{ cm}^3) = 2.68\text{ g/cm}^3$$

8. Determine the porosity computed from density.

$$n = 100\left(1 - \frac{\rho b}{\rho m}\right)\% = 100\left(1 - \frac{1.91\text{ g/cm}^3}{2.68\text{ g/cm}^3}\right)\% = 28.7\%$$

Question 2: A confined aquifer is 25 ft thick. The potentiometric heads were measured in monitoring wells A and B, as depth to water, at 23 and 31 ft. Monitoring wells A and B have the same reference point at 175 ft above mean sea level (AMSL). The two wells are along the same flow line, and the distance between these two wells is 285 ft. The hydraulic conductivity is 356 ft/day, and effective porosity is 0.23.

1. What is the hydraulic gradient between these two wells? Which well is upgradient, and which well is downgradient?
2. How many cubic feet per day are moving through a strip of the aquifer that is 20 ft wide?
3. What is the average linear velocity (seepage velocity)?

Answer 2:

1. $H_1 = 175\text{ ft} - 23\text{ ft} = 152\text{ ft amsl}$

   $H_2 = 175\text{ ft} - 31\text{ ft} = 144\text{ ft amsl}$

   $dh = 152 - 144 = 8\text{ ft}$

   $i = dh/dl = 8\text{ ft}/285\text{ ft} = 0.028$

   A–upgradient; B–downgradient

2. $A = b \times w = 25\text{ ft} \times 20\text{ ft} = 500\text{ ft}^2$

   $Q = -KiA = 356\ \text{ft/day} \times 7.48\ \text{gallon/ft}^3 \times 0.028 \times 500\ \text{ft}^2 = 37{,}280\ gpd = 4984\ \text{ft}^3/\text{day}$

3. $V = K\,i/n = 356\ \text{ft/day} \times 0.028/0.23 = 43.34\ \text{ft/day}$

Question 3: A well in a confined aquifer was pumped at a rate of 220 gal/min for 500 min. The aquifer is 48 ft thick. Time-drawdown data from an observation well located 824 ft away are given in the following table. Find the values for transmissivity (T), hydraulic conductivity (K), and storativity (S) of the tested aquifer.

Time after pumping started (min)	Drawdown (ft)
3	0.3
5	0.7
8	1.3
12	2.1
20	3.2
24	3.6
30	4.1
38	4.7
47	5.1
50	5.3
60	5.7
70	6.1
80	6.3
90	6.7
10	7.0
130	7.5
160	8.3
200	8.5
260	9.2
320	9.7
380	10.2
500	10.9

Answer 3:

The data are plotted in the following figure. The drawdown per log cycle is 5.5 ft, and t_0 is 5.2 min. Q is 220 gpm.

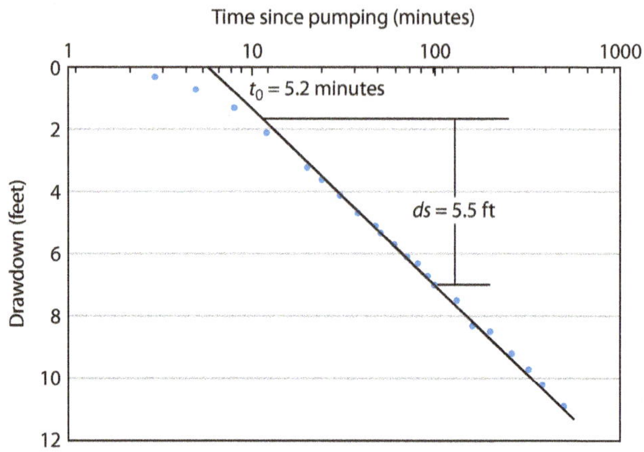

$$T = 35Q/ds = 35 \times 220/5.5 = 1400 \text{ ft}^2/\text{day}$$

$$K = T/b = 1400 \text{ ft}^2/\text{day}/48 \text{ ft} = 29.17 \text{ ft/day}$$

$$S = T \cdot t_o/640r^2 = \left(1400 \text{ ft}^2/\text{day} \times 5.2 \text{ min}\right)/\left(640 \times 824 \times 824\right) = 0.000017$$

Question 4: A well pumping at 77,000 ft³/day has observation wells located 10, 40, 150, and 400 ft away. After 0.14 day of pumping, the following drawdowns were observed:

Distance (ft)	Drawdown (ft)
10	15.1
40	9.4
150	4.4
300	1.7
400	0.25

Find the values for transmissivity (T) and storativity (S) of the tested aquifer.

Answer 4:

The data are plotted in the following figure. The drawdown per log cycle is 8.8 ft, and r_0 is 460 ft. Q is 77,000 ft³/day, which is 400 gpm (1 gpm = 192.5 ft³/day).

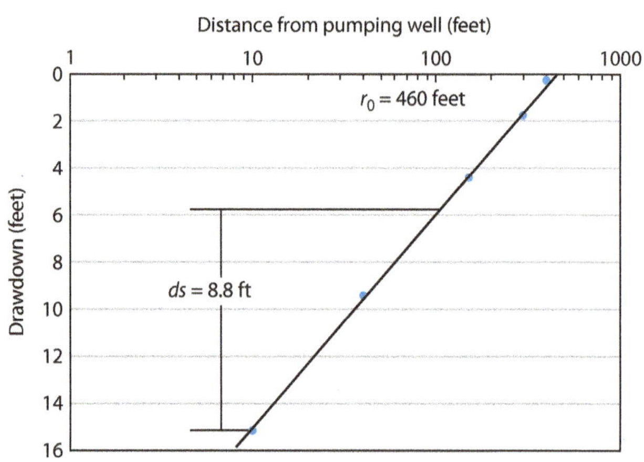

$$T = 70Q/ds = (70 \times 400 \text{ gpm})/8.8 = 3182 \text{ ft}^2/\text{day}$$

$$S = \frac{T\ t_e}{640 r_e^2} = \left(3182 \text{ft}^2/\text{day} \times 0.14 \text{ day} \times 1440 \text{min/day}\right) / \left(640 \times 460 \times 460\right)$$

$$= 0.00474$$

REFERENCES

American Society for Testing and Materials (A.S.T.M.), 1989, *Standard Test Method for Classification of Soils for Engineering Purposes*, D2487-85. Annual Book of A.S.T.M. Standards, Vol. 4.08, Philadelphia, PA, pp. 288–297.

American Petroleum Institute (API), 2000, No-Purge Groundwater Sampling, an Approach for Long-Term Monitoring, Newell, C.J., R.S. Lee, and A.H. Spexet, Groundwater Services, Inc. October.

Casagrande, A., 1948, Classification and Identification of Soils, *Transactions of American Society of Civil Engineers*, Vol. 113, pp. 901–992.

Darcy, H. P. G., 1856, *Les fontaines publiques de la ville de Dijon*. Paris: Victor Dalmont.

ITRC, 2004, *Technical and Regulatory Guidance for Using Polyethylene Diffusion Bag Samplers to Monitor Volatile Organic Compounds in Groundwater*. February.

ITRC, 2007a, *Protocol for Use of Five Passive Samplers to Sample for a Variety of Contaminants in Groundwater*. February.

ITRC, 2007b, *Vapor Intrusion Pathway: A Practical Guide*. VI-1. Washington, DC: Interstate Technology & Regulatory Council, Vapor Intrusion Team January.Jury, W. A., W. R. Gardner, and W. H. Gardner, 1991, *Soil Physics*. Wiley: New York, NY.

Kaminski, D., 2003, Go with the Flow. Low-flow sampling techniques improve sample quality and reduce monitoring program costs, *Water Well Journal*, June, pp. 24–28.

Meinzer, O. E., 1942, *Hydrology*, Dover, New York, NY.

McCarthy, D. F., 1993, *Essentials of Soil Mechanics and Foundations*, 4th edn, Regents/Prentice Hall: Englewood Cliffs, NJ, p. 618.

Morris, D. A. and A. I. Johnson, 1967, *Summary of Hydraulic and Physical Properties of Rock and Soil Materials, as analyzed by the Hydrologic Laboratories of the U.S. Geological Survey, 1948-60*, Geological Survey Water-Supply Paper 1839-D.

Nielsen, D. M., 2006, *Practical Handbook of Environmental Site Characterization and Ground-Water Monitoring*, 2nd edn, CRC Press, Taylor & Francis Group, Boca Raton, FL.

Puls, R. W. and M. J. Barcelona, 1996, Low-Flow (Minimal Drawdown) Ground-Water Sampling Procedures. EPA Ground Water Issue. EPA/540/S-95/504. April.

Yeskis, D. and Zavala, B. (2002), Ground-Water Sampling Guidance for Superfund and RCRA Project Managers. EPA Ground Water Forum Issue Paper. EPA/540/S-02/001, May.

5 Petroleum Hydrocarbon Environmental Forensics and Remedial Site Investigation

Jun Lu and Michael J. Wade

CONTENTS

5.1 INTRODUCTION

Environmental forensics is a relatively recent term for what has been classic environmental science studying contaminant releases. Almost all of the common definitions of the word *forensics* involve entry into some type of formalized debate that leads ultimately to some type of legal setting. One author defined environmental forensics as "the systematic examination of environmental information used in litigation." Subsequently, Sullivan et al. (2001) in the preface to their book on practical environmental forensics reference the legal destination (i.e., the courtroom) where the forensic process leads. However, in their book *Introduction to Environmental Forensics*, Murphy and Morrison (2002) refer to the fact that the word *forensic* is related to the Latin word *forum* and "refers to any public discussion or debate." They state that the term *forensic* is often related in the United States to some type of legal proceeding. By the time their contaminant-specific guide (Morrison and Murphy 2006) was published, the authors recognized that the term *environmental forensics* had evolved into a process whereby the source and age of a contaminant release could be determined for both scientific and legal proceedings. The further evolution of the term *environmental forensics* now encapsulates the process of evaluating physical, chemical, biological, and geological evidence to elucidate the source and age of multiple environmental contaminants. The *Introduction to Environmental Forensics* has since been expanded and updated (Morrison and Murphy 2015).

Environmental forensics is evolving not only on the technical front but also in the field of applications. Through its history, the focus has been to resolve liability issues. Due to strict environmental regulations, the polluted environment needs to be cleaned up. The paradigm of "polluters pay" underscores the importance of environmental forensics. At sites where contaminant sources are disputed, the sources need to be identified and remedial costs allocated among multiple potential responsible parties (PRPs). Investigations must be broadly based. Following only one limited single line of evidence will not suffice; multiple lines of evidence must be established, all leading to the same conclusions regarding source responsibility and cost allocation.

Recently, practitioners in the field of conventional remedial site investigation realized that the science of environmental forensics is fundamentally about understanding contaminant sources, which is one of the critical components of conceptual site models (CSMs) for site characterization. At a contaminated site, contaminants originate at primary sources and migrate through geological media (i.e., pathways). The footprints of contaminants on the pathways serve as secondary sources, which control the distribution of currently impacted areas. Because the contaminant sources are the starting points of the migration pathways, by better understanding contaminant sources, we can do a better job of predicting and delineating currently impacted areas.

In this chapter, we focus on petroleum hydrocarbon environmental forensics, because it happens that today's society is hydrocarbon based, relying on the carbon deposits laid down eons ago and now being recovered and put to a variety of uses. As with any process, nothing is 100% efficient, and current petroleum engineering produces residual materials with concomitant environmental impacts. Further, petroleum products can never be 100% controlled by humans, and releases occur from a wide variety of sources, including crude oil prospecting and resource development, bulk transportation, refining and refinery storage, product transportation and storage, as well as daily product use in a wide variety of circumstances.

Due to the expansive nature of the subject, a substantial amount of technical terminology occurs throughout the chapter. It is assumed that readers of this chapter work in the environmental field or are familiar with site investigation. We have included extensive technical reference documentation of the various subjects using peer-reviewed literature. As such, readers are encouraged to review the references cited to clarify technical details and to begin to enter the various detailed aspects of environmental forensic geochemistry.

5.2 PETROLEUM HYDROCARBON ENVIRONMENTAL FORENSIC TECHNIQUES

5.2.1 Introduction

Rarely is the investigator standing at the exact spot of an initial petroleum release when it occurs. That is not to say that the timing and extent of releases from a bulk crude oil carrier running aground near a populated area cannot be well documented. They can be and have been (Smith 1968). However, in terms of numbers, dramatic catastrophic occurrences are the exception rather than the rule. Much more commonly, petroleum release or releases occur quietly with no fanfare and certainly no witnesses. In fact, it can be many weeks, months, or years before the release or releases are discovered. Once they are discovered, however, the tasks begin of determining specifically what material was released, when it was first released, and the extent of the release(s). Much work on the source identification and age dating of petroleum releases of various petroleum products has been published in the peer-reviewed literature to date. We will provide a summary of the available techniques along with an assessment of the benefits and risks of applying these techniques.

5.2.2 Techniques Based on Conventional Site Investigation Data

5.2.2.1 Site History

Historical knowledge of site operations is critical to forensic evaluation, because it provides a timeline of the on-site handling, use, and storage of contaminants of concern. Most industrial sites have experienced either real estate transfers or environmental audits, and so a Phase I Environmental Site Assessment (ESI), Transaction Screen, or similar reports may be available. A Phase I ESI provides a good source of existing historical information. Where a site investigation has been or is being conducted, release history is further explored and typically included as part of the investigative report. It should be noted that for sites where remediation has been or is being conducted, the remediation activities likely altered the nature, concentration, and/or distribution of contaminants, which create complications for understanding the sources.

5.2.2.2 Preferential Pathway Analysis

When petroleum releases occur, as a light non-aqueous phase liquid (LNAPL) it flows downward through the vadose zone due to gravity and capillary forces. Its downward paths preferentially follow the geological media with higher permeability. At contaminated sites, geological media can be broadly categorized into two groups: sedimentary rocks and fractured bedrocks. The characteristics of sedimentary rocks are determined primarily by the processes that eroded and transported the sediment from its source area and by the setting and manner in which the sediments were deposited (the depositional environment), while the successful prediction of fluid flow in the subsurface in fractured rocks requires an understanding of fracture orientations, density, magnitudes, permeability, and degree of interconnectivity. At sites where locations of original releases cannot be easily identified, geological pathways provide important information that may help to infer or identify the source areas (Lu 2015).

5.2.2.3 Groundwater Dynamics Analysis

Groundwater dynamics refers to the process in which the groundwater hydrologic characteristics (elevation, flow direction, and hydraulic gradient) change in response to local and regional hydrogeological settings. In forensic evaluation involving contaminants in groundwater, groundwater flow direction and seepage velocity are two of the key elements of hydrogeological information used to determine the source. The groundwater flow direction is typically inferred from groundwater level contours. As contaminants migrate in the direction of groundwater flow, the source of contaminants is typically searched on-site and then traced upgradient off-site.

Groundwater seepage velocity is a function of saturated geological media (i.e., hydraulic conductivity and effective porosity) and groundwater gradient, which is determined by water levels. Among the variables, hydraulic conductivity is one of the most important and yet most uncertain aquifer properties. In forensic terms, the migration of contaminants may be interpreted as being "slow" or "fast," which could lead to the conclusion that there was an "older" or "more recent" release, and in turn, different source(s).

5.2.2.4 Spatial and Temporal Analysis of Contaminant Concentration in Groundwater

Laboratory chemical analyses are probably the most important data derived from site investigation for understanding the nature and extent of contaminants. In investigative reports, these data are commonly presented as iso-concentration maps (i.e., spatial patterns) and graphs of temporal trends of a specific contaminant to evaluate change in the contaminant plume. At sites with multiple contaminants of concern, spatial patterns may be used to infer the direction of original releases/sources relative to the plumes. For example, at sites impacted by gasoline, plume contours of benzene, toluene, ethylbenzene, and xylenes (BTEX) and oxygenates are often available. As methyl tertiary-butyl ether (MtBE) is much more soluble and has much lower retardation in soils, it tends to migrate ahead of BTEX constituents (Wilson 2003). If MtBE and BTEX are both from a gasoline source, the source area will likely be in the direction of the BTEX area or further upgradient.

It should be noted that in iso-concentration maps, the highest concentrations may or may not correspond to original releases, depending on whether there is a continuous source or a spill, among other complications. A continuous source will yield a plume, whereas a spill will yield a slug that grows with time as it moves down the groundwater flow path (Fetter 1994).

In a site investigation, the temporal trends are often presented in the investigation reports. At a monitoring well, the temporal trend of contaminants is typically presented as a graph of concentrations versus time. At a site where there is no LNAPL source present, it is expected that the concentration will decrease over time. However, it is not uncommon for the concentration to fluctuate due to factors such as water level rise and fall, a slight shift of the center line of the plume, or migration of free product toward the well. Additionally, when abnormally high concentrations occur, the occurrence should be flagged, and additional data should be collected to confirm or rule out other source(s) (Lu 2015).

5.2.2.5 Phase Equilibrium Analysis

Concentration data from the site investigation can also be used to detect the presence of non-aqueous phase liquids (NAPLs). The general rule of thumb is that chemical concentrations in the percent range or >10,000 mg/kg in soil would be considered to indicate the presence of NAPL in the vadose zone, and chemical concentrations greater than 1% of their combined effective solubility are generally considered to indicate the presence of NAPL in the saturated zone.

To obtain a more reliable estimation, quantitative analysis should be used. Petroleum hydrocarbon contaminants in the subsurface can be found in up to four different phases: dissolved in water, sorbed onto matrix particles, volatized in soil vapors, or as NAPLs. Assuming that no NAPL is present, there would be a hypothetical maximum mass of chemical which can be contained in a sample of soil, pore water, and soil gas (Feenstra et al. 1991). If the measured concentration exceeds the maximum equilibrium concentration, free NAPL is likely to be present in the media. For data collection, in addition to chemical concentrations, site-specific soil information such as soil moisture content, porosity, sorption parameters, and physical and chemical properties of the chemicals of interest shall be collected and analyzed. The rationales and limitations of this approach are discussed by many authors, including Cline et al. (1991), Cohen and Mercer (1993), and Feenstra et al (1991).

5.2.2.6 Mass Flux and Mass Discharge Analysis

Mass flux and discharge studies are commonly conducted in sites under investigation to understand the strength of a source or solute plume to facilitate the evaluation of natural attenuation and the assessment of risks posed by contamination to downgradient receptors (ITRC 2010). From a forensic perspective, the data from such studies are valuable in identifying potential additional source(s). For example, one of the ways to measure mass flux is the transect method, in which individual monitoring points are used to integrate concentration and flow data. Assuming that there is only one source at the site, the mass discharge from the transect near the source will be equal to or higher than that farther away from the source. A significantly higher discharge encountered in the downgradient direction indicates that additional mass is introduced into the plume, suggesting the presence of another potential source.

5.2.2.7 Contaminant Fate and Transport Modeling

In some sites, numeric models for the saturated zone (i.e., groundwater) are available. There are two types of models: one is the groundwater flow model, and the other is the groundwater contaminant fate and transport model. In forensic evaluation, fate and transport modeling would be preferred; however, the travel time of contaminants estimated from the flow model based on particle tracking is also valuable when the fate and transport model is not conducted or is unavailable.

Unsaturated zone models are used for assessing vadose zone contaminant loading to groundwater. Because the groundwater of concern may be contaminated by releases locally from above or groundwater from upgradient, understanding contaminant fate and transport in the vadose zone is important for source identification. Depending on the contaminant of concern, its release history, and site conditions, the contaminant of concern in groundwater may be from soils above or completely depleted from the vadose zone over time, and therefore, the vadose zone may no longer be a source of the contaminant of concern to the groundwater in the uppermost aquifer. Available models include SESOIL, VLEACH, HELP, and so on.

5.2.3 Petroleum Hydrocarbon Chemical Fingerprinting

Focusing on the nature of the release(s), the prime analytical method for characterization of petroleum hydrocarbon distributions (i.e., chemical fingerprinting) is gas chromatography (GC). Since its introduction, the basic hydrocarbon fingerprinting techniques using GC methods have gone through an extensive evolutionary process (Wade and Stainken 2016), to the point that now, a substantial amount of information comes to the investigator using the various chromatographic techniques. Many different detectors can be interfaced with the GC, with the two most commonly employed in hydrocarbon studies being gas chromatography using a flame ionization detector (GC/FID) and combined gas chromatography interfaced with a mass selective detector of some type (GC/MS). In fact, combining the GC/FID and GC/MS analytical approaches into a Phase I, Phase II approach offers the investigator a technically defensible approach that is potentially very efficient in terms of time and cost to elucidate the type of petroleum hydrocarbon contamination as well as assembling the more detailed mass spectrometry–based hydrocarbon data for detailed source assessment and age dating. This section discusses the basic types of petroleum hydrocarbon product pattern recognition, the use of hydrocarbon diagnostic ratios or compounds, and the use of biomarker hydrocarbons to evaluate product stability and original source(s).

5.2.3.1 Pattern Recognition

Petroleum products are formulated according to boiling point specifically for the end use for which the product is intended. There is no other reason to formulate a petroleum product than to facilitate its use. Products that do not function as intended will not last long in the marketplace, and neither will their manufacturers. Accordingly, much emphasis has been placed on the development of standard product specifications that are used the world over. Each country may have its own set of specifications, but as the petroleum market is worldwide, so too are the specifications that govern product formulation. The fundamental basis for breaking crude oil down into its desired components is distillation, as refining improvements have evolved with time (Speight 1980, 1991,

1999, 2007, 2013; Meyers 1986, 1997, 2004). As such, manufactured petroleum-derived consumer products are formulated based on boiling point distributions (Figure 5.1). The resulting specifications, therefore, have nothing to do with the gas chromatographic hydrocarbon distribution of such products; only physical aspects such as initial boiling point, final boiling point, specific gravity, distillation behavior, and so on are specified in the standardized product specifications.

However, the forensic investigator can take advantage of all the standardization that exists in the industry by GC/FID and GC/MS analysis of samples alone. Accordingly, light petroleum hydrocarbons such as the various automobile gasolines from various manufacturers can be expected to have highly similar, but certainly not identical, GC/FID gas chromatograms (Figures 5.2 through 5.4). In a similar manner, lighter distillate petroleum products such as kerosene and various diesel fuels all have common GC/FID chromatographic traits (Figures 5.5 through 5.7), but certainly, individual product chromatograms will vary from manufacturer to manufacturer. And completing the end of the distillation curve shown in Figure 5.1, the crude oil, heavy gas oil, and lubricating oil will have general GC/FID chromatographic traits based on boiling points also, but their chromatograms will also vary from manufacturer to manufacturer (Figures 5.8 through 5.10). It is within this product variability that the utility of the GC/FID analysis technique proves itself over and over again. When it is coupled with quantitative analysis by GC/MS for individual petroleum hydrocarbons, various types of information about the composition of each type of petroleum product (Figure 5.1) can be obtained. Further, knowing what products have been involved in petroleum release(s) and what products have *not* been involved in the same release(s) serves as an initial aid to focusing the investigation on likely source(s) and PRPs.

One illustration can serve to illustrate a number of points regarding source assessment using GC/FID analyses. Figure 5.11 shows the GC/FID chromatogram of an LNAPL identified in a site well. Fingerprint matches with the LNAPL sample showed that the product was not one product but two; a leaded gasoline and a kersone. Further comparison of similar chromatograms collected in the surrounding unsaturated soils showed that some soil samples matched the pattern in Figure 5.11 and some did not. In this manner, the footprints of the two petroleum products were tracked back to their original separate sources. Further, degradation of the original chromatographic signature can be followed using changes in the normal alkane (n-alkane) distribution of each sample to validate product movement and direction.

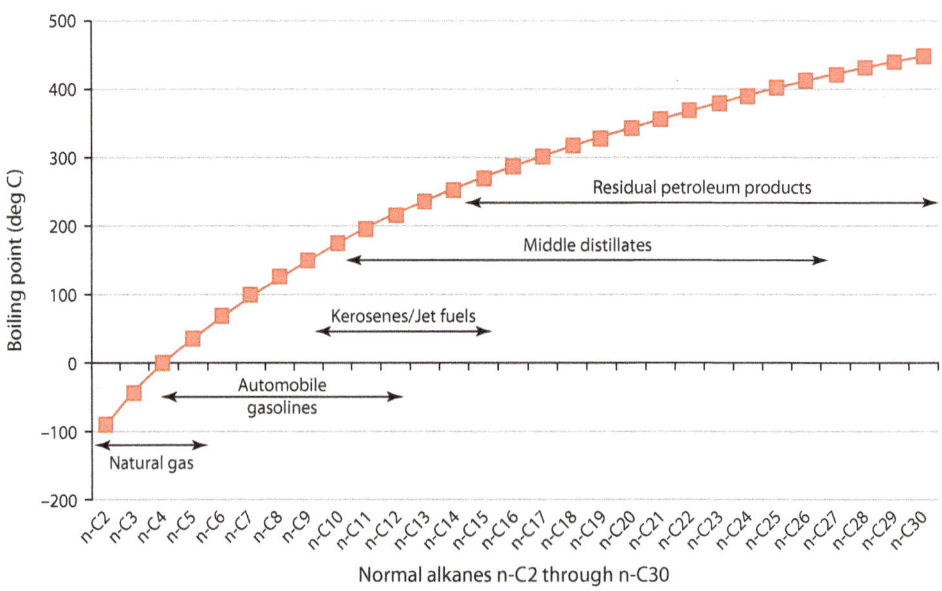

FIGURE 5.1 Boiling point distributions of general petroleum product ranges over the n-C2 to n-C30 interval.

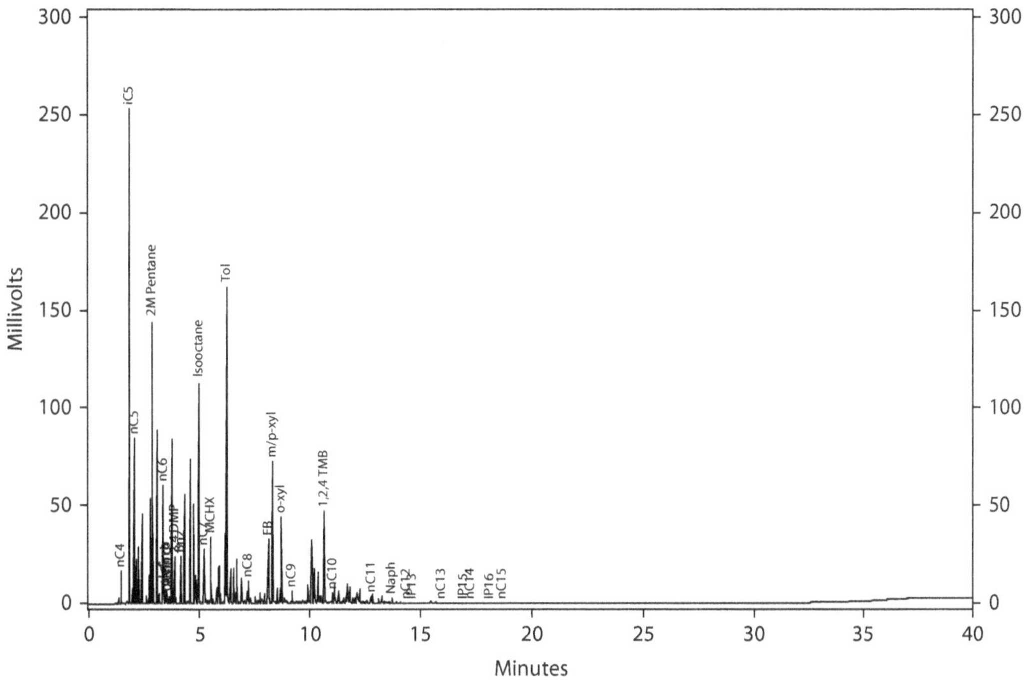

FIGURE 5.2 California Air Resource Board Phase 3 regular gasoline.

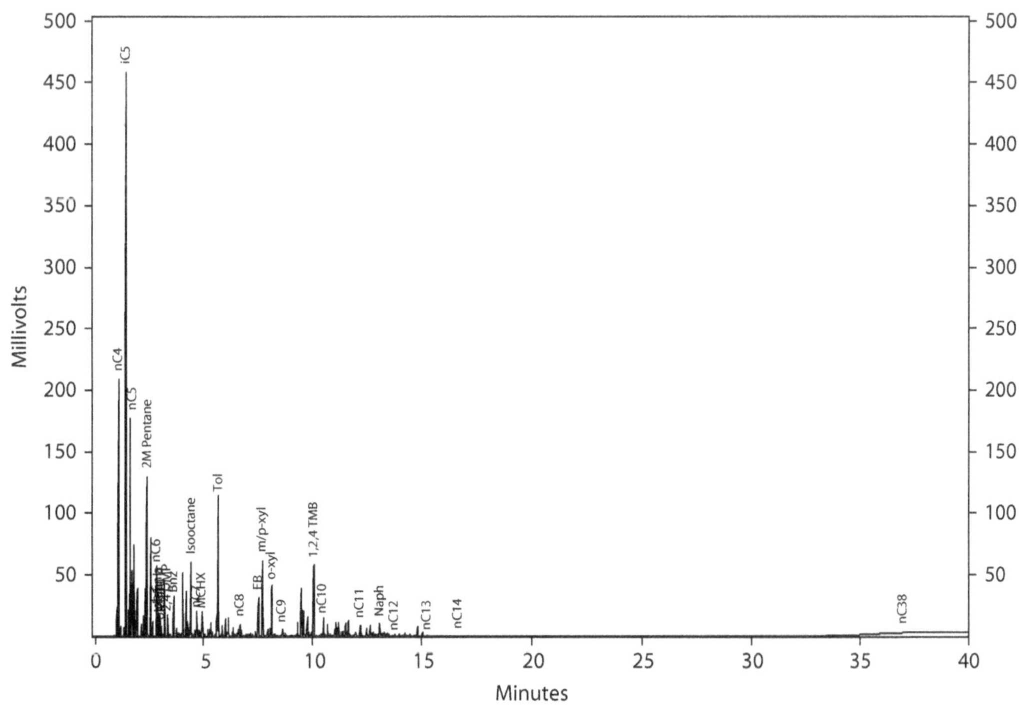

FIGURE 5.3 A type of regular gasoline.v

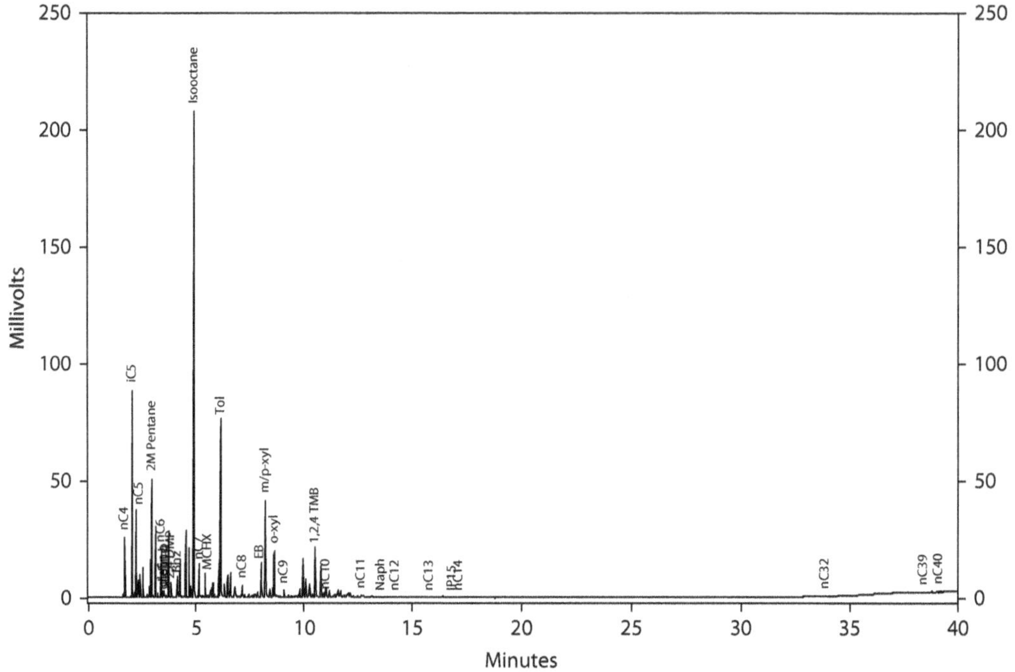

FIGURE 5.4 A type of high premium gasoline. Note the high concentration of isooctane.

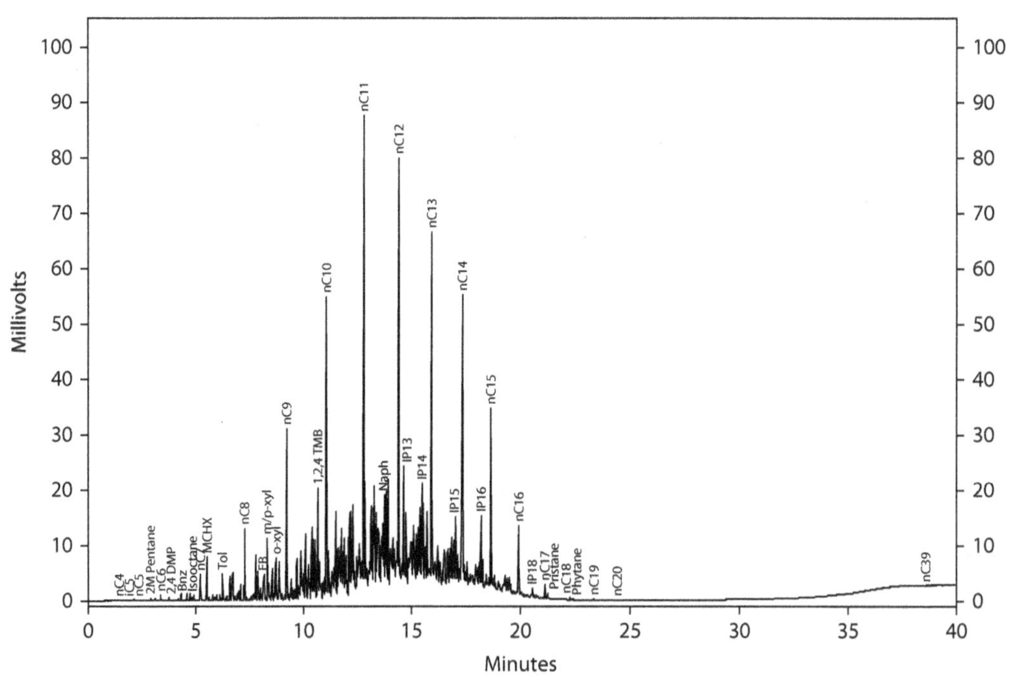

FIGURE 5.5 A type of kerosene fuel (Jet A).

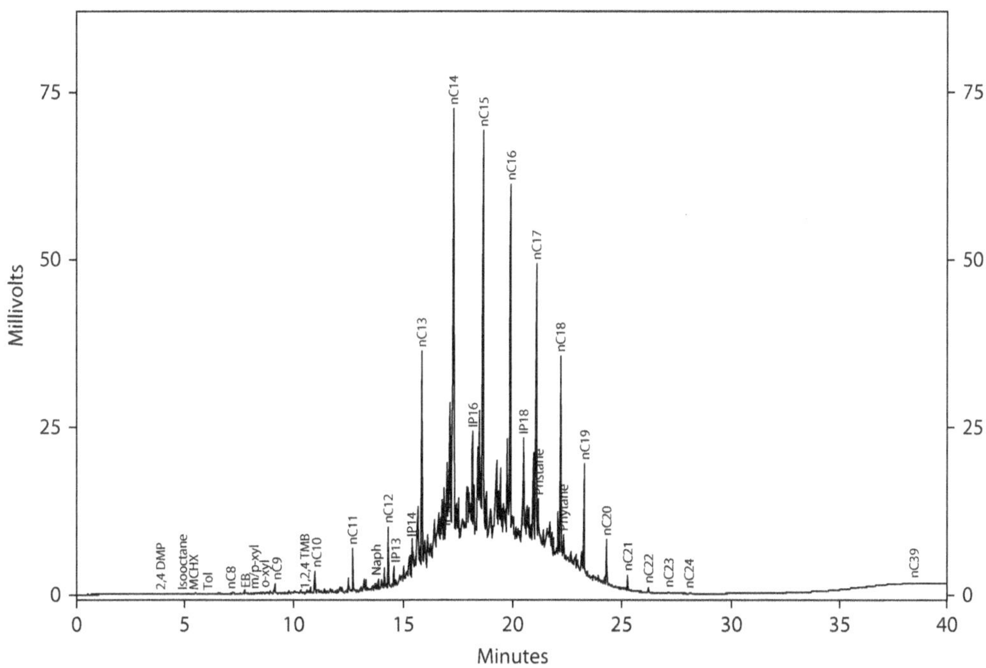

FIGURE 5.6 A type of lighter diesel fuel.

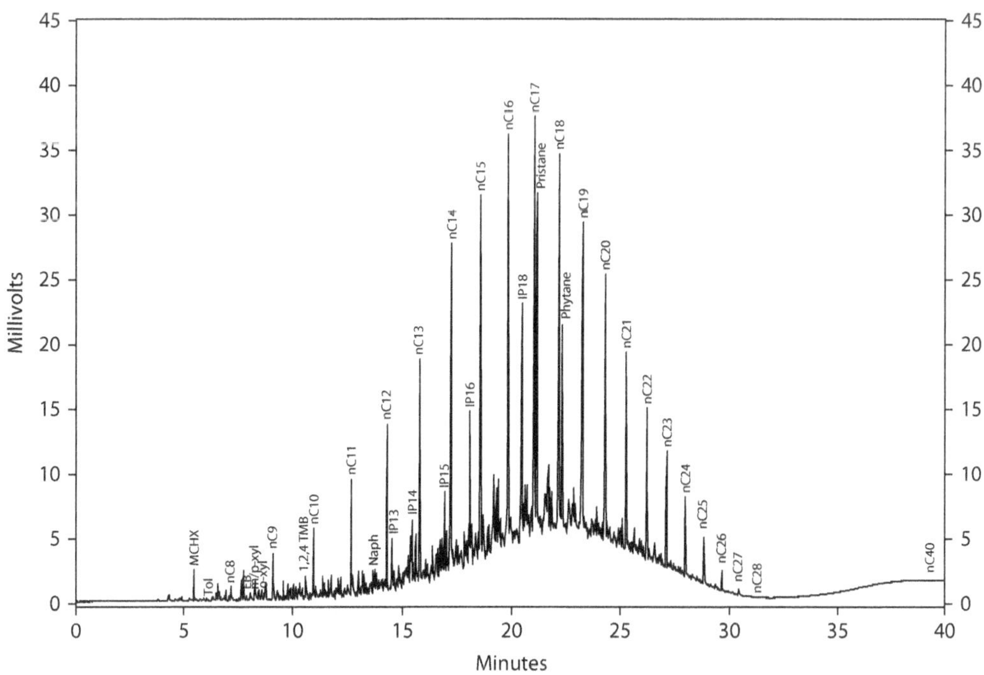

FIGURE 5.7 A type of heavier diesel fuel.

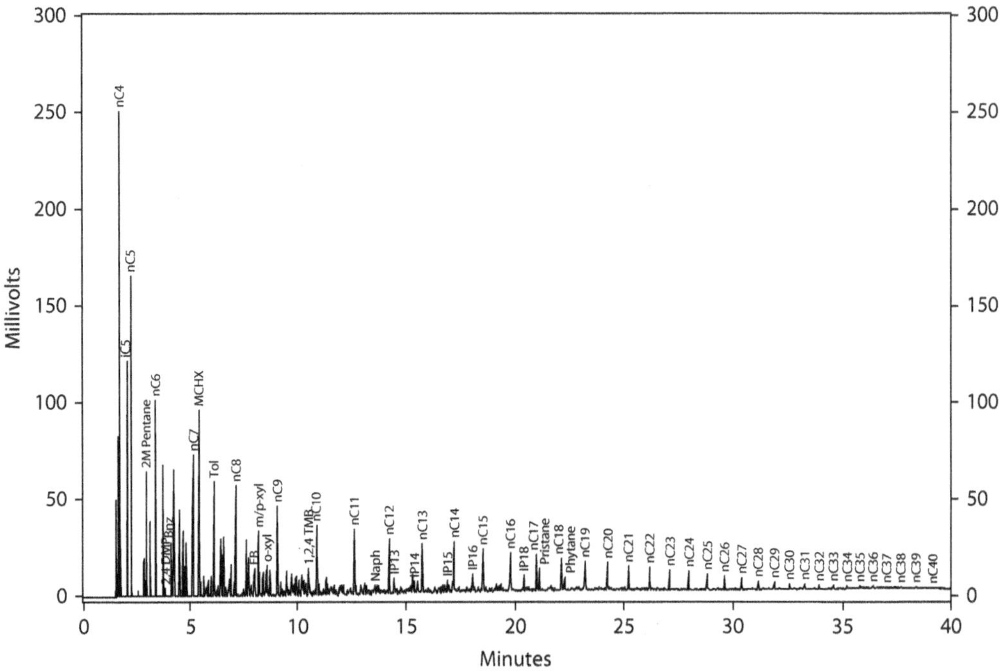

FIGURE 5.8 A type of California crude oil.

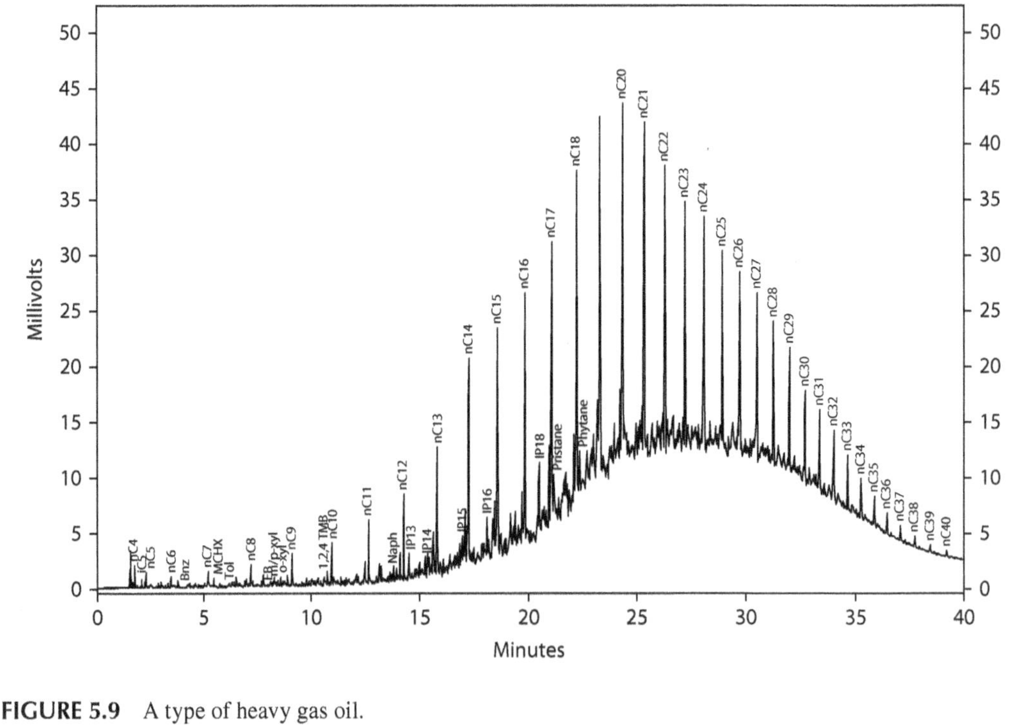

FIGURE 5.9 A type of heavy gas oil.

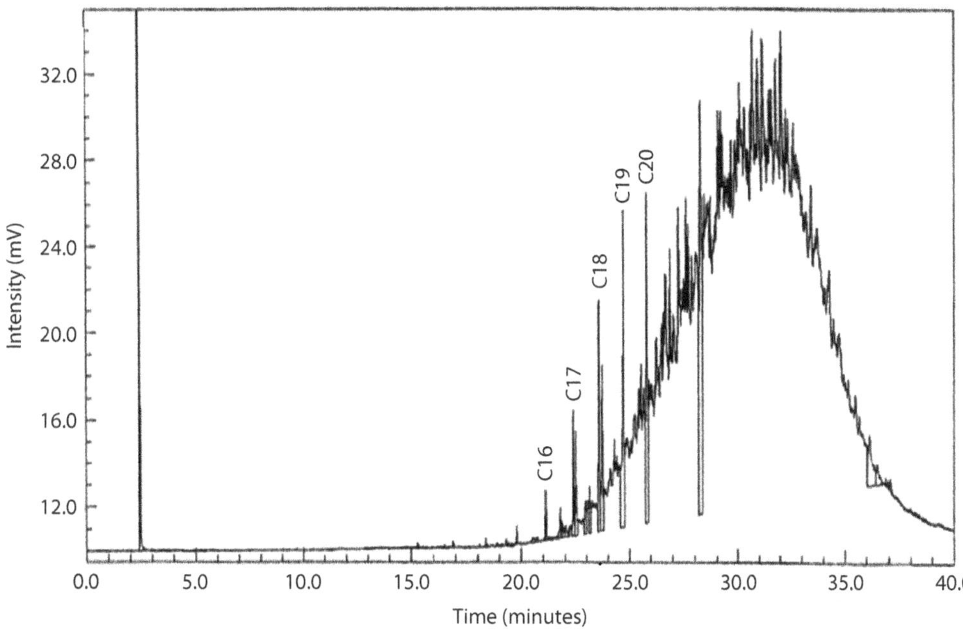

FIGURE 5.10 A type of lubricating oil.

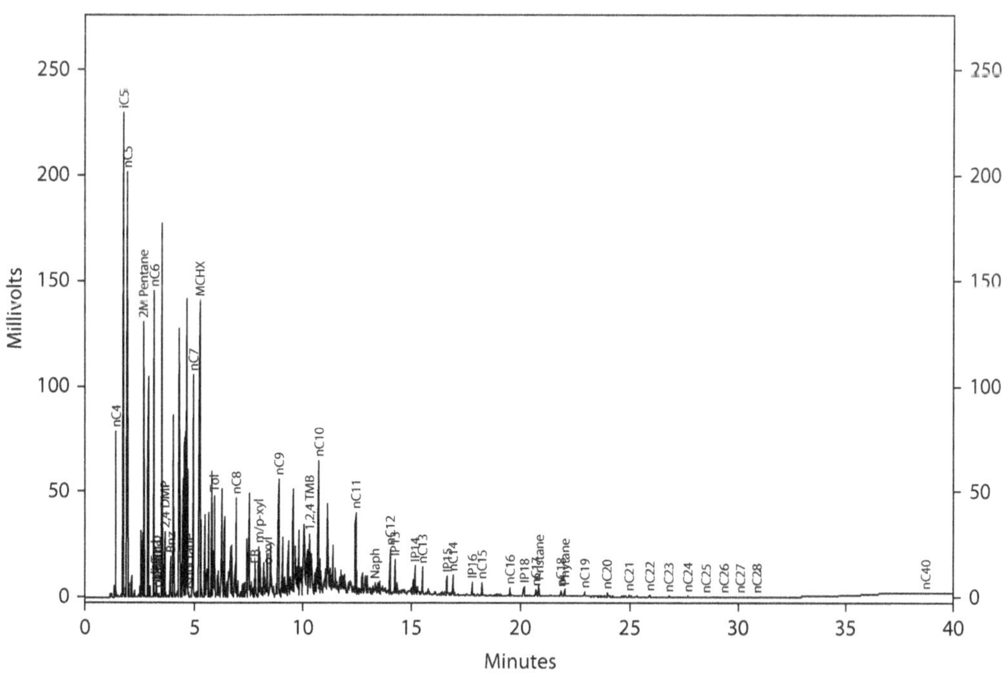

FIGURE 5.11 A product recovered from a monitoring well (a mixture of leaded gasoline product and kerosene product).

A modification of Environmental Protection Agency (EPA) Standard Method 8260B can be employed to support fingerprinting efforts in gasoline forensics investigations, age dating calculations, and evaluation of gasoline weathering. While the exact number of target analytes varies depending on the analytical laboratory, the method usually provides quantified data for 140–180 commonly identified volatile hydrocarbons. The analytes are representative of one of five gasoline compound classes: paraffins, isoparaffins, aromatics, naphthenes, and olefins (PIANO). Analysis of volatile gasoline range hydrocarbons is by purge-and-trap gas chromatography and mass spectrometry (GC/MS) operated in a full scan mode (Uhler et al. 2003; Stout et al. 2002; Douglas et al. 2007, 2015). Additional target compounds can be added if desired. These include contemporary gasoline oxygenate compounds such as tert-butyl alcohol (tBA), MtBE, di-isopropyl ether (DIPE), ethyl tert-butyl ether (EtBE), and tert-amyl methyl ether (tAME). Other commonly identified compounds include the lead scavengers 1,2-dichloroethane and 1,2-dibromoethane, the gasoline additive methylcyclopentadienyl manganese tricarbonyl (MMT), and volatile sulfur-containing hydrocarbons. Target compound concentrations are normally calculated using external five-point calibration data and reported in micrograms per liter for water samples, micrograms per kilogram dry weight for soil, and milligrams per kilogram for separate phase product or NAPL samples. Chromatograms should be provided for all samples. Also, separate method blank (MB), duplicate (SD), lab control (LCS), and lab control duplicate (LCSD) samples in each processed batch should be prepared and analyzed for quality control purposes.

Make no mistake about it; the instant a petroleum product is released to the environment, its chemical composition will begin to change. The agents of change are natural physical and chemical forces such as evaporation of the more volatile hydrocarbons, selective solubilization and therefore removal of the more water-soluble in preference to the less water-soluble hydrocarbons, adsorption/desorption phenomena with soil matrices, and, of course, biodegradation of some but not all hydrocarbons by ambient microorganisms, and so on. So, it should not come as any great surprise to investigators that the GC/FID and GC/MS chromatograms will change from their as-formulated state to an altered state. The overall process is described as *weathering* after the geological term for alteration of rock facies by physical and chemical reactions with the surrounding environment. Indeed, the term *weathering* was applied to petroleum products as early as the Marine Biological Association of the United Kingdom report on the *Torrey Canyon* oil spill in the late 1960s (Smith 1968); however, even earlier studies had focused on the environmental processing of spilled petroleum in the world's oceans (The Royal Society of London 1936) without actually applying the term *weathering*.

The degradation of petroleum hydrocarbons in weathering reactions is mechanism dependent and is governed greatly by the physical nature of the petroleum product and the chemical structure of individual hydrocarbon molecules. More environmentally labile hydrocarbons are selectively removed over the duration and mechanism of exposure; hence, the attractiveness of studying diagnostic compounds, ratio analyses, and biomarkers for fingerprinting purposes (discussed in the following sections). Figure 5.12 shows the before and after GC/FID chromatograms illustrative of degradation of a middle distillate over time. The same types of effects, although for different boiling point ranges of compounds, can be seen in gasolines, early distillates as well as later distillates, and crude oil releases. In fact, it is not reasonable to expect that the chromatograms of released oil will be comparable to those of laboratory standards or as-blended gasolines. Hydrocarbon weathering reactions are responsible for that phenomenon. However, it is critical for the investigator to collect and preserve for chemical analysis the suspected source material from tanks, pipelines, and so on if at all possible, so that a match between source and contaminant plume can be made.

5.2.3.2 Diagnostic Compounds and Hydrocarbon Ratios

The focus on identifying diagnostic compounds and subsequent calculations of selected ratios to elucidate trends or separate out multiple hydrocarbon sources has centered on the higher–molecular weight petroleum hydrocarbons, the polycyclic aromatic hydrocarbons (PAHs). The analytical

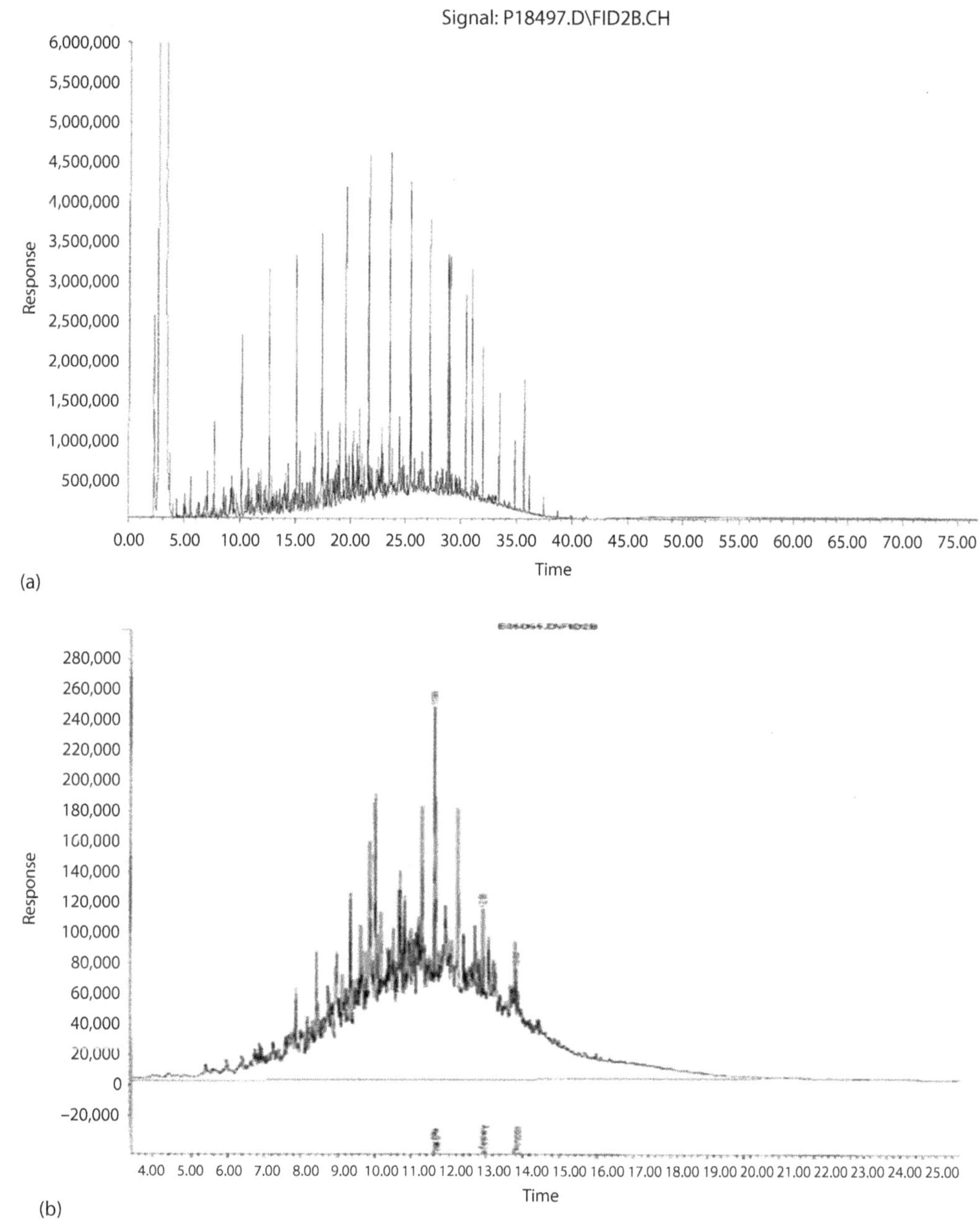

FIGURE 5.12 Example of (a) undegraded and (b) degraded middle distillate petroleum product GC/FID chromatograms

technique for identifying PAHs centered on a modification of the U.S. EPA Standard Method 8270 (U.S. EPA, SW 846). The EPA manual of standard analytical methods was developed in the 1980s to support EPA's work in implementation of the Resource Conservation and Recovery Act (RCRA). RCRA governs waste management and materials recovery and reuse, including the disposal of both hazardous and non-hazardous solid waste. RCRA created a data-based regulatory system, and standardized analytical methods had to be developed so that different analytical laboratories would use the same basic method development, chemical analyses, data collection, and quality assurance protocols, thereby making successful data comparisons among different sources of data (i.e., data from different analytical laboratories) more likely. In support of RCRA, EPA developed test methods for the analysis of various environmental media. The result was the preparation of the EPA publication *Test Methods for Evaluating Solid Waste: Physical/Chemical Methods*, also known as SW-846.

It did not take long for oil spill investigators to realize that for petroleum products, while the basic separation sciences still applied for semi-volatile hydrocarbons such as the PAHs, the method target analytes had to be changed to reflect typical petroleum product hydrocarbons (Sauer and Boehm 1991). The most commonly used SW-846 analytical method for the quantification of individual hydrocarbons in petroleum products is combined gas chromatography/mass spectrometry (GC/MS) based on Standard Method 8270, in either full scan mode (FS) or selected ion monitoring mode (SIM). In FS mode, the GC/MS data are gathered commonly over a scan range of 45–450 atomic mass units (amu). In SIM mode, the GC/MS data are gathered at specific amu typical for PAHs and other analytes of interest. In FS mode, detection limits tend to be comparable to typical Method 8270 results. However, when operated in the SIM mode, the GC/MS data acquisition approach helps reduce background mass spectrometric interferences, typically resulting in lower detection limits and cleaner peak shapes. Individual PAHs are identified based on their retention time and the specific SIM being scanned. GC/MS (SIM) should be employed when lower detection limits are required for ecological risk or hydrocarbon spill impact studies. Therefore, when dealing with complex petroleum matrices such as complex heavy distillate fuels, complex PAH-dominated wastes, and crude oils, a combination of GC/FID to identify components present followed by GC/MS SIM analysis for PAHs and other polycyclic hetero atom aromatic hydrocarbons to produce reliable quantitative detection limits for individual target analyses is recommended.

The GC/MS FS and SIM data acquisition approaches were applied in large oil spills (e.g., M/V *Exxon Valdez* and others) to distinguish crude oil–derived hydrocarbons from natural background hydrocarbon sources using a ratio analysis approach (Douglas et al. 1996), in the separation of PAH background sources from manufactured gas plant (MGP) PAH contamination (Costa et al. 2004; Saber et al. 2006; Wade 2015), in tracking the changes in hydrocarbon ratios in released Gulf of Mexico Crude Oil (GMCO) during resource damage assessment for the *Horizon* Gulf Oil spill (Stout 2016), and in numerous other applications (Boehm, Polycyclic Aromatic Hydrocarbons (PAHs), Chapter 15 in Morrison and Murphy 2006).

The work of Douglas et al. (1996) in studying crude oil chemistry represented the first wide-scale application of double ratio plots and is now considered typical of the application of alkylated PAH ratios to delineate petroleum source and weathering effects. The research used PAH ratios of alkylated dibenzothiophenes and phenanthrenes to track the alteration of spilled crude oils by weathering and biodegradation. The authors reported that the PAH ratios were useful source indicators even up to 98% depletion of total PAHs. The end result was that the PAH source and weathering ratios qualitatively described the depletion and sources of oil in subtidal sediment data from multiple oil spills from different parts of the world. Moreover, the authors also reported that other ratios of alkylated naphthalenes, phenanthrenes, and chrysenes allowed the assessment of the weathering an oil has undergone since initial release.

The uses of PAH ratios for other complex hydrocarbon products and matrices include the study of waste residues from MGP operations, which were common in the mid- to late nineteenth century and the early twentieth century in England and the United States. GC/MS (FS and SIM) and PAH ratio analyses were used to describe the separation of PAH background sources from MGP PAH contamination (Costa et al. 2004; Saber et al. 2006; Wade 2015). Costa et al. (2004) described the separation of PAH background and MGP-related impacts, illustrating the separation of separate PAH sources using double ratio plots employing several different sets of PAH ratios. While this method was successfully used by Costa et al. (2004) in certain instances, Wade (2015) cautions that focusing too narrowly on a limited number of PAH ratios to the exclusion of the presence of other individual PAH analytes might miss commonly encountered MGP-derived wastes such as naphthalene. Moreover, the figures used in Costa et al. (2004) were study specific, and great caution is urged in attempting to apply data assemblage illustrations to other instances. Saber et al. (2006), also working in the field of MGP source delineation, employed a higher-level analytical approach. Their work used a combination of GC/FID and GC/MS analyses combined with compound-specific isotope ratios (CSIRs) of carbon in PAHs, which were obtained using a gas chromatograph interfaced with an isotope ratio mass spectrometer (GC/IRMS). The more targeted and specific CSIRs of the sediment samples showed that the sample PAHs were not similar to those of the former MGP site soil samples. Their results indicated that the sediment PAHs originated from a mixture of sources, and some of the PAHs originated from a source other than the MGP operations. Wade (2015) employed similar GC/MS (FS and SIM) laboratory analysis coupled with statistical analyses of the entire suite of PAH compounds, showing that as a supplement to PAH ratio analyses, examination of the GC/MS (SIM) data as a total analytical suite might produce other useful results compared with those the simpler individual ratio or double ratio plot analyses were capable of providing.

Stout (2016) reported on the PAH and biomarker results comparing nearly 1600 oily matrix field-collected samples with the spilled Macondo oil in investigation of the Gulf of Mexico *Deepwater Horizon* (DWH) oil spill. Multiple sources of petroleum hydrocarbons included the DWH-spilled Macondo oil, natural seeps of crude oils, and other smaller oil spills in the Gulf of Mexico. Complicating factors included the rapid weathering of the Macondo oil and the fact that the DWH oil was chemically similar to the other sources of crude oil in the Gulf. The work described a three-tiered oil spill fingerprinting method, which included hydrocarbon ratio comparisons. The stability of various hydrocarbon ratios was followed over time. Numerous common ratios were shown to be stable, but not all ratios proved to be as stable as previously believed. The author reported that most (i.e., 22 of 29) but not all of the diagnostic ratios exhibited relative standard deviation values of less than 5% or 10%, a range of values that was termed acceptable for oil spill identification under short-term conditions and excellent under long-term conditions when different laboratory personnel and different batches were involved in data production. For the remaining seven diagnostic ratios, the relative standard deviation values were more variable and less reproducible, with over 10% variability over longer-term environmental exposure. Stout (2016) attributed the variability of the diagnostic ratio results to the effects of long term hydrocarbon weathering in the Gulf of Mexico under conditions optimized to produce hydrocarbon degradation.

Boehm (2006) provides a ready reference on the subject of polycyclic aromatic hydrocarbons (PAHs), including the application of investigatory techniques for hydrocarbon source identification (i.e., PAHs derived from spilled oil, background PAHs, etc.) plus mixing effects, illustrating the use of using PAH and alkylated PAH family data.

In summary, a GC/MS technique to gather accurate PAH data for complex petroleum-derived products and matrices is of great utility and is the preferred method for the study of such situations. However, great attention must be paid to the application of peer-reviewed information in specific circumstances regarding the stability of ratio indicators and the use of single and double plot data analyses.

5.2.3.3 Biomarkers

The use of molecular biomarkers for environmental forensic purposes takes advantage of previous geochemical research specifically developed for petroleum resource discovery and development. Considerable research applying this work to environmental forensics has shown that reliable source identifications can be made using biomarkers alone or in conjunction with PAH ratios for various petroleum products ranging from gasolines to crude oils (Prince et al. 1994; Bence et al. 1996; Boehm et al. 1997; Wang et al. 1999; Stout et al. 2001; Page et al. 2002; Wang and Fingas 2003; Wang et al. 2006; Stout and Douglas (2004); Stout et al. 2005; Wang et al. 2006a; Wang and Christensen 2006; Wang and Stout 2007; Stout and Wang 2008; Stout 2016.). The classes of biomarker to be selected for tracking sources of petroleum contamination are governed by the boiling point distribution of the petroleum contamination, lighter–molecular weight biomarkers for the lighter petroleum products and the heavier–molecular weight biomarkers for heavier petroleum products and crude oils. Diamondoid and drimane-based bicyclic sesquiterpane hydrocarbons have

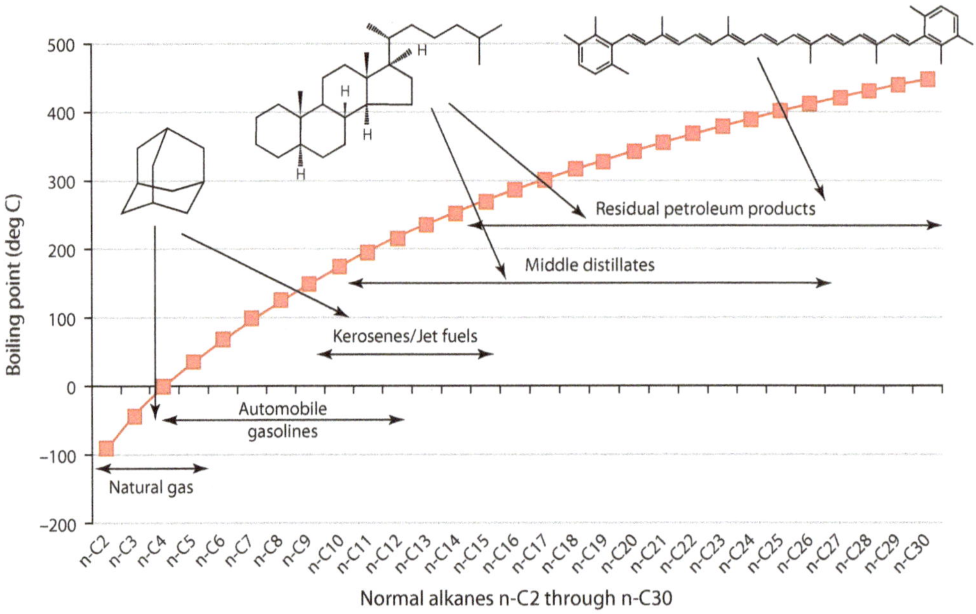

FIGURE 5.13 Structural formulas for general biomarker structures for diamondoid, bicyclic sesquiterpanes and steranes, triterpanes, and triaromatic steranes versus carbon range n-C2 through n-C30.

been used for lighter petroleum products such as gasoline, natural gas condensate, and middle distillates. Higher–molecular weight hydrocarbons such as steranes, triterpanes, and triaromatic steranes have been used for heavier petroleum products such as heavier fuel oils, including Bunker C oils and crude oils, often in conjunction with parent and alkyl PAH homologues.

Structural formulas for general biomarker structures are given in Figure 5.13 for diamondoids and bicyclic sesquiterpanes for lighter products and steranes, triterpanes, and triaromatic steranes for the heavier products. Table 5.1 gives a listing of common sterane, triterpane, and triaromatic sterane biomarkers used for fingerprinting purposes. Stout (2016) provides a listing of diagnostic ratios using biomarkers.

5.2.4 Stable Isotopic Analysis

Isotopes are atoms whose nuclei contain the same number of protons but a different number of neutrons. The stable isotopic composition of an element (e.g., carbon) is normally reported as delta (δ) values in parts per thousand (denoted as ‰) enrichments or depletions relative to a standard of known composition. δ values are calculated by

$$(\delta \text{ in ‰}) = \left(R_{\text{sample}} / R_{\text{standard}} - 1\right)1000 \tag{5.1}$$

where R is the ratio of the heavy to the light isotope in the sample or standard.

During chemical reactions, the partitioning of isotopes between two substances with different isotope ratios is called *isotope fractionation* (Hoefs 1987). The phenomenon of isotope fractionation has found applications in many scientific fields. In environmental site characterization and remediation at organic contaminant sites, one of the key elements is to evaluate the biodegradation process. Because the degradation of organic contaminants often results in a change in the ratio of stable isotopes, stable isotopes, especially compound-specific isotope analysis (CSIA), can provide an in-depth understanding of biodegradation or abiotic transformation processes in contaminated

TABLE 5.1
Biomarker Compounds for Oil Spill Studies

Peak	Compound Name	Code	Chemical Formula	GC/MS Target ions
Sesquiterpanes (BicyclicTerpanes)				
Drimanes and drimane-structure terpanes		–	C14H26, C15H28, C16H30	123
Diamondoids				
Adamantanes		–	C10H16, alkyl-C10H15	136, 135, 149, 163, 177
Diamantanes		–	C14H20, alkyl-C14H19	188, 187, 201, 215, 229
Terpanes				
1	C19 tricyclic terpane	TR19	C19H34	191
2	C20 tricyclic terpane	TR20	C20H36	191
3	C21 tricyclic terpane	TR21	C21H38	191
4	C22 tricyclic terpane	TR22	C22H40	191
5	C23 tricyclic terpane	TR23	C23H42	191
6	C24 tricyclic terpane	TR24	C24H44	191
7	C25 tricyclic terpane (a)	TR25A	C25H46	191
8	C25 tricyclic terpane (b) triplet: C24 tetracyclic terpane +	TR25B	C25H46	191
9	C26 (S + R) tricyclic terpanes		TET24+TR26A+TR26B C24H42 + C26H48	191
10	C28 tricyclic terpane (a)	TR28A	C28H52	191
11	C28 tricyclic terpane (b)	TR28B	C28H52	191
12	C29 tricyclic terpane (a)	TR29A	C29H54	191
13	C29 tricyclic terpane (b)	TR29B	C29H54	191
14	Ts: 18!(H),21"(H)-22,29,30-trisnorhopane	Ts	C27H46	191
15	17!(H),18!(H),21"(H)-25,28,30-trisnorhopane	TH27	C27H46	191, 177
16	Tm: 17!(H),21"(H)-22,29,30-trisnorhopane	Tm	C27H46	191
17	C30 tricyclic terpane 1	TR30A	C30H52	191
18	C30 tricyclic terpane 2	TR30B	C30H52	191
19	17!(H),18!(H),21"(H)-28,30-bisnorhopane	H28	C28H48	191

(Continued)

TABLE 5.1 (CONTINUED)
Biomarker Compounds for Oil Spill Studies

Peak	Compound Name	Code	Chemical Formula	GC/MS Target ions
20	17!(H),21"(H)-25-norhopane	NOR25H C29H50	191, 177	
21	17!(H),21"(H)-30-norhopane	H29	C29H50	191
22	18!(H),21"(H)-30-norneohopane (C29Ts)	C29Ts	C29H50	191
23	17!(H)-diahopane	DH30	C30H52	191
24	17!(H),21"(H)-30-norhopane (normoretane)	M29	C29H50	191
25	18!(H) and 18"(H)-oleanane	OL	C30H52	191, 412
26	17!(H),21"(H)-hopane	H30	C30H52	191
27	17!(H)-30-nor-29-homohopane	NOR30H C30H52	191	
28	17"(H),21!(H)-hopane (moretane)	M30	C30H52	191
29	22S-17!(H),21"(H)-30-homohopane	H31S	C31H54	191
30	22R-17!(H),21"(H)-30-homohopane	H31R	C31H54	191
31	Gammacerane	GAM	C30H52	191, 412
32	17"(H),21"(H)-hopane	I.S.	Lab STD	191
33	22S-17!(H),21"(H)-30,31-bishomohopane	H32S	C32H56	191
34	22R-17!(H),21"(H)-30,31-bishomohopane	H32R	C32H56	191
35	22S-17!(H),21"(H)-30,31,32-trishomohopane	H33S	C33H58	191
36	22R-17!(H),21"(H)-30,31,32-trishomohopane	H33R	C33H58	191
37	22S-17!(H),21"(H)-30,31,32,33-tetrakishomohopane	H314S	C34H60	191
38	22R-17!(H),21"(H)-30,31,32,33-tetrakishomohopane	H34R	C34H60	191
39	22S-17!(H),21"(H)-30,31,32,33,34-pentakishomohopane	H35S	C35H62	191
40	22R-17!(H),21"(H)-30,31,32,33,34-pentakishomohopane	H35R	C35H62	191
Steranes				
41	C20 5!(H),14!(H),17!(H)-sterane	S20	C20H34	217, 218
42	C21 5!(H),14"(H),17"(H)-sterane	S21	C21H36	217, 218
43	C22 5!(H),14"(H),17"(H)-sterane	S22	C22H38	217, 218
44	C27 20S – 13"(H),17!(H)-diasterane	DIA27S	C27 H48	217, 218, 259
45	C27 20R – 13"(H),17!(H)-diasterane	DIA27R	C27 H48	217, 218, 259

(Continued)

TABLE 5.1 (CONTINUED)
Biomarker Compounds for Oil Spill Studies

Peak	Compound Name	Code	Chemical Formula	GC/MS Target ions
46	C27 20S – 13!(H),17"(H)-diasterane	DIA27S2	$C_{27}H_{48}$	217, 218, 259
47	C27 20R – 13!(H),17"(H)-diasterane	DIA27R2	$C_{27}H_{48}$	217, 218, 259
48	C28 20S – 13"(H),17!(H)-diasterane	DIA28S	$C_{28}H_{50}$	217, 218, 259
49	C28 20R – 13"(H),17!(H)-diasterane	DIA28R	$C_{28}H_{50}$	217, 218, 259
50	C29 20S – 13"(H),17!(H)-diasterane	DIA29S	$C_{29}H_{52}$	217, 218, 259
51	C29 20R – 13!(H),17"(H)-diasterane	DIA29R	$C_{29}H_{52}$	217, 218, 259
52	C27 20S – 5!(H),14!(H),17!(H)-cholestane	C27S	$C_{27}H_{48}$	217, 218
53	C27 20R – 5!(H),14"(H),17"(H)-cholestane	C27""R	$C_{27}H_{48}$	217, 218
54	C27 20S – 5!(H),14"(H),17"(H)-cholestane	C27""S	$C_{27}H_{48}$	217, 218
55	C27 20R – 5!(H),14!(H),17!(H)-cholestane	C27R	$C_{27}H_{48}$	217, 218
56	C28 20S – 5!(H),14!(H),17!(H)-ergostane	C28S	$C_{28}H_{50}$	217, 218
57	C28 20R – 5!(H),14"(H),17"(H)-ergostane	C28""R	$C_{28}H_{50}$	217, 218
58	C28 20S – 5!(H),14"(H),17"(H)-ergostane	C28""S	$C_{28}H_{50}$	217, 218
59	C28 20R – 5!(H),14!(H),17!(H)-ergostane	C28R	$C_{28}H_{50}$	217, 218
60	C29 20S – 5!(H),14!(H),17!(H)-stigmastane	C29S	$C_{29}H_{52}$	217, 218
61	C29 20R – 5!(H),14"(H),17"(H)-stigmastane	C29""R	$C_{29}H_{52}$	217, 218
62	C29 20S – 5!(H),14"(H),17"(H)-stigmastane	C29""S	$C_{29}H_{52}$	217, 218
63	C29 20R – 5!(H),14!(H),17!(H)-stigmastane	C29R	$C_{29}H_{52}$	217, 218
64	C30 steranes	C30S	$C_{30}H_{54}$	217, 218
Monoaromatic Steranes				253
Triaromatic Steranes				231

Source: From Wang et al., *Environmental Forensics*, 7, 105–146, 2005.

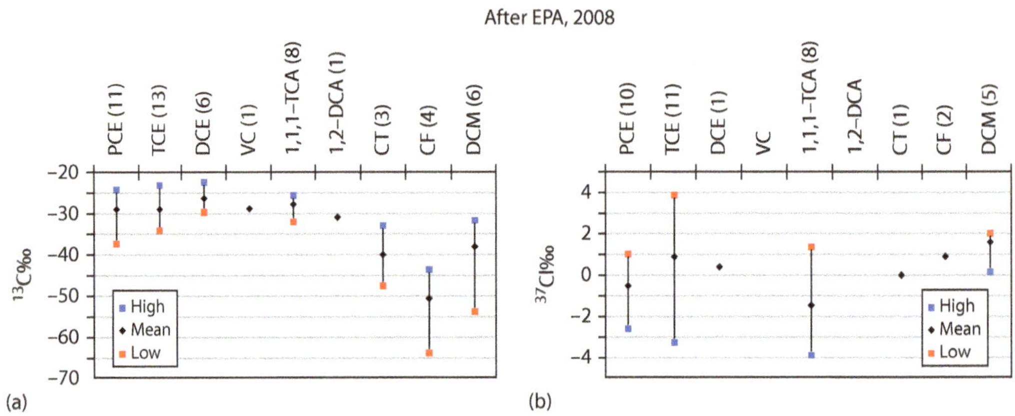

FIGURE 5.14 Minimum, maximum, and (a) mean carbon and (b) chlorine isotope ratio of chlorinated hydrocarbons from different manufacturers and production batches measured to date. The number in parentheses following each compound name indicates the number of samples analyzed for that compound. PCE is tetrachloroethylene, TCE is trichloroethylene, DCE is dichloroethylene, 1,1,1-TCA is 1,1,1-trichloroethane, 1,2-DCA is 1,2-dichloroethane, CT is carbon tetrachloride (tetrachloromethane), CF is chloroform (trichloromethane), and DCM is dichloromethane. (Data compiled from Beneteau, K. M., Aravena, R., and Frape, S.K., *Organic Geochemistry*, 30, 739–753, 1999; Holt et al. 1997; Hunkeler and Aravena 2000c; Jendrzejewski et al. 1997, 2001; Shouakar-Stash et al. 2003; van Warmerdam et al. 1995; Zwank et al. 2003.)

aquifers (EPA 2008). This better understanding helps in monitoring the progress of natural attenuation and selecting appropriate remedies.

From a forensic perspective, CSIA has become an emerging tool for source identification, especially for organic contaminants. The fundamental basis is the variability of isotope ratios of commercial products between manufacturers or between different production batches produced by the same manufacturer (EPA 2008). Figure 5.14 summarizes the data on variation in the isotopes of carbon and chlorine in chlorinated solvents and chlorinated production chemicals. As can be observed, some of the compounds have large differences in isotopic ratios, especially the carbon isotopes.

The interpretation of CSIA data is complicated by release scenarios and biotic and abiotic degradation. At a site where the history of release is not known, multiple interpretations may be possible even without degradation. In some cases, an apparent contiguous source of groundwater contamination may be heterogeneous with respect to its isotopic composition as a result of several different spill events over time and/or spilled compound from different sources with different isotope compositions. In other cases, different spills in different locations may have the same isotopic composition. Degradation (both biotic and abiotic) adds another level of complexity to CSIA data interpretation. Because differences in isotope ratios of different sources are commonly in the order of several percentages, the change in the isotope ratio due to biotic and/or abiotic degradation can rapidly become more important than variations due to different sources (EPA 2008). As such, for sites with CSIA data, the data interpretation needs to be in the context of other site information, as described throughout this chapter.

5.2.5 Statistics

Statistics is the science of data. It is one of the indispensable tools in environmental data quality assessment (EPA 2006), in waste management (EPA 1990), in groundwater monitoring (ITRC 2013), and in many other disciplines, because environmental forensics investigation can involve large amounts of data, and the application of statistics to extract source information from seemly "chaotic" data is only logical. Common applications include comparison of sample means, linear regression, and multivariate analysis.

Comparison of sample means may be conducted to differentiate contaminants of concern between natural occurrence (i.e., background) and manmade contamination at a site of interest. Many statistical tests are available for the analysis, including Student's t-test, paired t-test, and Wilcoxon's Rank Sum Test (Gauthier and Murphy 2003). Student's t-test is probably the most common test for comparing samples. Because it is a parametric technique, normal distribution of the sample is assumed. When paired data sets are available, examples of paired data include upwind/downwind, upstream/downstream, shallow/deep soil, and prior/post an event of interest. Wilcoxon's Rank Sum Test is a non-parametric technique and is used when data sets cannot be transformed to approximate a normal distribution.

In many environmental investigations, revealing trends (both spatial and temporal) in the data is of interest. At a contaminated site with a large network of groundwater monitoring wells, it is important to know whether a contaminant of concern increases or decreases over time and/or distance. Linear regression is a powerful tool for such applications. In forensic investigation, in addition to the applications described, linear regression is also used frequently for age dating, which will be discussed in a later section.

Multivariate analysis examines multiple variables to see how they relate to each other. Many methods are available, including principal component analysis (PCA), partial least squares (PLS), and polytopic vector analysis (PVA). A common objective of a multivariate approach is to identify a small number of uncorrelated variables from a large set of data. In a forensic context, the methods are used to determine the number of fingerprints present in the system, the multivariate chemical composition of each fingerprint, and the relative contribution of each fingerprint in each collected sample. Among these methods, PCA is probably most commonly used. In this analysis, reduction of dimension is achieved by transforming the data to a set of mutually orthogonal reference variables called *principal components* (PCs). If nearly all variability between samples can be accounted for by a small number of PCs, the relationship between multivariate samples may be assessed by simple inspection of a two- or three-dimensional plot: a PCs score plot (Johnson 2015). The PCA and other similar multivariate methods have been widely used to identify contaminant sources, to determine the contribution of each source in case of multiple sources, and to allocate the costs associated with environmental damage.

5.3 AGE DATING OF PETROLEUM RELEASES

Available age dating approaches offer a fairly simple approach to determining when releases of gasoline and distillate fuels may have occurred, but the various approaches are subject to much potential error in application. It is incumbent on the investigator to read the original peer-reviewed work carefully, with understanding, and then apply the appropriate constraints to individual case assignments. It is with this very strong caveat that we will discuss multiple approaches to age dating gasoline using alkyllead concentrations and the inherent uncertainties of such an approach (Stout et al. 2012; Wade 2014), oxygenate additives (Powers et al. 2001; Oudijk 2010), formulation variability using toluene to n-C8 ratios (Schmidt et al. 2002), and gasoline octane assessments (Kaplan et al. 1996; Schmidt et al. 2003). Due to the inherent interrelationships among leaded and unleaded gasoline additives, age dating based on any single approach is to be avoided. In a similar manner, we will discuss the current state of controversy involving age dating middle distillate fuels such as No. 2 fuel oil/diesel fuels (Christensen and Larsen 1993; Hurst and Schmidt 2005; Oudijk and Wade 2011; Oudijk 2012). Again, within the forensic geochemical community, there is no single age dating approach that is universally accepted by all investigators.

5.3.1 Leaded Gasolines

The story of the discovery and subsequent use of lead as an anti-knock additive is long and not without its disappointments. Very early in the life of the "horseless carriage" (i.e., the automobile), various ways of powering the vehicles were employed, including electricity, steam, and alcohol-based fuels, all developed in the later part of the nineteenth and the very early twentieth century in the United States. It was not until the early part of the twentieth century in the United States that petroleum products, which had heretofore been considered unwanted waste products of kerosene manufacturing, were considered for automobile fuels, leading to the development and refinement of the internal combustion engine. Even after the internal combustion engine was developed, alcohols were considered as potential fuels because of the unfortunate tendency for petroleum-based fuel to develop a knocking behavior in the engines. The knocking problem was eventually solved with the discovery of the moderating effects of lead alkylates, leading to the development of leaded gasolines starting in the mid-1920s in the United States and ending, finally, with the prohibition of the manufacture and sale of leaded automobile gasoline by the U.S. EPA after December 31, 1995; that is, on January 1, 1996. However, the age dating of leaded gasolines is not as simple as might be supposed (Wade 2014).

Age dating gasolines, automobile gasolines specifically, can be accomplished to a first approximation using the presence and concentrations of alkyllead additives. Using the presence/absence approach, some limits of gasoline manufacture can be placed on leaded gasoline contamination for age dating purposes for LNAPL samples only. For soil samples, things get much more complicated due to their environmental stability (Rhue et al. 1992) compared with the tendency of petroleum hydrocarbons to be degraded by a combination of processes collectively known as *weathering*. In dealing with lead concentrations, however, as soon as the grams per gallon values are calculated, and attempts are made to use calculated Pb/gal values to establish a more precise age estimate, severe problems arise. Unfortunately, there are simplistic modeling approaches using published regulatory total lead per gallon EPA limits that simply do not reflect the actual data collected in the leaded gasoline marketplace by the U.S. government (Wade 2014). Lead levels were shown to be highly variable in the semiannual leaded gasoline marketplace survey data gathered by the U.S. Department of Energy and the American Petroleum Association. The variability clearly shows that simple assumptions regarding regulatory guidelines put in place at various times for the U.S. automobile gasoline market are wildly inaccurate (Wade 2014). Great caution is urged if total lead level calculations are attempted in age dating leaded gasoline.

Further, age dating leaded gasolines in the presence of a used motor oil co-contaminant is not possible. Simply assuming that gasoline range hydrocarbons found in soils or groundwater when used motor oil is also present (analysis of samples by GC/FID will test for the presence of motor oil) came from the site under investigation is erroneous. Used motor oil has been shown to contain both simple and more complex aromatic hydrocarbons typically found in automobile gasolines resulting from the presence of partly burned gasolines, largely unburned gasolines, and chemically rearranged polycyclic aromatic hydrocarbons(PAH) that have been synthesized during the internal combustion process (Pruell and Quinn 1988; Wong and Wang 2001; Dominguez-Rosado and Pichtel 2003; California Environmental Protection Agency 2007). Similarly, the presence of diesel fuel hydrocarbons in used motor oil is highly likely. It is clear that gasoline range hydrocarbons found in used motor oil could originate from any brand of automobile gasoline sold in any gasoline market area throughout the country. Therefore, simply assuming that gasoline contamination found at a service station property that is contaminated with used motor oil throughout the soil column and/or floating on the groundwater actually came from gasoline being sold at that service station is a major error (Wade 2016). Again, if used motor oil contamination is present at the site, great caution is urged in assigning the responsibility for gasoline contamination to the property owner/operator alone. Vehicle repair operations and illegal disposal of used motor oil must be considered when used motor oil is discovered in soils or groundwater at a site.

5.3.2 Oxygenate Gasoline Additives

The history of oxygenate additives (i.e., alcohols and ethers) in automobile gasolines is long, variable, and somewhat exasperating; it is inexorably linked with the development of leaded gasoline additives. It was during the search to solve the "knocking problem" that researchers requested supplies of alcohol (ethanol) to test but were turned down due to the then current Prohibition Era ban on the availability and use of ethyl alcohol (the 18th Amendment to the U.S. Constitution was ratified on January 29, 1919). Therefore, ethyl alcohol (ethanol) was never tested for use as an anti-knock additive until much later in the life of the automobile. And in a very real sense, the subsequent problems of failed additives such as leaded gasoline and ether-treated gasolines might have been avoided completely had it not been for the failed sociological experiment of alcohol prohibition in the United States in the 1920s and early 1930s, which kept alcohol out of the hands of researchers. It was not until February 1933 that the U.S. Congress adopted a resolution proposing a 21st Amendment to the Constitution that would repeal Prohibition; it was finally ratified in December 1933, and the prohibition on ethanol sales was finally removed. However, it was not until the mid- to late-1970's that alcohols and ethers began to be used as gasoline additives during the U.S. EPA-mandated phase-down of leaded gasoline use in the United States.

Numerous oxygenate additives have been used in the U.S. automobile gasoline market, including alcohols such as methanol (methyl alcohol), ethanol (ethyl alcohol), and tert-butyl alcohol (tBA), and ethers such as methyl tert butyl ether (MtBE), ethyl tert-butyl ether (EtBE), diisopropyl ether (DIPE), and tert-amyl ether (tAME) have all been marketed by U.S. gasoline manufacturers at times. Of these, there have been significant markets in only two: MtBE, which faded over time, and ethanol, which has proved to be a lasting oxygenate additive.

The story of MtBE provides an opportunity for some age dating possibilities, but the variability of manufacture date and company-specific timing of use has made such age dating problematic, and investigators are urged to use caution when age dating via an oxygenate additive. MtBE was the first ether oxygenate approved for use in automobile gasoline by the U.S. EPA. Previous oxygenates approved for gasoline use by the U.S. EPA centered on alcohols such as tert-butyl alcohol. Historically, in the U.S. petroleum refining industry, the chemical MtBE has been put to two uses. First, soon after its approval for use, MtBE was used as an octane-enhancing additive, replacing lead alkylates, which had been used as octane-enhancing chemicals since the 1920s due to their impressive anti-knock characteristics. Second, beginning in the late 1980s and increasingly in the 1990s, MtBE was used as a gasoline oxygenate to manufacture reformulated gasolines (RFG) in compliance with the ozone nonattainment provisions of the U.S. Clean Air Act Amendments of 1990.

MtBE was in widespread commercial use as a gasoline additive in Europe in the early to mid-1970s. MtBE first began to be used as an automobile gasoline additive within the United States once its use was approved by the U.S. EPA in March 1979 (U.S. Federal Register Vol. 44, No. 45, page 12242). While the use of this particular additive was first approved by the U.S. EPA in March 1979, not all gasoline manufacturers began using MtBE in their branded gasoline at that time; moreover, MtBE was introduced into various grades of gasoline at various times by a single refiner. Thus, the analysis of the historical use of MtBE as an automobile gasoline additive grows complicated very quickly over time among different refiners and even for gasolines from different refineries owned by the same company.

U.S. gasoline refiners began the manufacture of MtBE for use in automobile gasolines at various times as pre-engineering, manufacturing, and total supply issues were addressed. The U.S. MtBE marketplace started with the first commercial manufacture of MtBE in the United States by Phillips Petroleum Company beginning in April 1979 (*Oil and Gas Journal*, May 21, 1979) (U.S. Federal Register Vol. 44, No. 45, Tuesday, March 6, 1979).

The first use of the additive MtBE in automobile gasoline was followed by ARCO Petroleum beginning sales of its own brand of MtBE-treated gasoline (ARCO Unleaded Supreme) in the fall of 1980 (Atlantic Richfield Corporation Annual Report for 1980). Subsequently, other gasoline manufacturers first began selling their own branded and MtBE-treated gasoline at various times during the 1980s, continuing into the mid- to late-1980's. MtBE was required to be added into gasoline by the U.S. government in 1992. However, different gasoline refiners and marketers began to introduce MtBE-treated gasolines into the marketplace at different times. At first, it was common industry practice for the gasoline refiners to add MtBE only to their unleaded premium gasolines. The single formulation use in a premium unleaded gasoline was due to two factors: first, in an effort to maximize profit margins from high octane gasolines during the early 1980s, emphasis was placed on marketing premium unleaded gasolines; second, in the United States during the early to mid-1980s, there was a limited number of MtBE manufacturing facilities, leading to a manufacturing bottleneck for MtBE. The manufacturing bottleneck led to a limit on the supply of MtBE in the automobile gasoline industry. The result was that not all unleaded gasolines sold in the 1980s were treated with MtBE. Further, those gasolines that did contain MtBE were blended initially at relatively low percentages (less than/up to 1% by volume) compared with today's modern reformulated unleaded gasolines, which can contain as much as 12% by volume. After some time, however, lower octane grades were blended with MtBE also. Therefore, not only are there timing issues with the additive MtBE among different refiners, as well as within one refiner at different refineries, but there are also timing issues for MtBE usage among different grades of gasoline within single refineries owned by a gasoline refiner. A further complicating factor, however, in determining which areas of the United States are ozone non-attainment areas is that some cities voluntarily declared themselves to be ozone non-attainment areas, requiring the distribution of RFG during the winter months to combat air pollution problems in their cities.

By the mid-2000s, the difficulties encountered in controlling MtBE as an environmental contaminant became so pervasive that individual states began banning MtBE in their gasolines, and by the mid-2000s, the U.S. EPA was forced to remove the requirements for RFG by banning the use of MtBE in RFG. However, just because the U.S. EPA banned MtBE in RFG, this does not mean that MtBE has been banned in all states in the United States. Not all states have banned MtBE gasolines. Some states still permit the use of MtBE in their automobile gasolines, and there is a consistent production of MtBE annually in the United States, reported by the U.S. Energy Information Administration (Figure 5.15). The investigator is urged to contact the state regulatory agency to

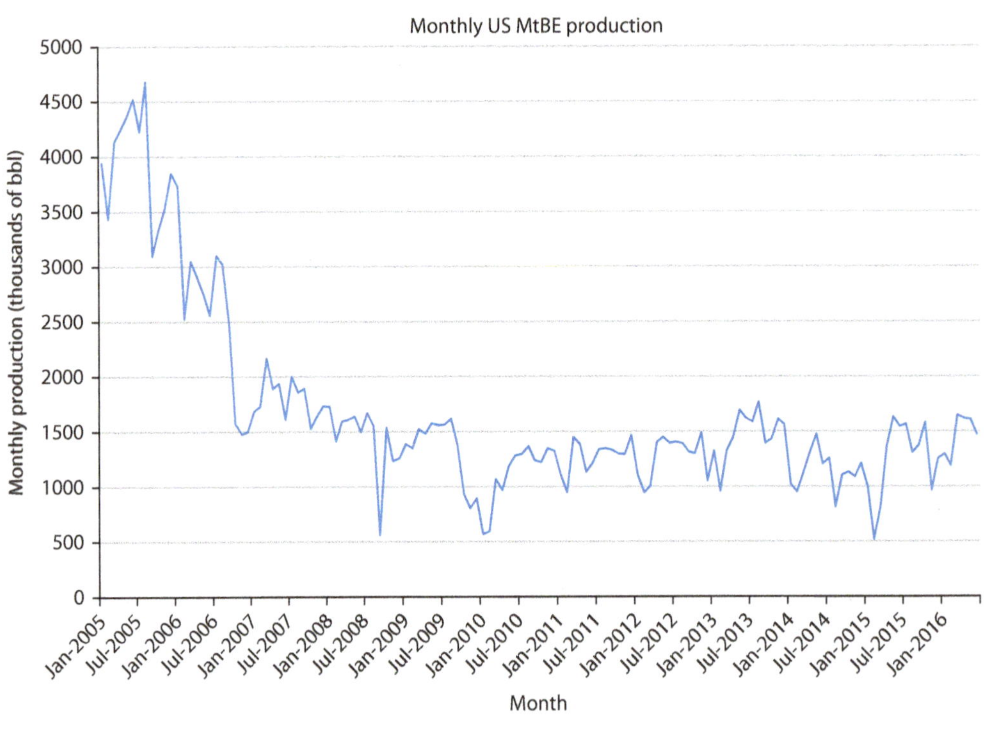

FIGURE 5.15 Annual production of MtBE in the United States as reported by the U.S. Energy Information Administration between January 2005 and January 2016.

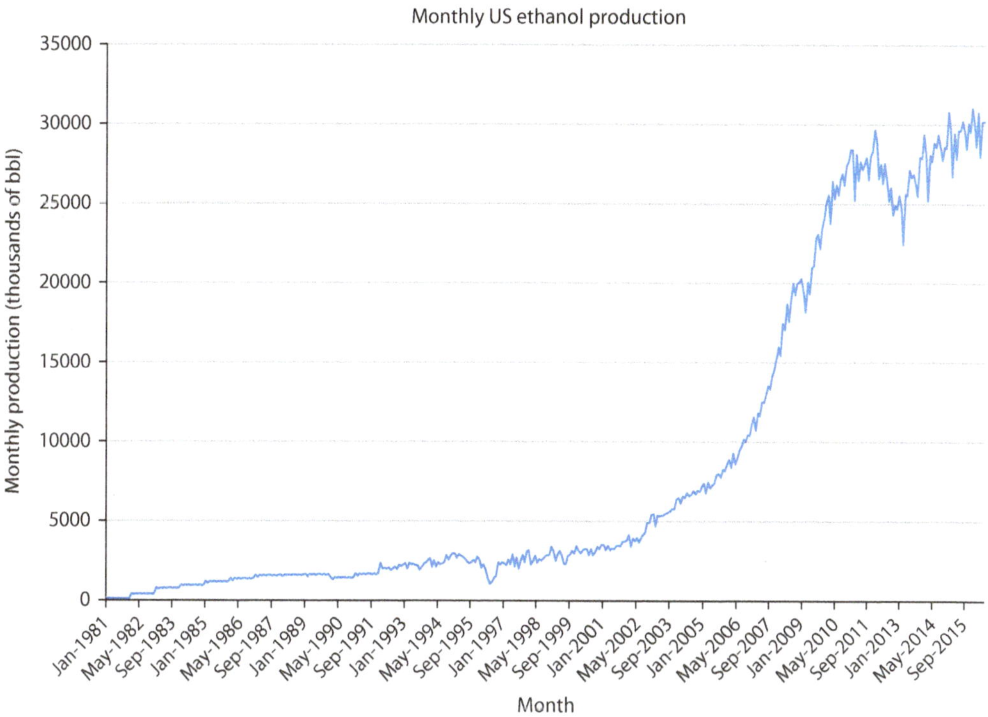

FIGURE 5.16 Annual production of ethanol in the United States as reported by the U.S. Energy Information Administration between January 1991 and January 2015.

determine the current state of MtBE ban in the area in which they are working if they are contemplating the use of MtBE in environmental matrices to age date gasoline contamination.

While the story of MtBE is complicated by high use/no use variability and start/stop complications of regulatory oversight, the story of ethanol as a gasoline additive is not complicated at all. Ethanol has emerged as the preferred automobile gasoline additive. Its environmental issues do not seem to be nearly as severe as those of MtBE (Powers et al. 2001). Moreover, the very low persistence of ethanol in groundwater makes it difficult to identify at normal laboratory detection limits. Annual ethanol usage, tracked by the U.S. Energy Information Administration, appears to have reached close to market saturation (Figure 5.16). Reports of persistence of ethanol in groundwater are rare. In the absence of unanticipated consequences of the use of ethanol as an automobile gasoline additive, the anticipated use of alcohol such as ethanol seems to have few negative environmental consequences.

5.3.3 Gasoline Formulation Characteristics and Forensics Information

When petroleum products, automobile gasolines or heavier distillate fuels, are manufactured, they leave the refineries with formulation characteristics that conform to the product specifications specific for each refiner. In addition, the finished gasolines of all types are formulated for specific driving seasons and specific regions of the United States. Because petroleum refineries are composed of complex engineering systems (Speight 1980, 1991, 1999, 2007, 2013; Meyers 1986, 1997, 2004), the key to their successful operation is consistency. Change, either deliberate or unintentional, offers challenges that must be met. Deliberate changes to specific engineering processes often involve manufacturing improvements, usually driven by newly engineered automobile engine performance concerns, competition among brands, or government regulatory requirements. Unintentional changes are usually catastrophic, and while they are almost always avoided, process accidents occur or systems fail. During normal refinery operations, therefore, change in any single product formulation is not good.

Petroleum products such as gasoline are engineered to produce a consistent physical formulation (boiling point range, vapor pressure, etc.). Also, the manufactured gasoline will have a certain basic chemical composition from each refinery, which can be viewed as a basic hydrocarbon backbone (Wade 2009) to which various additives such as ethanol or additions of octane-enhancing hydrocarbons will be added to control the octane (grade) of the gasolines manufactured. Other company proprietary performance-enhancing compounds are added as well. Moreover, additional more volatile hydrocarbons may be added or removed to control product vapor pressure (i.e., summer versus winter grade fuels). Octane-regulating hydrocarbons include branched chain saturated hydrocarbons such as isopentane and 2,2,4-methyl pentane (isooctane with its various reaction isomers), while the aromatic hydrocarbons added might include more toluene and/or xylene isomers or higher branched chain alkyl benzenoid hydrocarbons than are already in the backbone from the distillation process(es). Because differences exist between hydrocarbon backbones and octane-driven formulations, forensic research has shown that formulation era (Schmidt et al. 2002) and original as-formulated octane (Schmidt et al. 2003) estimates are possible. Information on the as-formulated age of a gasoline and/or its grade may provide significant forensic evidence, either individually or when dealing with multiple product mixed at a single site.

Schmidt et al. (2002) published a technique for estimating the age of regular/mid-grade gasolines released to the subsurface from the early 1970s up to 2001 that relied on formulation- and engineering-induced changes in the basic hydrocarbon backbone of gasolines. The authors examined chromatograms from 133 individual samples of fresh (from service stations, storage tanks, distribution terminals, or individual refineries) unreleased gasolines, representing 13 major refiners selling gasoline in the United States. To their very great credit, the authors identified the refiner, the source, the date, and the geographic location of each gasoline sample used in their data analysis. The results of their examination demonstrated that the toluene/n-C8 ratio could be used to estimate the era of initial gasoline formulation of the unreleased product (Figure 5.17). The data showed that for regular and mid-grade gasolines, the toluene/n-C8 ratio was normally distributed and fell within three general groups: 1973–1983 (43 samples), 1984–1993 (30 samples), and 1994–2001 (60 samples). Moreover, the authors also explored the effects of evaporation on the stability of the toluene/n-C8 ratio and reported that reliable results could still be obtained if a correction factor was applied up

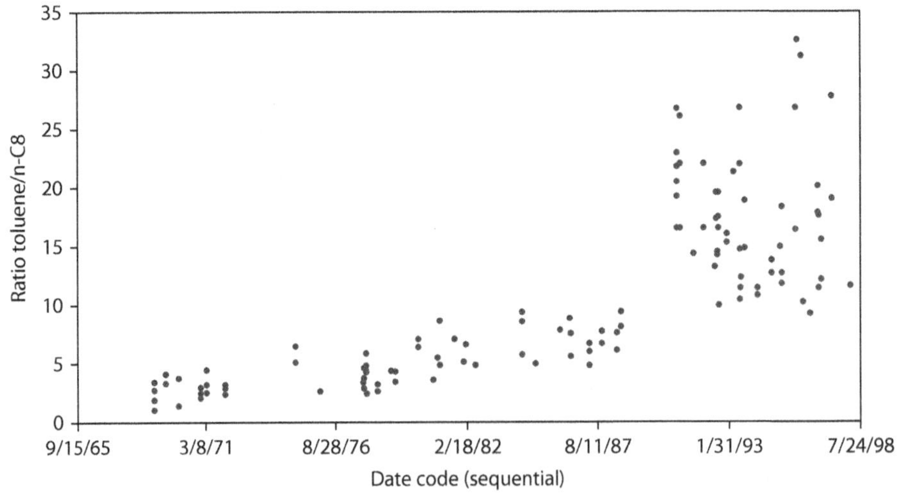

FIGURE 5.17 Toluene/n-C8 ratio versus date of initial gasoline formulation of unreleased regular and mid-grade gasolines. (Data from Schmidt, G., Beckmann D., and Torkelson, B.A, *Environmental Forensics*, 3, 145–162, 2002.)

to 50% evaporation, adjusting the age of initial formulation up to 25%. However, the authors also reported that premium grades could not be characterized in this manner, as the data were too variable to fall into any consistent date range.

There are several issues of note for this work. Initially, it was realized that there was a distinct geographical bias to the source of samples, with a dominant number of product samples coming from a single individual major Midwest refiner and gasoline marketer. Therefore, while the data set appears to represent an industry-wide phenomenon, something that has been well chronicled in each company's annual reports starting in the early 1970s, still we are left with the question … is this relationship applicable to all gasolines and all situations in all circumstances across the United States? And, obviously, the answer to that question has to be … no. Reviewing individual refiner annual reports (most are publicly owned) shows that the use of highly alkylated hydrocarbons in gasolines remained a highly specific company issue in the period of 1970–2000, and that the degree of use of alkylated hydrocarbons to influence octane varied by company. Therefore, industry-wide generalizations for age dating gasolines do not rest on a solid technical foundation in all circumstances, and the individual investigator is cautioned to verify that the data set supports the region of the United States where the investigation centers or risk challenge.

As a follow-on from their initial work, Schmidt et al. (2003) published a gas chromatography technique that could be used to estimate the OI of fresh and weathered gasolines released to soils and groundwater. The OI was defined as the ratio of the sum of isooctane and toluene to the sum of n-C7 and n-C8 (Equation 5.2). The degree of weathering of the samples was checked for qualitative compliance, with the finding that the technique is applicable provided that less than 75% of the sample has been lost to evaporation.

$$\text{Octane index}\left(\text{OI}\right)=\left(\text{isooctane}+\text{toluene}\right)/\left(\text{n-C7}+\text{n-C8}\right) \quad (5.2)$$

The authors quite rightly reported that this technique may not be appropriate if adulteration of the gasoline sample by distillate fuels such as No. 2 fuel oil/diesel fuel is suspected, if the sample shows strong indications of weathering by water washing or biodegradation, or if the gasoline sample being tested is not a neat sample; that is, it was not unreleased to the environment. The authors stressed that the OI technique has never been tested in other matrices other than a neat gasoline, such as gasoline hydrocarbons adsorbed to soils. Regardless of the authors' cautions, since the paper was published, our experience has shown that attempts have been made, without the benefit of peer-reviewed literature support, to extend the application of the technique to other such matrices. As pointed out by Wade (2016), the unsupported application of such age dating techniques or the use of any techniques not supported by the peer-reviewed literature is unwise and not scientifically reliable.

5.3.4 Age Dating Middle Distillate Fuels such as No. 2 Fuel Oil/Diesel Fuels

Christensen and Larsen (1993) published a paper on a method for determining the age of diesel oil spills in the soil. What followed was nothing less than the creation of a wide-ranging debate among forensic geochemists regarding whether, when, and how the Christensen and Larsen (C&L) approach might be applied to age date middle distillate fuels in the United States and other countries (Wade 2001, 2002; Hurst and Schmidt 2005; Kaplan and Galperin 1996; Kaplan et al. 1997; Galperin and Kaplan (2008a); Oudijk and Wade 2011). The original C&L paper dealt with GC/FID data gathered during the investigation of 12 diesel fuel releases over a period of 20+ years in Denmark and the Netherlands. The results showed that there was a statistically significant (95% probability) relationship between the age of the release and the n-C17/pristine ratio (Figure 5.18). Within the paper, the authors established a very clear list of criteria for the approach to be applied: (1) the diesel oil concentration in the samples had to exceed 100 mg/kg; (2) the samples had to be collected at least 1 m below ground level; (3) the samples had to be collected from the unsaturated zone beneath a paved area; and (4) the analyses had to include information on single components

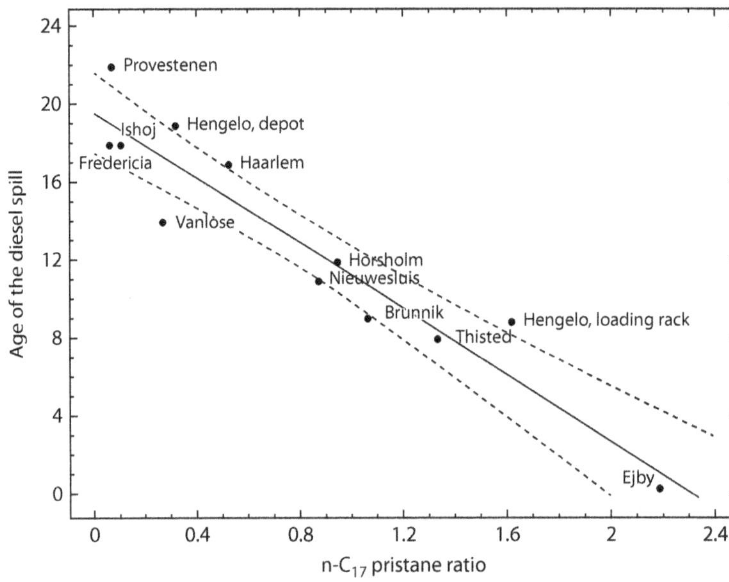

FIGURE 5.18 Relationship between the n-C17/pristine ratio and date of initial diesel fuel release. (Data from Christensen, J.B. and Larsen, T.H, *Fall 1993 Groundwater Monitoring and Remediation*, 142–129, 1993.)

such as n-C17, n-C18, pristine, and phytane. Also, the authors explicitly specified the exclusion of any samples where any petroleum product other than diesel fuel (for example, gasoline) was present. With the restrictions put in place within the C&L relationship, the exposure environment being studied was limited by the authors themselves, in that it was "to a large degree protected from direct factors increasing the rate or microbial activity" and in the "vadose zone containing lighter-than-water Non-Aqueous Phase Liquids (LNAPL) present in concentrations up to residual saturation" (Christensen and Larsen 1993).

Examination of the applicability of the C&L technique in the United States began with Wade (2001). Wade (2002) and others commented on the C&L approach shortly after the publication of Wade (2001) , and discussion began almost immediately on the applicability of the C&L technique within the U.S. peer-reviewed geochemical literature. Kaplan (2003), writing about the age dating of environmental contamination, included the C&L relationship for discussion. Hurst and Schmidt (2005) envisioned a wider application of the n-C17/pristine ratio to all circumstances involving release of distillate fuels, regardless of their location. Galperin and Kaplan (2008b) provided a derivation of the empirical C&L relationship from first principles, concluding that the C&L relationship adequately described the expected microbial effects of hydrocarbon degradation. By the mid-2000s, the forensic geochemical community was deeply polarized between those investigators who advocated wide use of the C&L approach and those who saw distinct problems with its use regardless of circumstances. Oudijk and Wade (2011) made a case for reasonable scientific thought within the forensic geochemical community, cautioning both extremes of the community not to be overly aggressive with application or dismissal but to evaluate the data on a case-by-case basis and use such results as one line of evidence but not the only line of evidence to be considered. Since that time, the use of the C&L approach has become one of a number of lines of evidence pursued in age dating distillate fuel releases. Moreover, subsequent to the C&L paper and related discussion being published, attempts have been made to apply the approach to such petroleum products as Jet A/kerosene, No. 2 fuel oil or diesel fuels, or No. 3 and 4 fuel oils, with varying degrees of success. Current forensic geochemical thought is that the C&L age dating approach has been used and that due consideration should be given to all factors and all data in any age dating assessment, including the C&L approach as published. However, it should be stressed that the C&L age dating approach for distillate fuels such as No. 2 fuel oil, diesel fuel, or others is not universally accepted within the forensic scientific community and that its use should be tempered with reasonableness, caution, and strict adherence to all aspects of the peer-reviewed published literature on the subject of petroleum age dating.

5.3.5 Forensic Dendroecology

Dendroecology is the study of tree ring growth to elucidate ecological information regarding the environment. It is an ongoing and established scientific procedure used to gather a time series of data regarding the ecology of the tree's environment. It is based on the principle that nearly all species of trees lay down one ring of growth per year; the rings are accurately counted, and additional information is gathered. Forensic dendroecology, on the other hand, is a newer development that takes advantage of the established dendroecology procedures and adds in data measurements to quantify selected contaminants such as metals, petroleum hydrocarbons, and chlorinated hydrocarbons. This technology has been applied in petroleum geochemical forensics to track the release of leaded gasolines, petroleum hydrocarbons, and sulfur-containing hydrocarbons in fuel oils/diesel fuel to the surrounding environment, with the aim of establishing financial liability for the release of selected contaminants (leaded gasolines or middle distillates) to the environment. Under simple circumstances, where a single release has occurred at a site, and it can be documented that no other sources are impacting the site, such an approach can provide additional information to other site investigation results using the methods discussed in this chapter. Attention must be paid to documenting the complete list of petroleum products released at the site by GC/FID before undertaking forensic dendroecological studies. In the event that multiple gasolines from multiple sources are in evidence or that used motor oil has been released at the site, forensic dendroecology results will be subject to serious technical challenge, and its potential addition to the results from the other forensic tools arrayed at a site needs to be considered carefully. However, under favorable circumstances, the principles of forensic dendroecology can be considered.

5.4 APPLICATIONS TO REMEDIAL SITE INVESTIGATION

As described in previous sections, the technical core of environmental forensics is fundamentally source identification. In remedial site investigations, the understanding of contaminant sources is one of the key components of CSMs. This section provides an overview of applications of chemical fingerprinting, arguably one of the most critical petroleum hydrocarbon forensic techniques, in remedial site investigations (Lu 2016). The first scenario is to use GC/FID chromatograms to enhance the interpretation of laser-induced fluorescence (LIF) high-resolution site characterization data so that LNAPL at a petroleum refinery site can be mapped in three dimensions more accurately. The second scenario is about determining LNAPL migration pathways so that a sounder remedial strategy can be developed. The third scenario is about the application of chemical fingerprinting in natural source zone evaluation. The final scenario is about avoiding unnecessary site investigations through proving the non-existence of perceived or suspected contaminant sources. With each scenario, an example is provided to illustrate how the chemical fingerprinting technique can be used to effectively characterize the petroleum hydrocarbons at petroleum-impacted sites.

5.4.1 Enhancement of Laser-Induced Fluorescence Data Interpretation to Facilitate LNAPL Mapping

LIF is a technique using UV or visible laser light delivered via direct push boring tools to excite (put energy into) PAH molecules present in NAPLs. The resulting fluorescence below the surface is measured. The intensity of fluorescence provides a means of semi-quantitatively estimating the saturation of NAPL (i.e., magnitude of impact) within the subsurface. Because this technique provides a means of characterizing LNAPL presence within the subsurface at high resolution, it has been used widely in the investigation of petroleum-impacted sites.

At sites with single or known product(s), a qualitative analysis of the observed fluorescence spectrum and fluorescence lifetime (the time between laser light absorption and fluorescence) may be used to identify different types of NAPL at different areas of the site or at different sites. However, at sites with multiple and/or unknown types of releases, it has become difficult or impossible to determine the types of LNAPL in subsurface with LIF data alone. Because the risk is dependent on LNAPL types (e.g., gasoline product has more environmental risk than heavier products such as diesel or lubricate oil), the identification of LNAPL types becomes critical in site investigation. The following case study illustrates how chemical fingerprint data may be used to achieve this.

The site is a large petroleum facility (Lu et al. 2013). LIF investigation was conducted throughout the site primarily to delineate the extent of the released LNAPL. In some areas where a comingled LNAPL plume was identified or suspected, understanding how the different petroleum hydrocarbons are distributed in the soil column is important, as soluble and toxic compounds (e.g., BTEX) in groundwater are more likely sourced from the products immediately in contact with groundwater (i.e., near the water table).

Figure 5.19a shows a ROST boring location relative to two monitoring wells (MW-1 and MW-2) where measurable LNAPL is present. MW-1 was screened between 35′ and 45′ below ground surface (bgs), and MW-2 was screened from 26′ to 66′ bgs, spanning a longer interval. GC/FID chromatograms are available for the LNAPL samples collected from both wells (Figures 5.19b,c). Based on the chromatograms, LNAPL from MW-2 contains a mixture of diesel, gasoline, and possibly kerosene, and LNAPL from MW-1 contains only diesel, which appears identical to the diesel found in MW-2.

Figure 5.19d shows the ROST profile from the push location described in the previous paragraph. LNAPL detected using LIF occurs in three distinct intervals (i.e., 22′–32′ bgs; 40′–43′ bgs; 58′–60′ bgs). Based on the distinctive fill colors (i.e., orange/brown at the upper interval; green in the middle interval; and blue in the lower interval), the NAPLs at the three intervals appear different. The orange/brown response (i.e., the middle interval) occurs between 40′ and 43′ bgs and correlates with the screen interval of MW-2, which contains diesel NAPL. The lower response is a dominantly blue waveform, which is typical of kerosene fluorescence characteristics. The upper response cannot be associated with a specific product based on fluorescence characteristics alone (i.e., fill color and waveform), likely due to complications of environmental alteration and soil matrix. However, based on the presence of gasoline in MW-2, which was screened across the entire vertical interval in which the three responses are detected in the ROST profile, the upper response is likely weathered gasoline.

The information derived from ROST is valuable in correlating and therefore mapping LNAPL occurrence across the site. In this case, the NAPLs in contact with groundwater are determined to be kerosene and/or diesel with high confidence. This finding, combined with the fact that the area is in the interior of the facility, provides decision makers with critical information in prioritizing the areas for remediation and selecting effective remedial technologies.

5.4.2 Developing Remedial Strategy by Mapping Release Migration Pathways

As described previously, CSMs are indispensable road maps for remedial site investigations. In any CSM, the two most critical components are contaminant source and migration pathways. During remedial site investigations, there are always data gaps in understanding sources and migration pathways, and therefore, the information on these two components is closely related in the process of interpreting or determining the extent of contaminants. In other words, different interpretations of migration pathways would lead to different delineations of the extent of contaminants, which

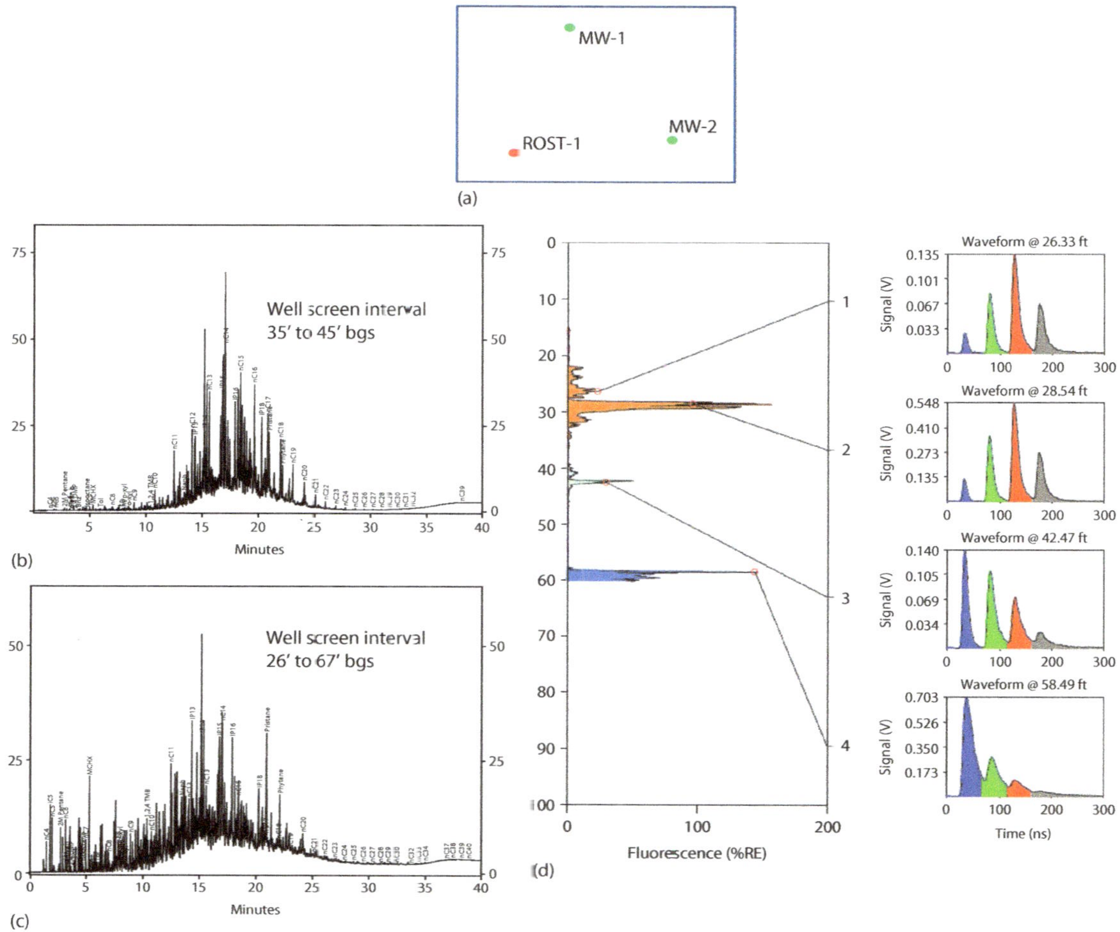

FIGURE 5.19 (a) The ROST boring location relative to two monitoring wells (MW-1 and MW-2); (b) GC/FID chromatogram for the LNAPL sample collected from MW-1; (c) GC/FID chromatogram for the LNAPL sample collected from MW-2. (d) the ROST profile from the push location in (a) (*x*-axis is the fluorescence intensity and *y*-axis represents the depth from ground surface in feet).

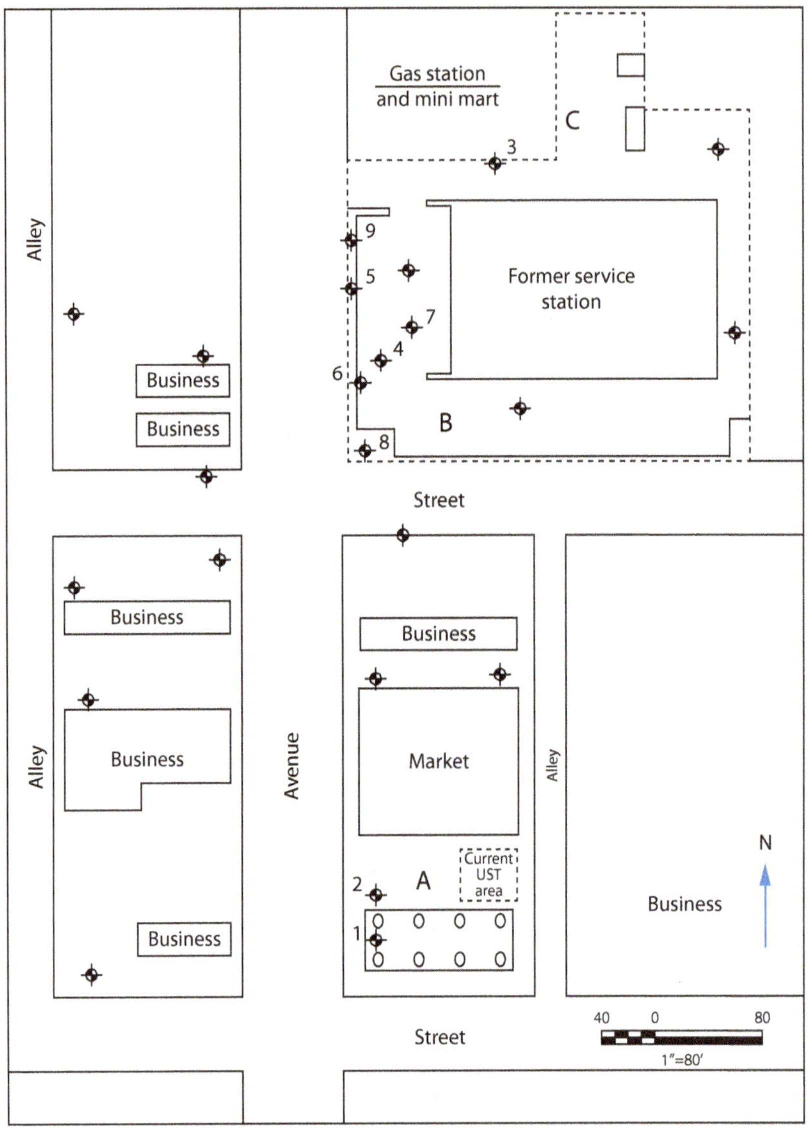

FIGURE 5.20 Site map. (From Mohler, R. and Lu, J. Use of statistic tools to improve confidence in analytical conclusions, Environmental Forensics: Proceedings of 2009 Annual Conference, Royal Society of Chemistry Special Publication No. 327, 2009.)

directly affects the development of a remedial strategy or design. In this regard, chemical fingerprint data may be used to enhance understanding of the migration pathways. A case study follows (Mohler and Lu 2009).

The subject site is located in a well-developed business district in a city in southern California (Figure 5.20) and is currently developed as a service station (A). To the north, there were two former service stations, one approximately 100 ft from the site (B) and another 200 ft from the site (C, further north). Groundwater monitoring data from wells at the site indicated that shallow groundwater was present at a depth of approximately 35–38 ft below ground surface. The interpreted groundwater flow direction was from (C), to (B), to (A) in general. Among the monitoring wells shown at the site, most of them contain petroleum free product(s). At the site, the liability associated cost allocation was not an issue, as the B and C properties had been acquired by the owner of

A property. However, understanding the source relationship of LNAPL is important in determining the migration pathways, which in turn helped in the development of a remedial strategy.

In resolving "source" and comingling issues, free product samples were taken from nine monitoring wells for forensics analysis. Among the nine samples, one was taken from near the former C station, six from near the former B station, and two from the subject site (A). The hydrocarbon fingerprint data were generated using a gas chromatograph with flame ionization detection (GC-FID). Lead speciation of the product samples was performed using a gas chromatograph coupled to an electron capture detector (GC-ECD). Total lead was determined using gas chromatography inductively coupled plasma spectroscopy (GC-ICP).

Detailed analysis can be found in Mohler and Lu (2009). Multivariate analysis results for lead and hydrocarbons are slightly different, but both revealed that Samples 1, 2, and 3 are similar, suggesting similar source. From a perspective of remedial investigation, the similarity of Samples 1, 2, and 3 suggests that the LNAPL plume composed of the LNAPL products spreads from Site C through A. Further delineation would have been necessary to map out the extent of contamination between the two locations.

5.4.3 Evaluating Natural Source Zone Depletion at Sites with LNAPL

At sites where LNAPL is present, natural source zone depletion (NSZD) emerges as an objective benchmark by which to compare the relative effectiveness of different remedial alternatives. NSZD is a combination of processes, including dissolution of LNAPL constituents into groundwater, volatilization of LNAPL constituents into the vadose zone, and subsequent biodegradation of the LNAPL constituents in both aqueous phase and vapor phase (ITRC 2009a,b). The depleted mass of the LNAPL constituents can be assessed both qualitatively and quantitatively for the saturated zone and the unsaturated zone. In current practice, the assessment of the depleted mass is primarily based on the concentrations of hydrocarbons in the dissolved phase (in the saturated zone) and the vapor phase (in the unsaturated zone); the chemical characteristics of LNAPL are not considered as part of the assessment process.

At sites with a long history of monitoring and compliance, LNAPL samples may have been collected to obtain chemical fingerprints for various purposes. When available, these data provide important information to understand the occurrence of mass distribution from the source zone to the groundwater and vadose zones and to identify NSZD processes. For example, if evaporation is the dominant process, the preferentially depleted constituents will be those with lower boiling points or molecular weights. Figure 5.21a,b are two GC/FID chromatograms of a type of crude oil collected from a storage tank (fresh) and from a pond for processed fluid (crude oil and water). As can be observed from Figure 5.21b, the constituents of lower molecular weight (i.e., retention time up to approximately 10 min) were almost completely depleted due primarily to evaporation.

In the case of a dissolution-dominant process, the preferentially depleted constituents will be those with higher solubility, such as BTEX. Figure 5.22a,b are chromatograms from LNAPL collected from the same well. In the LNAPL collected in 2006, toluene was the predominant component and dwarfed other products (gasoline range and heavy products) on the chromatogram. However, in 2012, the LNAPL appears very different. It was apparent that preferential loss of toluene to groundwater, largely through dissolution, made other components show up on the chromatogram more prominently.

In the case of a biodegradation-dominant process, the chromatogram pattern of LNAPL will change, but the hydrocarbon range may not be affected significantly, as the most pronounced loss would be alkanes, which occur throughout the entire hydrocarbon range. Figures 5.23a,b are the chromatograms of diesel range LNAPL collected from the same well in 2002 and 2006. In 2002, the alkanes were prominent across the entire hydrocarbon range; however, in 2006, the alkanes were noticeably depleted. Note that in 2002, the peak of nC17 was much higher than pristine's (pristine is an isoprenoid, which is relatively resistant to biodegradation), and the peak of nC18 was

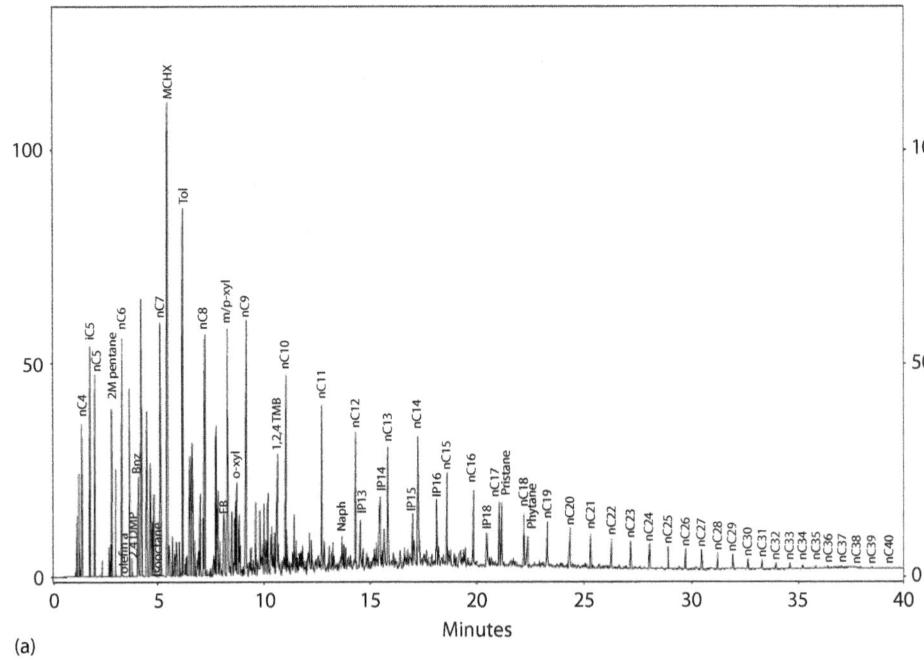

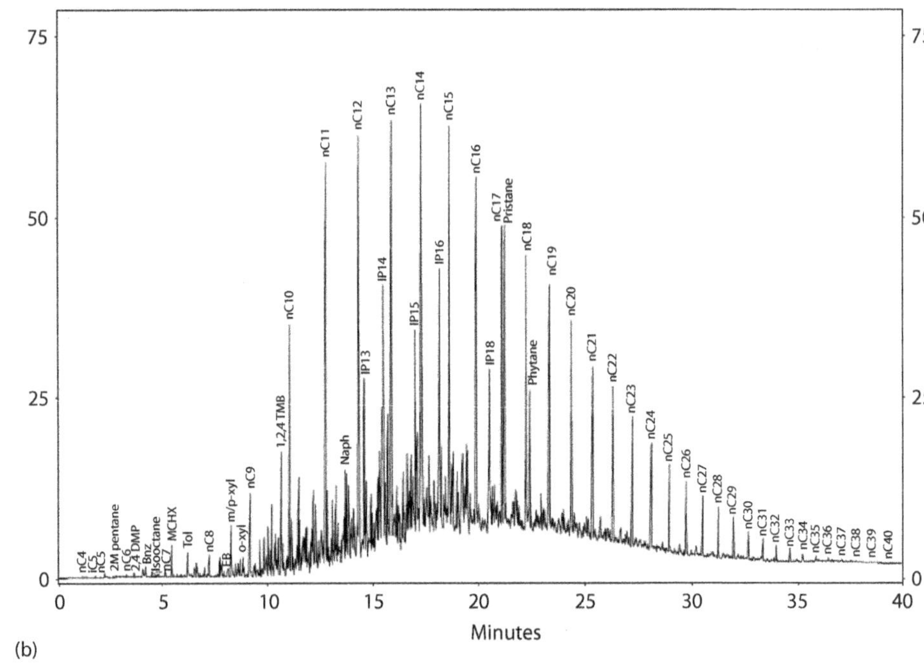

FIGURE 5.21 GC/FID chromatogram of a type of crude collected (a) from a storage tank (fresh) and (b) from a pond for processed fluid (crude and water).

much higher than phytane's (phytane is another isoprenoid, which is similar to pristine). In 2006, the situation reversed; in other words, the peaks of the alkanes were much lower than those of their neighboring isoprenoid compounds.

Based on the examples provided in the preceding paragraphs, by examining the characteristics of the chromatograms and/or comparing the chemical fingerprints of LNAPL collected at different

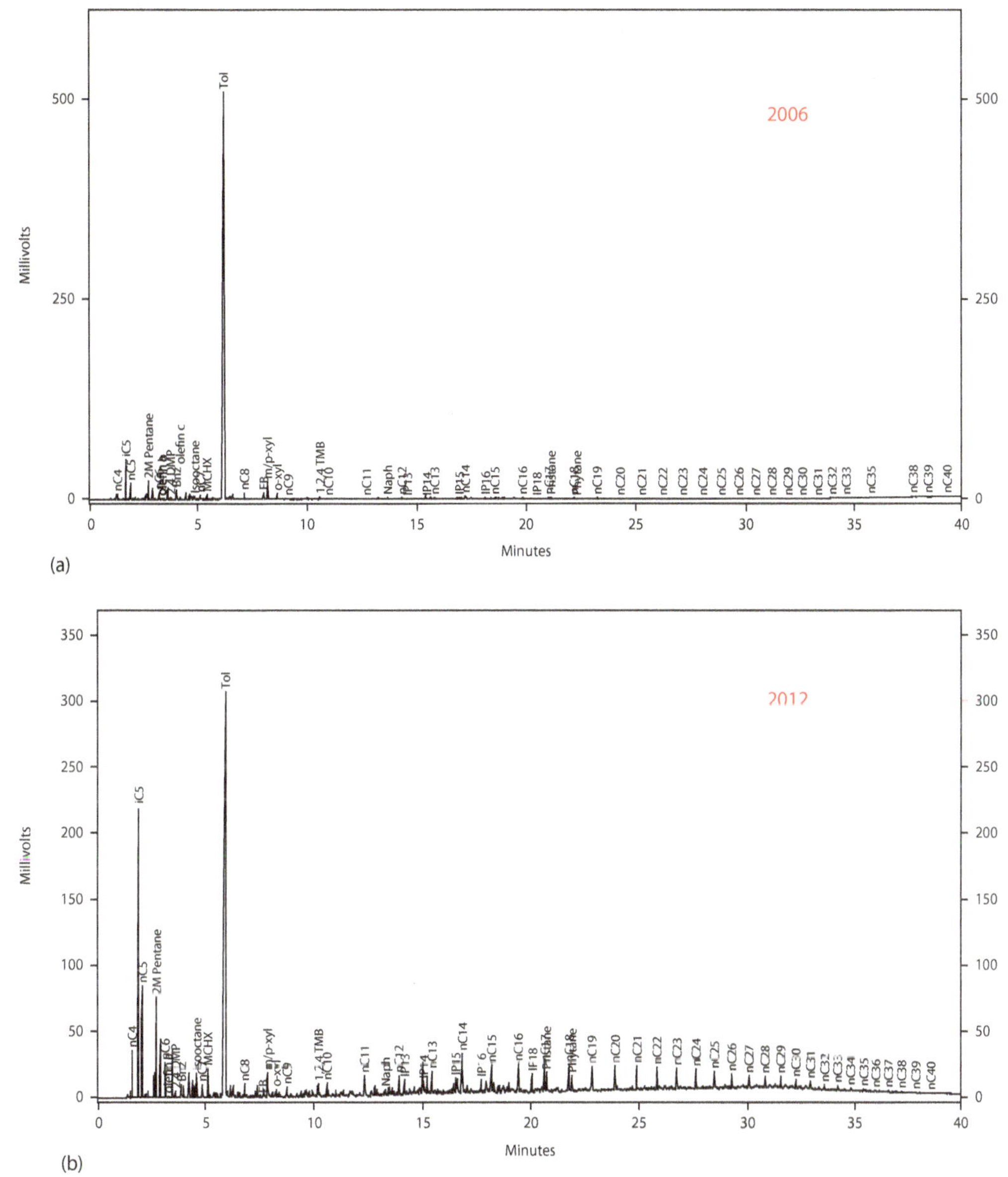

FIGURE 5.22 GC/FID chromatograms from LNAPL collected from a monitoring well (a) in 2006 and (b) in 2012.

times, the dominant process and its magnitude can be evaluated to complement the quantitative assessment of the depleted mass based on the hydrocarbons of interest in the vapor and dissolved phases.

5.4.4 Avoiding Unnecessary Site Investigations by Eliminating Perceived On-Going Contaminant Sources

At active petroleum facilities such as refineries and tank farms, release prevention is a key component of operation. For obvious releases, their sources should be identified and measures taken to stop releases. However, there are cases in which, when suspected releases cannot be ruled out, vigorous

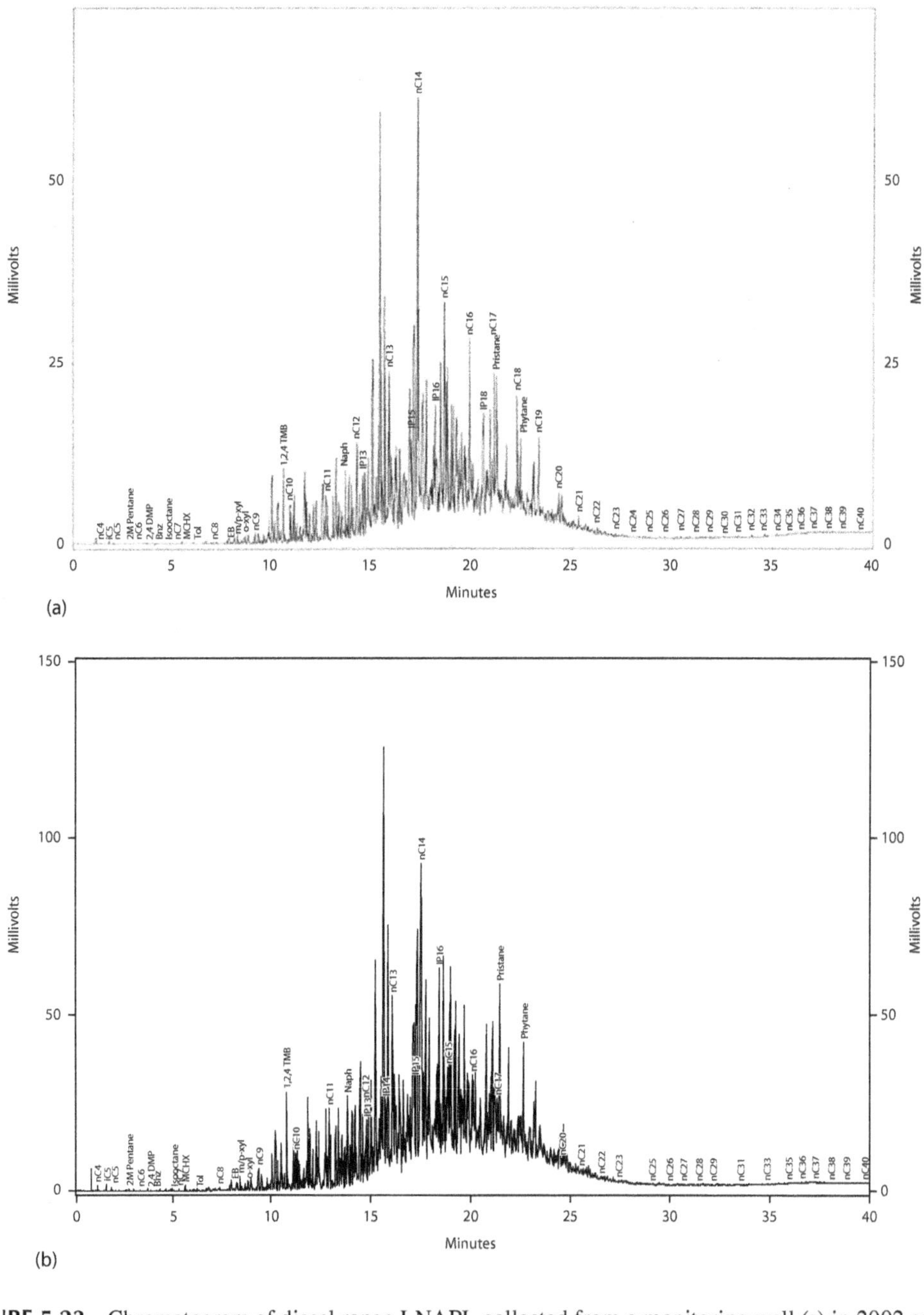

FIGURE 5.23 Chromatogram of diesel range LNAPL collected from a monitoring well (a) in 2002 and (b) in 2006.

investigation must be conducted to identify the "sources." This can be a waste of resources when it is found that the release did not occur. In this regard, environmental forensics has a role to play in eliminating or minimizing such incidents. There follows a summary of a case study illustrating this (Lu and Kowalewski 2016).

During a routine inspection in a petroleum product terminal, hydrocarbon staining was found in soil along the perimeter of a diesel fuel tank (Figure 5.24). As the observation was made a few days after a magnitude 5.1 earthquake in the general area of the terminal, the hydrocarbon staining in the

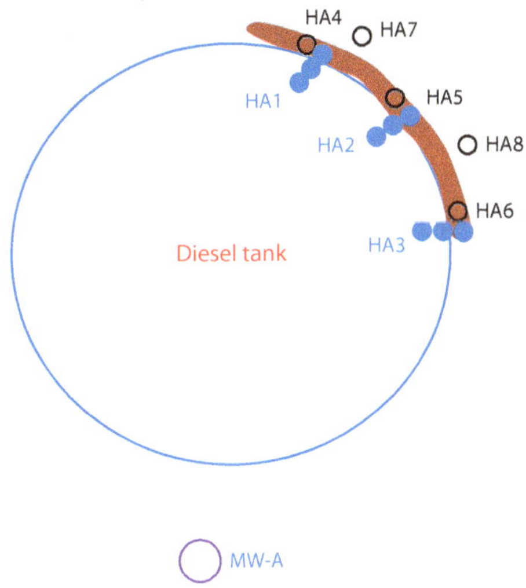

FIGURE 5.24 Sample locations relative to the suspected leaking tank. Area in red is hydrocarbon-stained soil. Black circles are for vertical hand auger sampling boreholes, and blue dots are slanted sampling boreholes beneath the tank. MW-A is the monitoring well with LNAPL. (From Lu, J. and Kowalewski, D., *Journal of Environmental Protection*, 7(3), 2016.)

soil was suspected to be the result of a leak from the tank. The observation triggered an immediate tank inspection, which included examination of the tank exterior, piping, and connections; pressure testing; wall thickness gauging; and an inventory check. The results of the initial inspection suggested that there were no integrity issues with the tank.

Before more vigorous and more disruptive inspections were conducted, chemical fingerprinting was performed to assess the nature of soil staining and the potential connection with the previous inspections. Chemical fingerprinting was conducted to assess the nature of soil staining, phase liquid (LNAPL) sample from the monitoring well in the immediate vicinity, and representative soil samples from the perimeter of and beneath the tank. A tiered laboratory analytical approach was taken. Initial GC/FID carbon chain analysis was conducted for all samples, and the results were used to screen out the "stained" soil samples that were not related to the fuel in the tank. For a few stained samples that could not be ruled out with the first-tier analysis, GC/MS was conducted for detailed fingerprint analyses.

With the fingerprint data collected, GC/MS full scan m/z 85 ion profile, associated diagnostic compounds, and sesquiterpene, terpane, and sterane biomarkers (Figures 5.25 and 5.26) provided critical lines of evidence for differentiation between the hydrocarbons in the soil and the fuel in the tank. In conjunction with other information, it was determined that the soil staining was not a result of suspected tank release but remobilization of hydrocarbons from the historically impacted soil beneath the tank to the soil along the perimeter of the tank. This case study demonstrated the elimination of unnecessary remedial investigation through diesel source identification and correlation using knowledge of LNAPL mobility in response to earthquake seismic waves and chemical fingerprinting.

5.5 SUMMARY

The work now referred to as *forensic geochemistry* has been carried out for decades under names such as *resource assessment*, *impact assessment*, *classic environmental science*, *studying contaminant releases*, and the like. Since the 1990s, when the name became linked with environmental investigations of crude oil spill investigations in support of Natural Resource Damage Assessments (NRDA) or law suits engendering various levels of legal hearings, including U.S. federal or various state court cases, such work has been referred to as *forensic geochemistry*. However, other

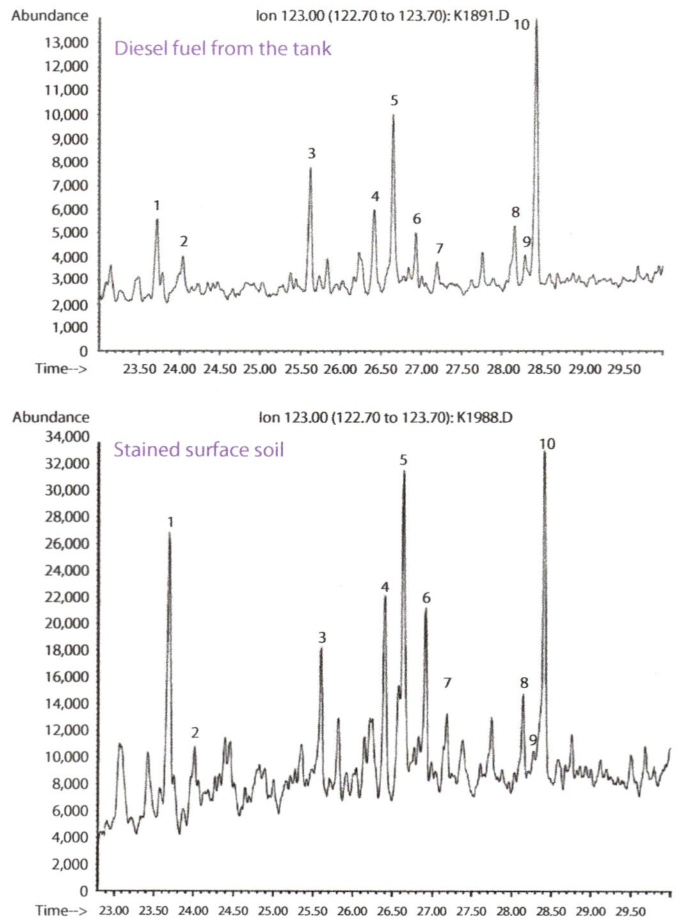

Ratios of Sesquiterpane ratios

Ratios	Tank diesel fuel	Stained surface soil	Difference*
SES1 / SES2	2.60	4.37	68%
SES1 / SES3	0.81	2.26	180%
SES1 / SES5	0.57	1.04	81%
SES3 / SES5	0.71	0.46	35%
SES3 / SES10	0.47	0.49	3%
SES4 / SES5	0.56	0.62	12%
SES5 / SES10	0.67	1.07	60%
SES6 / SES5	0.31	0.45	43%
SES8 / SES10	0.22	0.27	25%

*Difference between the soil sample and tank fuel sample

FIGURE 5.25 Sesquiterpane profiles and ratios. (From Lu, J. and Kowalewski, D., *Journal of Environmental Protection*, 7(3), 2016.)

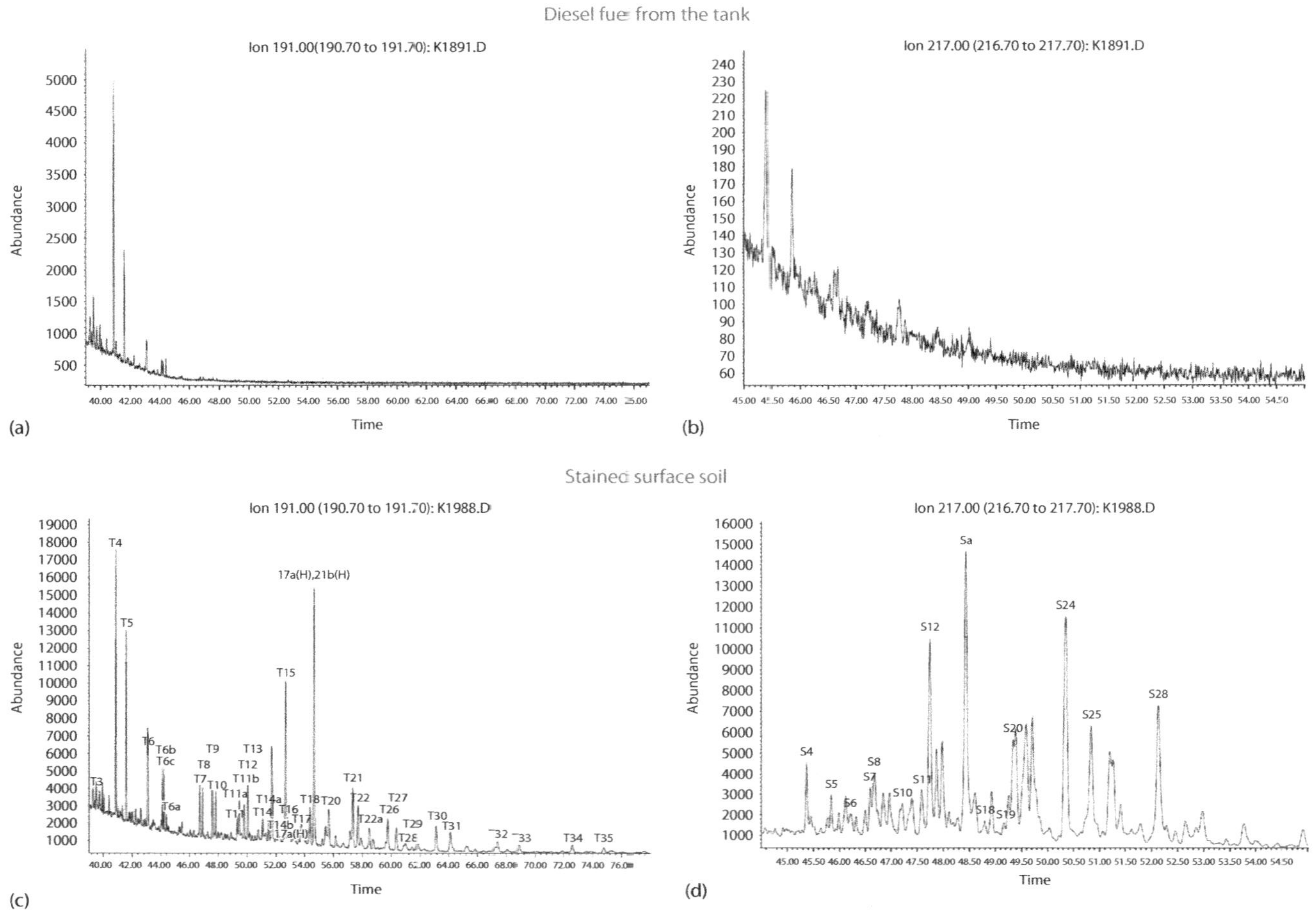

FIGURE 5.26 m/z 191 and 217 ion profiles of the selected samples of interest: (a) terpane biomarkers of diesel fuel from the tank; (b) sterane biomarkers of product sample from the monitoring well nearby the tank; (c) terpane biomarkers of stained surface soil; and (d) sterane biomarkers of soil sample from beneath the tank (at 9 ft bgs). (From Lu, J. and Kowalewski, D., *Journal of Environmental Protection*, 7(3), 2016.)

environmental needs always existed, and now it has been realized that there are wider needs and that these forensic geochemical techniques can be applied to a much wider audience. It must be remembered that the techniques have been documented in a substantial number of publications in the peer-reviewed technical literature. The scientific bases for these techniques exist and must be followed faithfully in the conduct of such investigations going forward. In the final analysis, the work completed must conform to that already widely accepted within the forensics and legal communities. Furthermore, the forensic geochemical community is continually being renewed with advancements in scholarly research in this field. As such, when using such techniques, it will be wise to remain current in the literature support for your work as well.

PRACTICE AND EXERCISES

Question 1: What is environmental forensics? How does it evolve?

Answer 1: Environmental forensics is best defined as the formalized study of environmental science of contaminant releases leading to entry into some type of formalized debate, which leads ultimately to some type of legal setting. It has been defined as the systematic examination of environmental data used in litigation, and where the litigation occurs in the United States is the courtroom. However, the word *forensic* is related to the Latin word *forum* and "refers to any public discussion or debate"; thus, the final destination being the courtroom often is not realized. But the process is conducted as though the courtroom were the final destination.

Question 2: How is environmental forensics related to remedial site investigations?

Answer 2: The science of environmental forensics is fundamentally about understanding contaminant sources, which is one of the critical components of a conceptual site model for site characterization. At a contaminated site, contaminants originate at primary sources and migrate through geological media (i.e., pathways). The footprints of contaminants on the pathways serve as secondary sources, which control the distribution of currently impacted areas. Because the contaminant sources are the starting points of the migration pathways, by better understanding contaminant sources, we can do a better job of predicting and delineating currently impacted areas, which is the major objective of remedial site investigations.

Question 3: What approach would a subject matter expert take to an environmental forensic investigation?

Answer 3: A subject matter expert would typically take a tiered approach, starting with gathering general information (i.e., from conventional site investigation) and then conducting more focused or specialized studies including chemical fingerprinting, age dating, stable isotope analysis, statistics, and so on.

Question 4: What forensic information can you extract from conventional site investigation?

Answer 4: A substantial amount of information related to contaminant sources may be obtained from conventional site investigations. Documents with source information may include site history, geological cross sections, hydrogeological reports, groundwater flow and fate and transport model reports, and so on.

Question 5: What is chemical fingerprinting of petroleum products? What are the common techniques?

Answer 5: Chemical fingerprinting is the detailed examination of the composition of petroleum hydrocarbon-derived fuel products. It takes advantage of state-of-the-art analytical chemical techniques such as gas chromatography and combined gas chromatography/mass spectrometry, using capillary column separation techniques to develop such detailed chemical data on a compound-by-compound basis for some products and for general classification of hydrocarbons with the more complex petroleum products. Using the detailed hydrocarbon composition data, the composition of the hydrocarbon backbone of each sample can be examined and changes understood.

Question 6: What is the fundamental basis of stable isotope analysis for source identification and correlation?

Answer 6: The fundamental basis is the variability of isotope ratios of commercial products between manufacturers, or between different production batches produced by the same manufacturer.

Question 7: What is age dating? Why it is important?

Answer 7: The approaches to age dating have been developed as an aid to identifying when contamination from petroleum product may have begun at a site. The importance of such information is related to the need for such age-related information for the determination of financial responsibilities placed on the responsible party for the initial release of petroleum products to the environment. Various age dating approaches have been described in the peer-reviewed literature, and it is here that the student must go to understand the various issues surrounding age dating. The simple explanation is that available age dating approaches must NOT be simply blindly applied in each and every situation; rather, the literature must be consulted to understand the limitations and assumptions behind each and every age dating paper in the literature. Failure to do so presents a great risk to the scientist and to the scientist's client that the results will be inapplicable to their current site. Age dating of gasolines and distillate fuels and age dating using stable isotopes and radioisotopes have been completed successfully under a variety of circumstances by following such published peer-reviewed technical literature strictures.

Question 8: Provide two examples of applications of environmental forensics to remedial site investigations.

REFERENCES

Atlantic Richfield Corporation, 1980. Annual Report. Atlantic Richfield Corporation (ARCO), Los Angeles, CA.

Bence, A.E., Kvenvolclen, K.A., and Kennicutt, II, M.C. 1996. Organic geochemistry applied to environmental assessments of Prince William Sound, Alaska, after the Exxon Valdez oil spill. A review. *Organic Geochemistry*. 24: 7–42.

Beneteau, K.M., Aravena, R., and Frape, S.K. 1999. Isotopic characterization of chlorinated solvents laboratory and field results. *Organic Geochemistry*. 30: 739–753.

Boehm, P.D. 2006. Polycyclic aromatic hydrocarbons (PAHs). In: *Environmental Forensics – Contaminant Specific Guide*, R.D. Morrison and B.L. Murphy, Eds., Academic Press, San Diego, pp. 313–334.

Boehm, P.D., Douglas, G.S., Burns, W.A., Mankiewiez, P.J., Page, D.S., and Burns, A.R. 1997. Application of petroleum hydrocarbon chemical fingerprinting and allocation techniques after the Exxon Valdez oil spill. *Marine Pollution Bulletin*. 34: 599–613.

California Environmental Protection Agency. 2007. *A Review of The Potential Human and Environmental Health Impacts of Synthetic Motor Oils*. Report prepared by California Environmental Protection Agency, Office of Environmental Health Hazard Assessment, Integrated Risk Assessment Branch. 16 Pages. Report dated April 2007.

Christensen, J.B. and Larsen, T.H. 1993. Method for determining the age of diesel oil spills in the soil. *Fall 1993 Groundwater Monitoring and Remediation*. Fall 1993: 142–129.

Cline, P.V., Delfino, J.J., Rao, P., and Suresh, C. 1991. Partitioning of aromatic constituents into water from gasoline and other complex solvent mixtures. *Environmental Science and Technology*. 25: 914–920.

Cohen, R.M. and Mercer, J.W. 1993. *DNAPL Site Evaluation*. C.K. Smoley, Boca Raton.

Costa, H.J., White, H.A., and Ruspantini, J.J. 2004. Distinguishing PAH backgrounds and MGP residues in sediments of a freshwater creek. *Environmental Forensics*. 5: 171–182. ISSN: 1527–5922 print/1527–5930 online.

Crimi, M. 2012. The Geochemical Effects of In Situ Chemical Oxidation: Advantages and Challenges, AECOM webinar, October 25, 2012.

Dominguez-Rosado, E. and Pichtel, J. 2003. Chemical characterization of fresh, used and weathered motor oil via GC/MS, NMR and FTIR techniques. *Proceedings of the Indiana Academy of Science*. 112(2): 109–116.

Douglas, G.S., Bence, A.E., Prince, R.C., McMillen, S.J., and Butler, E.L. 1996. Environmental stability of selected petroleum hydrocarbon source and weathering ratio. *Environmental Science and Technology*. 30: 2332–2389.

Douglas, G.S., Emsbo-Mattingly, S.D., Stout, S.A., Uhler, A.D., and McCarthy, K.J. 2007. Chemical fingerprinting methods. In: *Introduction to Environmental Forensics*, 2nd edn., B.L. Murphy and R.D. Morrison, Eds., Academic Press, New York, pp. 311–454.

Douglas, G.S., Emsbo-Mattingly, S.D., Stout, S.A., Uhler, A.D., and McCarthy, K.J. 2015. Hydrocarbon fingerprinting methods. In: *Introduction to Environmental Forensics*, 3rd edn., B.L. Murphy and R.D. Morrison, Eds., Academic Press, San Diego.

Feenstra, S., Mackay, D.M., and Cherry, J.A. 1991. A method for assessing residual NAPL based on organic chemical concentrations in soil samples. *Ground Water Monitoring and Remediation*, Spring 1991.

Fetter, C.W. 1994. *Applied Hydrogeology*. Prentice-Hall, Upper Saddle River, NJ.

Galperin, Y. and Kaplan, I.R. 2008a. Age significance of n-C17/Pr ratios in forensic investigations of refined product and crude oil releases: Discussion. *Environmental Geosciences*. 15(2): 85–86.

Galperin, Y. and Kaplan, I.R. 2008b. Zero-order kinetics model for the Christensen-Larsen method for fugitive fuel age estimates. *Ground Water Monitoring and Remediation*. 28(2) Spring: 1–4.

Gauthier, T.D. and Murphy, B.L. 2003. Age dating groundwater plumes based on ratios of 1,1-dichloroethylene to 1,1,1-trichloroethane: An uncertainty analysis. *Environmental Forensics*. 4: 205–213.

Hoefs, J. 1987. *Stable Isotope Geochemistry*. Springer, New York, NY.

Holt, B. D., Sturchio, N. C., Abrajano, T., and Heraty, L. J., 1997. Conversion of chlorinated volatile organic compounds to carbon dioxide and methyl chloride for isotopic analysis of carbon and chlorine. Analytical Chemistry 69: 2727–2733.

Hunkeler, D. and Aravena, R., 2000a. Evidence of substantial carbon isotope fractionation among substrate, inorganic carbon, and biomass during aerobic mineralization of 1,2-dichloroethane by Xanthobacter autotrophicus. Applied and Environmental Microbiology 66: 4870–4876.

Hunkeler, D., Meckenstock, R.U., Lollar, B., Schmidt, T.C., and Wilson, J.T., 2009. A Guide for Assessing Biodegradation and Source Identification of Organic Groundwater Contaminants Using Compound Specific Isotope Analysis (CSIA). U.S. Environmental Protection Agency, Washington, D.C., EPA/600/R-08/148, 2009.

Hurst, R.W. and Schmidt, G.W. 2005. Age significance of nC17/Pr ratios in forensic investigations of refined product and crude oil releases. *Environmental Geosciences*. 12(3): ISSN 1075-9565. 177–192.

Interstate Technology & Regulatory Council. 2009a. *Evaluating Natural Source Zone Depletion at Sites with LNAPL*. Interstate Technology & Regulatory Council, LNAPLs Team, Washington, DC. www.itrcweb.org.

Interstate Technology & Regulatory Council. 2009b. *Evaluating LNAPL Remedial Technologies for Achieving Project Goals*. Interstate Technology & Regulatory Council, LNAPLs Team, Washington, DC. www.itrcweb.org.

Interstate Technology & Regulatory Council. 2010. *Technology Overview—Use and Measurement of Mass Flux and Mass Discharge*. Interstate Technology & Regulatory Council, Integrated DNAPL Site Strategy Team, Washington, DC. www.itrcweb.org.

Jendrzejewski, N., Eggenkamp, H.G.M., and Coleman, M.L., 1997. Sequential determination of chlorine and carbon isotopic composition in single microliter samples of chlorinated solvents. Analytical Chemistry 69: 4259–4266.

Jendrzejewski, N., Eggenkamp, H.G.M., and Coleman, M. L., 2001. Characterization of chlorinated hydrocarbons from chlorine and carbon isotopic compositions: scope of application to environmental problems. Applied Geochemistry 16: 1021–1031.

Johnson, G.W. 2015. Principal components analysis and receptor models in environmental forensics. In: *Introduction to Environmental Forensics*, 3rd edn. B.L. Murphy and R.D. Morrison, Eds., Academic Press, San Diego.

Interstate Technology & Regulatory Council, 2013. *Groundwater statistics and monitoring compliance*. Interstate Technology & Regulatory Council, Groundwater statistics and monitoring complianceTeam, Washington, DC.

Kaplan, I.R. 2003. Age dating of environmental organic residues. *Environmental Forensics*. 4: 95–141.

Kaplan, I.R. and Galperin, Y. 1996. How to recognize a hydrocarbon fuel in the environment and estimate its age of release. In: *Groundwater and Soil Contamination: Technical Preparation and Litigation Management*, T.J. Bois and B.J. Luther, Eds., Wiley Law Publishers, New York, NY, pp. 145–199.

Kaplan, I.R., Galperin, Y., Lu, S.-T., and Lee, R.-P. 1996. Patterns of chemical changes during environmental alteration of hydrocarbon fuels. *Groundwater Monitoring and Remediation, Fall*. Fall 1996: 113–124.

Kaplan, I.R., Galperin, Y., Lu, S.-T., and Lee, R.P. 1997. Forensic environmental geochemistry: Differentiation of fuel-types, their sources and release time. *Organic Geochemistry*. 27 (5–6): 289–317.

Lu, J. 2015. Identification of forensic information from existing conventional site investigation data. In: *Introduction to Environmental Forensics*, B.L. Murphy and R.D. Morrison, Eds., ISBN: 978012406962, Elsevier/Academic Press, San Diego, CA.

Lu, J. 2016. Use of chemical fingerprint data to enhance remedial site investigations at petroleum impacted sites. *Environmental Forensics*. 17(1): 19–26.

Lu, J. and Kowalewski, D. 2016. Resolution of suspected tank release of diesel fuel using diagnostic compounds and biomarkers. *Journal of Environmental Protection*. 7(3): 473–480.

Lu, J., Germain, R., and Andrews, T. 2013. NAPL source identification utilizing data from laser induced fluorescence (LIF) screening tools. In: *Proceedings of 2013 Annual Conference, Royal Society of Chemistry Special Publication,* Cambridge, UK.

Meyers, R.A. 1986. *Handbook of Petroleum Refining Processes*. Chemical Process Technology Handbook Series. McGraw-Hill, New York, NY.

Meyers, R.A. 1997. *Handbook of Petroleum Refining Processes*. 2nd edn. Chemical Process Technology Handbook Series. McGraw-Hill, New York, NY.

Meyers, R.A. 2004. *Handbook of Petroleum Refining Processes*. 3rd edn. Chemical Process Technology Handbook Series. McGraw-Hill, New York, NY.

Mohler, R. and Lu, J. 2009. Use of statistical tools to improve confidence in analytical conclusions. In: *Environmental Forensics: Proceedings of 2009 Annual Conference*, Royal Society of Chemistry Special Publication No. 327, Cambridge, UK.

Morrison, R.D. and Murphy, B.L. 2006. Environmental Forensics – Contaminants Specific Guide. Academic Press, San Diego.

Murphy, B.L. and Morrison, R.D. 2002. *Introduction to Environmental Forensics*. 1st edn. Academic Press, San Diego.

Murphy, B.L. and Morrison, R.D. 2015. *Introduction to Environmental Forensics*. 3rd edn. Academic Press, San Diego.

Oil and Gas Journal, 1979. Published by the PennWell Corporation, May 21, 1979 Issue, Houston, TX.

Oudijk, G. 2010. The rise and fall of organometallic additives in automobile gasolines. *Environmental Forensics Journal*. 11(1): 17–49.

Oudijk, G. 2012. Age dating of middle-distillate fuels released to the subsurface environment. In: *Earth Sciences*, Dr. Imran Ahmad Dar, Ed., ISBN: 978-9533-307-861-8, In Tech. Available from www.intechopen.com/.

Oudijk, G. and Wade, M.J. 2011. Editorial commentaries and responses. On age-dating distillate fuels: A commentary on the methods of Hurst & Schmidt (2005) and Galperin & Kaplan (2008). *Environmental Forensics*. 12: 1–2. ISSN: 1527–5922 print/1527–5930 online.

Page, D.S., Bence, A.E., Burns, W.A., Boehm, P.D., Brown, J.S., and Douglas, G.S. 2002. Holistic approach to hydrocarbon source allocation in the subtidal sediments of Prince William Sound Embayments. *Journal of Environmental Forensics*. 3: 331 340.

Powers, S.E., Rice, D., Dooher, B., and Alvarez, P.J. J. 2001. Will ethanol-blended gasoline affect groundwater quality? *Environmental Science and Technology*. 35(1): 24A–30A.

Prince, R.C., Elmendorf, D.L., Lute, J.R., Hsu, C.S., Haith, C.E., Senius, J.D., Dechert, G.J., et al. 1994. 17a(H),21b(H)-hopane as a conserved internal marker for estimating the biodegradation of crude oil. *Environmental Science and Technology*. 28(1): 142–145.

Pruell, R.J. and Quinn, J.G. 1988. Accumulation of polycyclic aromatic hydrocarbons in crankcase oil. *Environmental Pollution*. 49(1988): 89–97.

Rhue, R.D., Mansell, R.S., Ou, L.-T., Cox, R., Tang, S.R., and Ouyang, Y. 1992. The fate and behavior of lead alkyls in the environment: A review. *Critical Reviews in Environmental Control*. 22(3/4): 169–193.

Saber, D., Mauro, D., and Sirivedhin, T. 2006. Environmental forensics investigation in sediments near a former manufactured gas plant site. *Environmental Forensics*. 7: 65–75. ISSN: 1527–5922 print / 1527–5930 online.

Sauer, T.C. and Boehm, P.D. 1991. The use of defensible analytical chemical measurements for oil spill natural resource damage assessments. *Proceedings, 1991 International Oil Spill Conference*. American Petroleum Institute, Washington, DC, pp. 363–369.

Schmidt, G., Beckmann D., and Torkelson, B.A. 2002. Technique for estimating the age of regular/mid-grade gasolines released to the subsurface since the early 1970's. *Environmental Forensics*. 3: 145–162.

Schmidt, G.W., Beckmann, D.D., and Torkelson, B.E. 2003. Gas chromatography technique for estimating the octane rating of fresh and weathered gasolines released to soils and groundwater. *Environmental Forensics*. 4: 75–80. doi: 10.1080/15275920303484.

Smith, J.E., Ed. 1968. *"Torrey Canyon" Pollution and Marine Life. A report by the Plymouth Laboratory of the Marine Biological Association of the United Kingdom.* Marine Biological Association of the United Kingdom, University Printing House, Cambridge, UK. 196 pages.

Speight, J.G. 1980. *The Chemistry and Technology of Petroleum.* Chemical Industries Series 3. Marcel Dekker, New York, NY.

Speight, J.G. 1991. *The Chemistry and Technology of Petroleum.* 2nd edn., Revised and Expanded. Chemical Industries Series 44. Marcel Dekker,, New York, NY. 760 pages.

Speight, J.G. 1999. *The Chemistry and Technology of Petroleum.* 3rd edn., Revised and Expanded. Chemical Industries Series 76. Marcel Dekker,, New York, NY.

Speight, J.G. 2007. *The Chemistry and Technology of Petroleum.* 4th edn. CRC Press, Baton Rouge, LA.

Speight, J.G. 2013. *The Chemistry and Technology of Petroleum.* 5th edn. CRC Press, Baton Rouge, LA.

Stout, S.A. 2016. Oil spill fingerprinting method for oily matrices used in the Deepwater Horizon NRDA. *Environmental Forensics.* 17(3): 218–243. doi: 10.1080/15275922.2016.1177759.

Stout, S.A. and Douglas, G.S. 2004. Diamondoid hydrocarbons—application in the chemical fingerprinting of natural gas condensate and gasoline. *Environmental Forensics.* 5(4): 225–235.

Stout, S.A. and Wang, Z. 2008. Diagnostic compounds for fingerprinting petroleum in the environment. In: *Environmental Forensics*, R.E. Hester and R.M. Harrison, Eds., Special Publication No. 26. Royal Society of Chemistry, London, pp. 54–104.

Stout, S.A., Uhler, A.D., and Boehm, P.D. 2001. Recognition of and allocation among sources of PAH in urban sediments. *Environmental Claims Journal.* 13: 141–158.

Stout, S.A., Uhler, A.D., McCarthy, K.J., and Emsbo-Mattingly, S. 2002. Chemical fingerprinting of hydrocarbons (chapter 6). In: *Introduction to Environmental Forensics*, B.L. Murphy and R.D. Morrison, Eds., Academic Press, San Diego, CA.

Stout, S.A., Uhler, A.D., and McCarthy, K.J. 2005. Middle distillate fuel fingerprinting using drimane-based bicyclic sesquiterpanes. *Environmental Forensics.* 6: 241–251. ISSN: 1527–5922 print/1527–5930 online.

Stout, S.A., Douglas, G.S., and Uhler, A.D. 2012. Automotive gasoline. Chapter 18. In: *Environmental Forensic Contaminant Specific Guide*, R.D. Morrison and B.L. Murphy, Eds., Elsevier, New York, NY, 533 pp.

Sullivan, P.J., Agardy, F.J., and Traub, R.K. 2001. *Practical Environmental Forensics.* Environmental Engineering series, Nelson L. Nemerow Ed., Series Advisor, John Wiley & Sons, , New York, NY. 587 pages.

The Royal Society of London. 1936. *The Pollution of the Sea and Shore by Oil.* Report submitted to the Council of the Royal Society and printed for private circulation. Report prepared by N.K. Adam, F.R.S., The Sir William Ramsey Laboratories of Physical and Inorganic Chemistry, University College, London, WC1, 25 October, 1936.

Uhler, R.M., Healey, E.M., McCarthy, K.J., Uhler, A.D., and Stout, S.A. 2003. Molecular fingerprinting of gasoline by a modified EPA 8260 gas chromatography-mass spectrometry method. *International Journal of Environmental Analytical Chemistry.* 83(1): 1–20.

USEPA, 2006. Data quality assessment: statistical methods for practitioners. Office of Environmental Information, Publication EPA/240/B-06/003, February, Washington D.C.

U.S. Federal Register, 1979. U.S. Government Printing Office, Vol. 44(45): 12242, Washington, D.C.

Wade, M.J. 2001. Age-dating diesel fuel spills: Using the European empirical time-based model in the USA. *Environmental Forensics.* **2**: 347–359.

Wade, M.J. 2002. Commentaries and perspectives. Invited commentary on the Christensen and Larsen Technique. *Environmental Forensics.* **3**: 13.

Wade, M.J. 2009. Thoughts on manufacturing changes in the US Petroleum Industry: Implications for age-dating, calculating weathering indices and hydrocarbon fingerprinting. *International Journal of Soil, Sediment and Water.* 2(1), ISSN: 1940-3259. 8 pages.

Wade, M.J. 2014. Failures in the historical age dating of separate phase product leaded gasoline releases in the United States forensic geochemical community.. In: *Environmental Forensics: Proceedings of the 2013 INEF Annual Conference*, R.D. Morrison and G. O'Sullivan, Eds., Royal Society of Chemistry, London, pp. 37–66. Print ISBN: 978-1-84973-944-3. PDF eISBN: 978-1-78262-834-7.

Wade, M.J. 2015. Statistics of lost historical coal tar PAH contamination in soils and sediments. In: *Environmental Forensics: Proceedings of the 2014 INEF Conference*, G. O'Sullivan and D. Megson, Eds., Royal Society of Chemistry, London, pp. 104–129.

Wade, M.J. 2016. From Ockham's Razor to Rube Goldberg: Don't rely on forensic age-dating miracles. *Environmental Forensics.* 17(2): 131–135, contributed editorial.

Wade, M.J. and Stainken, D. 2016. The history of hydrocarbon analyses: Whence has forensic geochemical hydrocarbon fingerprinting come. *Journal of Environmental Protection.* 7: 303–311. Published online February 2016 in SciRes. www.scirp.org/journal/jep. doi: 10.4236/jep.2016.73026.

Wang, Z. and Christensen, J.H. 2006. Crude oil and refined product fingerprinting: Applications (chapter 17). In: *Environmental Forensics – Contaminants Specific Guide*, B.L. Murphy and R.D. Morrison, Eds., Academic Press, San Diego, CA.

Wang, Z. and Fingas, M.F. 2003. Development of oil hydrocarbon fingerprinting and identification techniques. *Marine Pollution Bulletin*. 47: 423–452.

Wang, Z. and Stout, S.A. 2007. *Oil Spill Environmental Forensics – Fingerprinting and Source Identification Environmental Forensics*. Academic Press, San Diego, CA.

Wang, Z., Fingas, M.F., and Page, S.D. 1999. Oil spill identification. *Journal of Chromatography A*. 843: 369–411.

Wang, Z., Stout, S.A., and Fingas, M.F. 2006. Forensic fingerprinting of biomarkers for oil spill characterization and source identification. *Environmental Forensics*. 7: 105–146. ISSN: 1527–5922 print / 1527–5930 online.

Wang, Z., Fingas, M., Yang C., and Christensen, J.H. 2006a. Crude oil and refined product fingerprinting: Principles (chapter 16). In: *Environmental Forensics – Contaminants Specific Guide*, R.D. Morrison and B.L. Murphy, Eds., Academic Press, San Diego, CA.

Wilson, J.T. 2003. Fate and transport of MTBE and other gasoline components. In: *MTBE Remediation Handbook*, E.E. Moyer and P.T. Kostecki, Eds., Amherst Scientific Publishers, Amherst, MA.

Wong, P.K. and Wang, J. 2001. The accumulation of polycyclic aromatic hydrocarbons in lubricating oil over time – A comparison of supercritical fluid and liquid-liquid extraction methods. *Environmental Pollution*. 112: 407–415.

6 Vapor Intrusion

Chawn Y. Jeng and Xihong Zhai

CONTENTS

6.1 INTRODUCTION

Vapor intrusion occurs when there is a migration of vapor-forming chemicals from subsurface sources into nearby buildings. It has received attention from the scientific and regulatory communities in the past two decades with an increasing awareness that volatile organic compounds (VOCs) in soil and groundwater could pose threats to indoor air quality via a vapor intrusion pathway similar to that of radon gas entering homes. Figure 6.1 is an illustration of various factors that control and affect the migration of vapors from contaminated soil and groundwater into buildings through cracks in the foundation and openings for utility lines. Atmospheric conditions and building ventilation are also shown to influence soil gas intrusion.

Historically, the environmental community considered that direct exposure to soil and groundwater would pose the greatest threat to human health from subsurface contamination. As long as people do

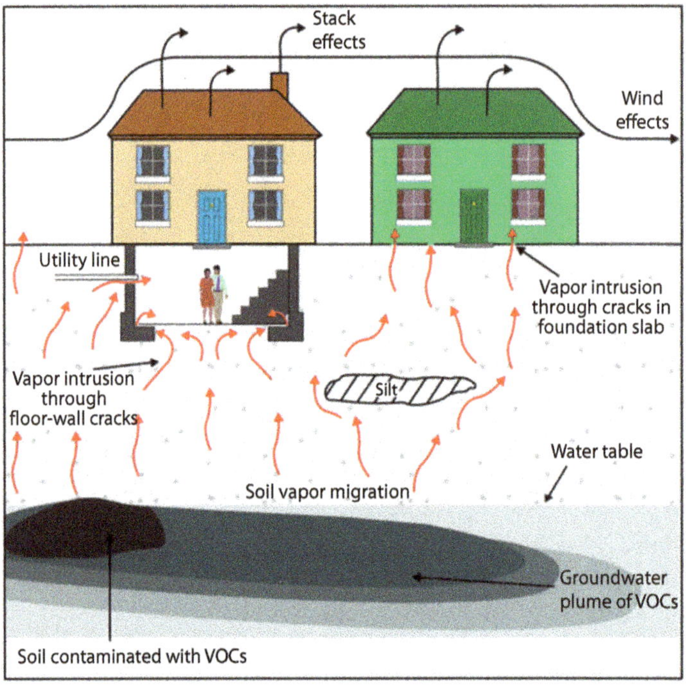

FIGURE 6.1 Migration of soil vapors to indoor air. (From www.epa.gov/vaporintrusion.)

not contact the impacted soil or drink the contaminated groundwater, the contaminants in the subsurface are unlikely to cause harm to humans. However, environmental professionals and regulators noticed the significance of vapor intrusion beginning in the late 1980s to early 1990s through some investigative and modeling work on the transport of contaminant vapors into houses with a focus on radon (Nazaroff et al., 1985, 1987; Loureiro, 1990; Johnson and Ettinger, 1991; Little et al., 1992). Since then, academics and industry leaders have made great efforts to understand the fundamentals of how subsurface vapors enter buildings and how to prevent or mitigate this pathway. Regulators have also requested that vapor intrusion be evaluated at sites where toxic volatile chemicals are present in the subsurface and have reopened closed cases to determine whether the vapor intrusion pathway is complete and thus needs to be reassessed.

This chapter is organized into the following sections: Section 6.2 presents some basic concepts for understanding subsurface vapor intrusion; Section 6.3 describes common investigation methods for assessing the vapor intrusion pathway; Section 6.4 provides an overview of the risk evaluation and management at vapor intrusion sites; and Section 6.5 discusses various technologies for mitigating the vapor intrusion pathway where it may pose threats to indoor air quality.

Numerous technical documents, publications, tools, and other resources are available for vapor intrusion assessment, evaluation, and mitigation methods. The following web pages provide useful information for readers who would like to learn more about these topics:

- www.epa.gov/vaporintrusion
- www.itrcweb.org/Team (Petroleum Vapor Intrusion and Vapor Intrusion)
- www.clu-in.org/issues/default.focus/sec/Vapor_Intrusion
- www.dtsc.ca.gov/SiteCleanup/Vapor_Intrusion.cfm

6.2 BASIC CONCEPTS FOR UNDERSTANDING VAPOR INTRUSION

Vapor intrusion is a potential concern at any building, existing or planned, located near soil or groundwater contaminated with toxic chemicals that can volatilize. Both diffusion and advection are mechanisms of transport of subsurface soil gas into the indoor air environment. Diffusion is the mechanism by which soil gas moves from high concentration to low concentration due to a concentration gradient. Advection is the transport mechanism by which soil gas moves due to differences in pressure. These pressure differences can be generated by atmospheric pressure changes, temperature changes creating natural convection in the soil, or forced pressure changes due to building ventilation systems. Once soil gases enter the "building zone of influence," they are generally swept into the building through foundation cracks by advection due to the indoor-outdoor building pressure differential. The reach of the "building zone of influence" on soil gas flow is usually less than a few feet vertically and horizontally.

The migration of volatile contaminants in the subsurface to indoor air is complex, and many factors may influence the vapor intrusion pathway, such as contaminant type, contaminant source and transport mechanisms, soil properties and degree of heterogeneity, building conditions and ventilation, meteorological conditions, and so on (ITRC, 2007; USEPA, 2012a). The following sections discuss how these key factors may affect soil vapor migration into buildings and the indoor air quality.

6.2.1 Contaminant Type

The USEPA defines volatile organic compounds (VOCs) as chemicals with a Henry's Law constant greater than 10^{-5} atm-m^3/mol at room temperature. Two classes of VOCs are commonly encountered at sites with soil and groundwater contamination:

- Petroleum hydrocarbons (PHCs) including gasoline, diesel, and jet fuel, and key individual constituents such as benzene, toluene, ethylbenzene, xylenes, and naphthalene.
- Chlorinated VOCs (CVOCs) used in dry cleaning and in industrial manufacturing as degreasing solvents, including tetrachloroethylene (PCE), trichloroethylene (TCE), 1,1,1-trichloroethane (TCA), and their degradation products.

The key difference between these vapors in the subsurface is that PHCs usually biodegrade readily under aerobic conditions, whereas CVOCs typically biodegrade under anaerobic conditions and at slower rates. Additionally, the biodegradation of chlorinated solvents can produce toxic degradation products, such as 1,2-dichloroethylene (1,2-DCE) and vinyl chloride (VC), whereas the degradation of petroleum hydrocarbons produces carbon dioxide and water and sometimes methane or other simple hydrocarbons under anaerobic conditions. Because petroleum degradation is relatively rapid when oxygen is present, it can limit the potential significance of vapor intrusion for PHCs under many conditions. In contrast, chlorinated chemicals are more difficult to degrade in soil vapor and groundwater, and thus, CVOC plumes often extend farther than petroleum plumes (USEPA, 2012b).

Other differences that may affect vapor intrusion potential include: (a) PHC liquids are less dense than water and can float on the groundwater table, whereas chlorinated solvents are heavier than water and sink through the groundwater aquifer, and (b) background PHC concentrations in air from indoor and outdoor sources (e.g., vehicle emissions) are more prevalent than CVOCs, which can make the interpretation of indoor air sampling results and vapor intrusion evaluation more difficult. These differences often lead to different strategies in addressing vapor intrusion–related human exposure to these two types of contaminants.

6.2.2 Contaminant Source and Transport

VOCs may be present beneath buildings directly as soil vapors move in the unsaturated/vadose zone through soil pores. They can also volatilize from dissolved groundwater plumes and non-aqueous phase liquid (NAPL). Thus, the key transport mechanisms controlling vapor migration into indoor air include (a) diffusion, that is, vapors move from areas of high concentration to areas of low concentration; (b) advection, that is, areas of high pressure to low pressure; (c) volatilization from a groundwater or NAPL source, which is governed by Henry's Law and Raoult's Law, respectively; and (d) air flow and mixing pattern inside the building, which are mainly affected by ventilation, wind, and stack effects. Figure 6.2 shows an example conceptual site model (CSM) of vapor intrusion from contaminated groundwater.

As shown in Figure 6.2, the migration of volatile contaminants in the subsurface into indoor air is complex and depends on many factors, such as physical properties of the compound (e.g., Henry's Law constant and partition coefficient K_{oc}), the subsurface conditions (e.g., depth to groundwater and infiltration), and the surface conditions (e.g., presence of buildings and impermeable surfaces vs. watered landscapes), as well as the other factors discussed in Section 6.2. For example, besides foundation cracks and openings in the wall of the building, vapors can enter into a building preferentially through other preferential entry points such as utility piping and floor drains (Section 6.2.4). Similarly, vapors can also migrate preferentially along subsurface features such as sewer lines and utility corridors. Thus, the prediction of contaminant movement by vapor intrusion is difficult, as the subsurface and building conditions vary from site to site. Nonetheless, modeling tools that incorporate the key transport and fate processes have been developed to aid decision-making (see Section 6.3.3).

Field data have been collected from sites with potential vapor intrusion conditions, which provide some general rules to identify sites with greater vapor intrusion potential. For example, the U.S. Environmental Protection Agency (USEPA) and some states such as California recommend that buildings within 100 ft (laterally or vertically) of subsurface contamination may be considered for potential risk from vapor intrusion. The 100-ft distance may not be appropriate in all cases; for example, if preferential pathways exist in the subsurface that link the contaminant plume to buildings. If the contaminant plume is not well defined, it may be necessary to evaluate potential

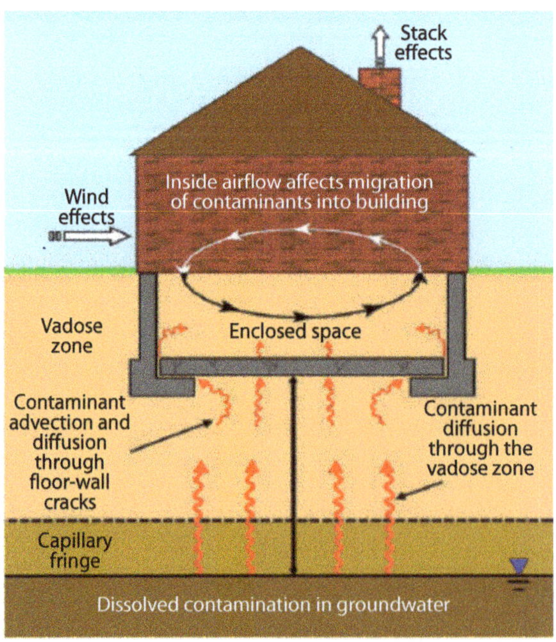

FIGURE 6.2 CSM of vapor intrusion from contaminated groundwater. (From Interstate Technology and Regulatory Council (ITRC), *Vapor Intrusion Pathway: A Practical Guideline* (www.itrcweb.org/Documents/VI-1.pdf), January, 2007. Figure 2.1.)

pathways from a distance greater than 100 ft. Some states, such as New Jersey (NJDEP, 2012), established different distance criteria based on the contaminant type (petroleum versus chlorinated hydrocarbons), as discussed in Section 6.2.1.

6.2.3 Soil Properties

The soil properties and degrees of heterogeneity between the contaminant source and a building are among the most important factors that influence soil vapor migration and thus, should be studied carefully during vapor intrusion evaluation. Specifically, the following soil physical parameters have been identified to affect vapor transport (diffusion and advection) in the soil column:

- Total porosity (n)
- Soil water-filled porosity (θw)
- Soil air-filled porosity (θa)
- Dry bulk density (ρs)
- Soil vapor permeability (kv)
- Soil and groundwater temperature (Ts)

These soil properties can be obtained by standard geotechnical methods in a laboratory or measured in the field. They have been used as input parameters for modeling vapor intrusion on a site-specific basis; for example, in the USEPA Vapor Intrusion Johnson and Ettinger Model (USEPA, 2017). Grain size (sand, silt, and clay) distributions can also be used to classify the soil type in each soil layer for modeling purposes. For example, the USEPA Vapor Intrusion Model provides default soil physical properties corresponding to 12 soil types under the United States Soil Conservation Service (SCS) classification.

The heterogeneity of soils in the lithologic units also affects contaminant migration. For example, lateral continuous dense soil (e.g., clay) layers may hinder contaminant upward migration, while highly permeable (e.g., sand) units may enhance contaminant movement. The effects of heterogeneous subsurface and geologic barriers on soil vapor distribution and indoor air concentration have been simulated by mathematical modeling and shown to be significant (USEPA, 2012a). Thus, a review of soil boring logs and soil characteristics, along with the history of site operations and spills/releases and the development of a CSM containing this and other relevant information, is critical for understanding and assessing the vapor intrusion pathway.

6.2.4 Building Conditions and Ventilation

Building types and conditions are also important factors in assessing the potential for subsurface vapor intrusion. Information regarding the construction and operation of a building should be used to assess the likelihood of subsurface soil gas entering the building through cracks and openings via diffusive and advective flow. Specific building-related factors that may affect vapor intrusion include building age, construction materials, foundation and slab type (e.g., slab-on-grade vs. basement), building dimension and layout, ventilation type (natural vs. mechanical), presence of potential vapor entry points (e.g., sump, drain, elevator shaft), and surface cover outside the building.

Air within a building is exchanged and mixed through three processes: (1) mechanical or forced ventilation, such as the operation of heating, ventilation and air conditioning (HVAC) systems and other devices such as exhaust fans, ceiling fans, and fireplaces; (2) natural ventilation, such as the opening and closing of doors and windows; and (3) infiltration of uncontrolled airflow through cracks and openings caused by indoor-outdoor temperature differences and barometric pressure changes. Consequently, the ventilation in a building directly affects vapor intrusion, as vapor is drawn into the building through openings mainly by the pressure difference between the soil and the interior of the building induced by building heating and mechanical ventilation as well as wind and stack effects (see Figure 6.2). On the other hand, if a building ventilation system generates positive air pressure within the structure relative to the subsurface, soil gas migration into the building may be minimized or eliminated.

Building ventilation also provides mixing of external (outdoor) air with the soil vapor flow entering into the building, resulting in dilution of contaminant concentrations in indoor air. This process is typically characterized by an indoor air exchange rate (AER), defined as the number of times that the total volume of air within a building is replaced by external air per hour. Scientific literature has reported that residential air exchange rates typically range from 0.2 to 2.0 air exchanges per hour in most homes in the United States and vary depending on building conditions and seasonal phenomena (e.g., wind/stack effect, pressure, and temperature; see Section 6.2.5). For commercial buildings, AERs also vary as a function of building construction, HVAC/energy efficiency, and climatic conditions, and they are generally higher than those in residences.

Finally, features inside buildings such as utility piping, bathroom/kitchen sinks, floor drains, sumps, and elevator shafts are likely vapor entry points and may create preferential pathways for vapor intrusion. In addition, surface pavements, permeable subsurface conditions (e.g., gravels and fractures), and underground utilities (e.g., sewer lines, manholes, and utility corridors) near buildings may significantly enhance vapor migration in the subsurface into buildings. As discussed in Section 6.2.2, the CSM should identify these known or potential preferential pathways for vapor intrusion so that they can be evaluated during site investigations. For example, instruments that provide real-time measurements of contaminants (e.g., hand-held photoionization detector [PID] or portable gas chromatograph/mass spectrometer [GC/MS]) are useful to investigate preferential vapor entry points and the spatial variability of chemical levels in air inside a building.

6.2.5 Meteorological Conditions

A variety of meteorological conditions, such as temperature, barometric pressure, wind, rainfall, and snow cover, can influence soil gas concentrations and vapor flux into a building. For example, temporal variations in the soil gas are typically within a factor of 2 during a season and up to a factor of 5 between seasons (ITRC, 2007). The importance of these variables is greater for soil gas closer to the surface; they are less significant for vapors at depths greater than 5 ft below the surface. The effects of certain meteorological-related parameters (e.g., pressure, wind load, and seasonal water table fluctuation) on soil vapor distribution and indoor air concentration have been simulated by mathematical modeling (USEPA, 2012a).

Indoor air may be more susceptible to both short-term (e.g., diurnal) and long-term (seasonal and year-to-year) variations in meteorological conditions. Other occupant-related activities, such as HVAC operations, often result in complex patterns for observed temporal trends of indoor air data. Figure 6.3 shows an example of temporal plots of weekly, biweekly, and monthly concentrations of PCE and chloroform inside a residence where USEPA conducted a vapor intrusion study (USEPA, 2012d).

As shown in Figure 6.3, the variations in indoor air samples were more than ten-fold between seasons, but the short-term (weekly to monthly) variability was less significant. Moreover, as the sampling duration increased from weekly to monthly, the variability generally decreased, because the meteorological and occupant-related influences tend to average out over longer intervals. Thus, collection of time-integrated samples over longer periods (e.g., 7–14 days) using passive samplers may be advantageous to reduce short-term temporal variability associated with changes in weather conditions. In addition, samples collected during unusual weather conditions (e.g., significant rainfall, high wind speed, and major storm events) are not representative of normal conditions and thus, may not be suitable for use in risk evaluation (see Section 6.4.2).

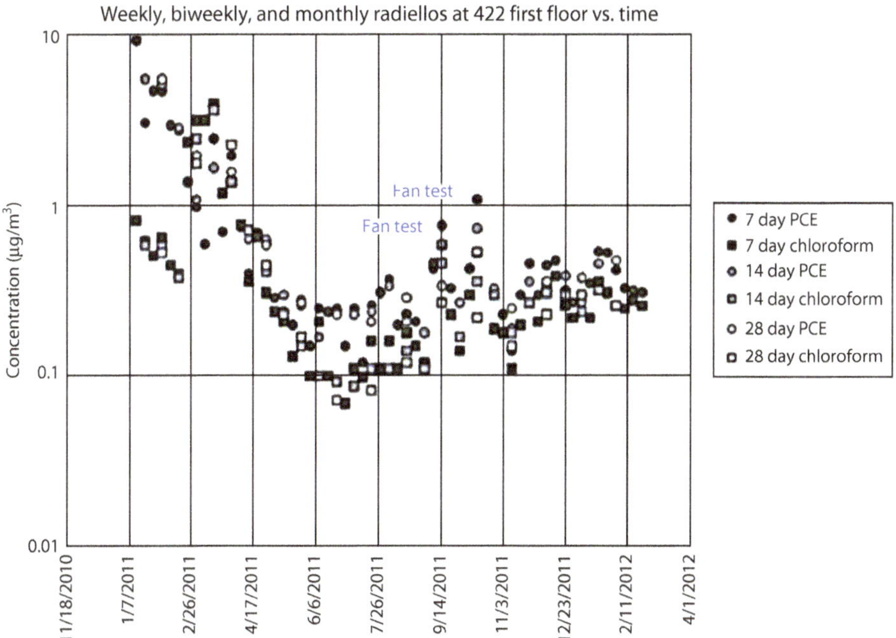

FIGURE 6.3 PCE and chloroform concentrations in indoor air with weekly, biweekly, and monthly duration. (From USEPA, *Fluctuation of Indoor Radon and VOC Concentrations due to Seasonal Variations.* EPA/600/R/12/673. September, 2012. Figure 5.4.)

6.3 INVESTIGATION OF VAPOR INTRUSION

As discussed in Section 6.2, one of the key steps in assessing the vapor intrusion pathway is the development of a CSM based on all available information on a contaminated site. If vapor intrusion is identified as a potentially complete pathway in the CSM, further investigation and evaluation are generally warranted. This section provides a step-by-step approach on how to conduct additional investigations for evaluating the vapor intrusion pathway properly. Because the CSM is a dynamic tool and should be updated as new information becomes available, this process may be repeated, as the updated CSM may identify additional data gaps following each stage of investigation, and additional data collection is needed to support the risk evaluation and management decision (see Section 6.4).

Typically, the investigation process for vapor intrusion is conducted in a phased approach, including a preliminary screening phase and a detailed investigation phase. The CSM should be used to guide both preliminary screening and detailed investigation phases, which are discussed further in Sections 6.3.1 and 6.3.2, respectively. The use of an attenuation factor (defined as the ratio of the indoor air and soil gas concentrations) as a tool for preliminary screening as well as site-specific evaluation is discussed in Section 6.3.3.

6.3.1 Preliminary Screening Evaluation

Most guidance documents recommend conducting a preliminary screening evaluation using available data before committing resources to a detailed site investigation. The following steps are commonly used during the preliminary screening phase to determine whether the collection of additional data is necessary (ITRC, 2007; DTSC, 2011a):

- Does the site represent an acute exposure concern (e.g., high methane level)?
- Are there sufficient characterization data to evaluate this pathway?

- Are any of the site COCs both volatile and toxic?
- Are buildings located in close proximity to volatile chemicals in the subsurface?
- What are the appropriate exposure scenarios and screening levels?
- Do the data exceed the appropriate generic screening levels?
- Does exceedance of the generic screening level warrant further investigation?

Figure 6.4 shows an example of how these steps are used in the preliminary screening phase for a decision-making flowchart.

Besides chronic health effects associated with most VOCs, vapor intrusion can also cause acute hazards to humans, such as odors that may cause physiological effects, and potential fire and explosion conditions for certain chemicals (e.g., methane and gasoline vapor). As indicated in Figure 6.4, vapor intrusion should be evaluated for acute exposure to determine whether immediate response

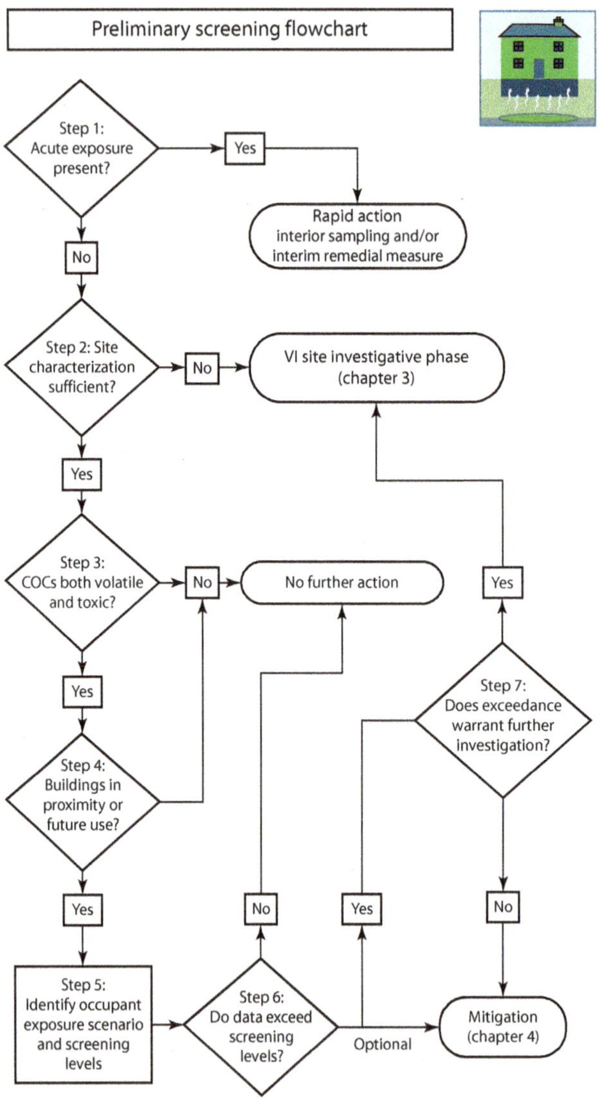

FIGURE 6.4 Example preliminary screening flowchart. (From Interstate Technology and Regulatory Council (ITRC), *Vapor Intrusion Pathway: A Practical Guideline* (www.itrcweb.org/Documents/VI-1.pdf), January, 2007. Figure 2.2.)

action is warranted (see Section 6.4.3). For example, the lower explosive limit (LEL) of methane is 5% or 50,000 parts per million by volume (ppmv), so emergency response (e.g., evacuation of buildings and contacting the local fire department) is typically recommended when the level in indoor air exceeds 10%–20% of the LEL.

6.3.2 Detailed Site Investigation

For a site that does not pass the preliminary screening evaluation, a detailed site investigation should be warranted to collect additional data for conducting a site-specific evaluation of vapor intrusion. Because the vapor intrusion pathway is complicated by the influences of many factors, as discussed in Section 6.2, a multiple-line-of-evidence (MLE) approach is usually recommended for both data collection and data evaluation to reduce uncertainties in decision-making.

Table 6.1 summarizes the approaches that are commonly taken for the investigation and evaluation of vapor intrusion. It also lists some of the limitations of each approach. Thus, several investigative approaches are frequently applied to the same site to gather multiple lines of evidence for supporting the decision-making.

Besides the investigative approaches listed in Table 6.1, other lines of evidence have also been collected pertaining to vapor intrusion, especially at complex sites:

- Background (outdoor and indoor air) concentrations
- Constituent ratios across media (e.g., soil gas and indoor air)
- Spatial and temporal trend/variability of data
- Site geology and soil types
- Building construction and current conditions (e.g., layout, use, preferential pathways)
- Known/suspected preferential transport pathways
- Predictive fate and transport modeling
- Other matrix (soil, NAPL, crawl space) data
- Other gas (e.g., radon) data as tracers
- Isotope analysis for source differentiation

TABLE 6.1
Common Approaches for Vapor Intrusion Investigation

Media Investigated	Evaluation Method	Principal Issues
Indoor air	Indoor air concentrations directly measured by canisters or passive samplers	Intrusive; background sources may interfere with sampling results; seasonal variations; cannot be used for future buildings
Sub-slab vapor[a]	Attenuation factor estimated or measured (e.g., using radon) to predict indoor air concentration	Intrusive; fewest pathway assumptions required, but attenuation factors may be conservative for many buildings
Soil gas[a]	Attenuation factor or modeling based on site-specific conditions used to predict indoor air concentration	Fewer pathway assumptions required, but more spatial and temporal variability than groundwater
Groundwater[a]	Attenuation factor or modeling based on site-specific conditions used to predict indoor air concentration	Most pathway assumptions required; generally less spatial and temporal variability than soil gas or sub-slab vapor

Source: Adapted from Interstate Technology and Regulatory Council (ITRC), *Vapor Intrusion Pathway: A Practical Guideline* (www.itrcweb.org/Documents/VI-1.pdf), January, 2007. Table 3.1.

[a] The attenuation factors for subsurface media (i.e., sub-slab vapor, soil gas, and groundwater) are discussed in Section 6.3.3.

Each line of evidence may be weighted differently depending on its nature and data quality. For example, soil matrix data, while providing valuable information for understanding the distribution of contaminant mass in the vadose zone, typically show more spatial variability than soil gas and groundwater data; thus, they may be less reliable for predicting vapor intrusion health risks than using soil gas and groundwater data. Additionally, some lines of evidence may conflict with the others, and professional judgment based on the CSM should be used in making a final decision. The use of an MLE approach to evaluate vapor intrusion is further discussed in Section 6.4.1.

Besides these common approaches, other innovative techniques have also been used to assess the vapor intrusion pathway, such as soil and indoor flux measurements and real-time continuous monitoring in soil gas and indoor air. Guidance documents (e.g., ITRC, 2007; DTSC, 2011a; USEPA, 2012b, 2015) are available for detailed discussion on development of the CSM and collection of reliable data using both conventional and innovative methods during the investigation for vapor intrusion evaluation, which is outside the scope of this chapter.

6.3.3 Attenuation Factor

As discussed in Section 6.3.2, an attenuation factor is required to predict the indoor air concentration using data collected from the subsurface media (i.e., sub-slab vapor, soil gas, and groundwater). The attenuation factor (alpha or α) is defined as the ratio between the air concentration inside a building and the vapor concentration in the subsurface (sub-slab, soil gas, or groundwater) for a given chemical:

$$\alpha = \frac{C_{\text{building}}}{C_{\text{subsurface}}}$$

where:

C_{building} is the building/indoor air concentration (μg/m^3)
$C_{\text{subsurface}}$ is the subsurface vapor concentration (μg/m^3)

Based on this definition, the attenuation factor is chemical specific, medium specific, and building specific. For example, USEPA compiled a database of the attenuation factors observed for vapor intrusion of chlorinated VOCs in residential buildings (EPA, 2012c). It consists of pairings of concentrations in indoor air and subsurface samples (groundwater, exterior soil gas, sub-slab soil gas, and crawlspace vapor) from actual sites in the United States and represents the most comprehensive vapor intrusion data available at this time. The database was analyzed and screened to reduce the impacts of background sources on indoor air concentrations and is considered representative of vapor intrusion of chlorinated VOCs from subsurface vapor sources into residential buildings for most conditions. Figure 6.5 shows the observed attenuation factors using paired indoor air and groundwater (vapor) concentrations from the USEPA database.

These empirically based distributions may be used to identify high-end (conservative) attenuation factors for the preliminary screening of potential vapor intrusion sites before deciding whether a site-specific evaluation is necessary. Table 6.2 shows examples of these high-end attenuation factors that are recommended as default attenuation factor values by USEPA and some state agencies for regulatory purposes.

While regulatory agencies may recommend different default attenuation factors (note that some of the differences are intended to reflect different building types), they generally represent reasonable ranges that are protective of most conditions where vapor intrusion of chlorinated VOCs may occur. As discussed in Section 6.2.1, vapor intrusion of petroleum hydrocarbons is limited by aerobic degradation under most subsurface environments where oxygen is present. Thus, these default attenuation factors tend to be even more conservative for the screening of potential vapor intrusion sites with petroleum hydrocarbons.

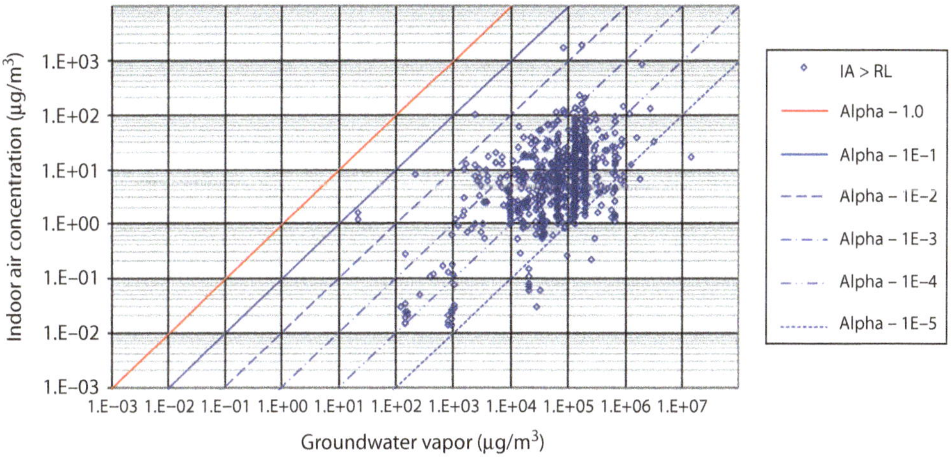

FIGURE 6.5 Sample plot of attenuation factor (α) for indoor air and groundwater vapor concentrations. (From USEPA, *Vapor Intrusion Database: Evaluation and Characterization of Attenuation Factors for Chlorinated Volatile Organic Compounds and Residential Buildings.* EPA 530-R-10-002. March 16, 2012. Figure 16c.)

TABLE 6.2
Example Default Attenuation Factors for Preliminary Screening

	USEPA	California (DTSC)	Wisconsin (WDNR)
Sub slab vapor	0.03	0.05 (residential/commercial)	0.1 (residential/small commercial) 0.01 (large commercial)
Soil gas	0.03	0.002/0.001 (residential-existing/future) 0.001/0.0005 (commercial-existing/future)	0.01 (residential/small commercial) 0.001 (large commercial)
Groundwater	0.001	Not available	0.001 (residential/small commercial) 0.0001 (large commercial)

Source: From USEPA, *Technical Guide for Assessing and Mitigating the Vapor Intrusion Pathway from Subsurface Vapor Sources to Indoor Air.* OSWER Publication 9200.2-154. June 2015.

Besides attenuation factors that are based on empirical data, default attenuation factors are sometimes derived using generic modeling with protective, generic assumptions (e.g., permeable soil and low air exchange rate). Additionally, modeling using site-specific soil and building parameters can be used to derive "refined" attenuation factors or predict indoor air concentrations, especially when a site fails the preliminary screening evaluation using the default attenuation factors. Several modeling tools are supported by regulatory agencies (USEPA, 2014, 2017; DTSC, 2011a) and published in the literature (Abreu and Johnson, 2005, 2006; API, 2012), and they can be used for such purposes.

6.4 RISK EVALUATION AND MANAGEMENT AT VAPOR INTRUSION SITES

This section provides an overview of the risk evaluation and management at vapor intrusion sites and how the health risk–based approach is applied to the decision-making process. Section 6.4.1 discusses the evaluation of MLE during the investigation phase with a case study. Section 6.4.2

describes the characterization of the health risks and other factors, such as background concentrations, building condition and controls, and land/building use restriction, to support the risk management decision. Consequently, the corresponding response action options for addressing vapor intrusion concerns are grouped into three general categories (immediate/urgent responses, interim mitigation measures, and source remediation) and discussed in Section 6.4.3.

6.4.1 Evaluation of Lines of Evidence

Section 6.3.2 describes the collection of data to be used as MLE for evaluating vapor intrusion. While not every line of evidence is needed to make decisions at a site, it is important to obtain as many lines of evidence as possible to reduce uncertainties in the risk evaluation and risk management decision. Some lines of evidence may be more critical than others, depending on the CSM and site conditions. In general, the closer the sampled medium is to the receptor (e.g., indoor air), the more relevant the corresponding line of evidence for assessing the current exposure and health risk. While all lines of evidence should be considered in the decision-making, individual lines of evidence may be inconsistent with other lines of evidence. In those cases, the CSM should be re-evaluated and adjusted, if necessary, to better reflect the weight of all available evidence.

When the available lines of evidence do not support a confident decision, collection of additional lines of evidence should be considered, such as multiple rounds of indoor air and subsurface vapor sampling to evaluate the spatial and temporal variability of the data, testing of indoor air using portable real-time instruments to diagnose indoor sources of volatile chemicals or identify preferential pathways, or site-specific modeling of vapor intrusion/attenuation (USEPA, 2015).

The following case study (Jeng et al., 2008) illustrates how the MLE approach is used to evaluate vapor intrusion exposure into a number of industrial and residential buildings adjacent to a refinery with contaminated soil gas and groundwater.

6.4.1.1 Case Study: MLE Approach to Assess the Significance of Subsurface Petroleum Vapor Intrusion

More than 200 soil gas samples were collected at 5 and 15 ft below the surface from 116 locations in and around residential and commercial buildings adjacent to a refinery tank farm located in southern California to determine the lateral and vertical extent of hazardous vapors in soil. All samples were analyzed for methane and VOCs using the USEPA Methods TO-3 and TO-15, respectively, by State-certified laboratories. The principal constituents detected in the soil gas samples were methane and benzene. The source of methane was likely the biodegradation of petroleum hydrocarbons in the soil, groundwater, and residual NAPL smear zone. Ethylbenzene, toluene, and xylenes were also detected but generally at lower concentrations than benzene. Table 6.3 shows the ranges of soil gas sampling results and comparison with the California Human Health Screening Levels (CHHSLs) for the key chemicals detected in soil gas.

Nine residential homes and four commercial buildings were selected for subsequent sampling and evaluation based on elevated shallow soil gas concentrations relative to the screening levels. Based on these soil gas sampling results, additional sub-slab vapor or indoor/ambient air data were collected in the commercial and residential buildings, respectively, as discussed in the following paragraphs.

A total of eight sub-slab soil vapor probes (two per building) were installed in the four commercial buildings that were selected for further study based on the soil gas data screening results. It should be noted that indoor air sampling was not recommended in this case, because industrial and laboratory chemicals containing VOCs were regularly used in these buildings. These VOC sources could not be removed prior to the sampling, thus introducing bias to indoor air sample collection and difficulty in interpretation of results. Table 6.4 summarizes the results of sub-slab vapor sampling for the key chemicals.

TABLE 6.3
Soil Vapor Monitoring Results

Chemical	Soil Gas Concentration	Residential CHHSL	Commercial CHHSL
Methane (ppmv)	ND–500,000	5,000[a]	5,000[a]
Benzene (ppbv)	ND–160,000	10	44
Ethylbenzene (ppbv)	ND–6,200	98	330
Toluene (ppbv)	ND–9,700	36,000	100,000
o-Xylene (ppbv)	ND–1,200	73,000	200,000
m/p-Xylene (ppbv)	ND–4,700	73,000	200,000

[a] 10% of methane LEL.
ND, Not detected.

TABLE 6.4
Sub-Slab Vapor Monitoring Results

Chemical	Sub-Slab Vapor Concentration	Screening Level
Methane (ppmv)	1–4.9	5,000[a]
Benzene (ppbv)	ND–1	4.3
Ethylbenzene (ppbv)	All ND	37
Toluene (ppbv)	ND–0.59	11,000
o-Xylene (ppbv)	ND–1.5	23,000
m/p-Xylene (ppbv)	ND–2.5	23,000

[a] 10% of methane LEL.
ND, Not detected.

The screening levels for sub-slab vapor samples were derived using the indoor air CHHSLs for commercial and industrial land use with a site-specific attenuation factor of 0.01. Because the sub-slab vapor concentrations beneath these commercial buildings were significantly lower than the screening levels, indicating that the potential for subsurface vapor intrusion was unlikely, no subsequent action was recommended.

The indoor air quality assessment of nine homes was performed in accordance with regulatory guidance. Up to four indoor air samples (including one duplicate sample) were collected from commonly occupied areas (i.e., central living area and bedroom) and potential VOC infiltration points (bathroom or kitchen) in each residence. At least one ambient air sample was also collected outside of each residence. Table 6.5 summarizes indoor and ambient air sampling results for the key chemicals.

For methane, toluene, and xylenes, the indoor air concentrations are all significantly lower than their respective indoor air CHHSLs for residential land use. Some benzene and ethylbenzene concentrations in indoor air are higher than the residential CHHSLs but are comparable to the range of ambient air concentrations. Because the key constituents present in soil gas were also found in most indoor air samples, additional data analyses were performed to evaluate the potential significance of subsurface petroleum vapor migration into these residences.

First, indoor air data from these homes were compared with regional and site-specific ambient air results for any significant difference from subsurface vapor intrusion or other factors. As shown in Table 6.5, the indoor air concentrations for all key chemicals are comparable to the range of ambient air concentrations measured outside these residences, except for toluene in the three samples

TABLE 6.5
Indoor and Ambient Air Sampling Results

Chemical	Indoor Air Concentration	Ambient Air Concentration	Residential CHHSL
Methane (ppmv)	ND–32	ND–87	500[a]
Benzene (ppbv)	ND–1.8	ND–1.5	0.03
Ethylbenzene (ppbv)	ND–2.2	ND–1.7	0.22
Toluene (ppbv)	1.4–13	0.97–2.9	81.8
o-Xylene (ppbv)	ND–0.85	ND	166
m/p-Xylene (ppbv)	ND–4.2	ND–3.1	166

[a] 1% of methane LEL (a CHHSL value is not available for methane).
ND, Not detected.

collected from one residence (11–13 ppbv). While the source of toluene in this case is unknown, the three samples also exhibited elevated levels of PCE (3.7–4 ppbv) even though no PCE was detected in shallow soil gas near this residence. It is possible that not all VOC sources were removed during the building preparation. Nonetheless, the toluene levels found in this home are significantly lower than the risk-based CHHSL and therefore, do not pose a significant health risk. The indoor air concentrations in these nine homes are also consistent with typical background levels of these chemicals found in indoor air in North America. Additionally, indoor air samples were collected from several homes outside the soil vapor plume area, and their results are also comparable to the indoor air concentrations of the nine homes within the soil vapor plume area. Based on these comparisons, the indoor air concentrations at these nine homes are consistent with ambient and indoor air quality in typical residences and do not appear to be affected to any significant level by the soil gas plume present in this area.

Second, indoor air, ambient air, and soil vapor concentrations collected from each residence were plotted to examine spatial relationship among the data sets. Figure 6.6a–f present the spatial data plots for the six key constituents found in soil gas.

As shown in these figures, soil gas concentrations are generally lower in the homes that are farther away from the suspected source area (near Residence 1), but the corresponding indoor air data do not follow a decreasing trend and are relatively constant among the nine homes for most chemicals. In fact, the indoor air concentrations typically match well with ambient air data, indicating that ambient air may have a strong influence on the indoor air quality at these homes. Slightly elevated levels of methane and/or benzene were detected in Residences 4, 6, and 9, which could be attributed to indoor sources (natural gas leaks from stoves and water heaters were reported in these homes).

Finally, ratios of measured indoor air and soil gas concentrations (i.e., "apparent attenuation factors") were calculated for the key constituents in several residences, and they are shown in Table 6.6.

It should be noted that these apparent attenuation factors are specific to individual residences (i.e., affected by differences in building structures and ventilation among the residences) but are expected to be comparable for the key constituents within the same residence, because these chemicals have similar physical and transport properties. For example, the attenuation factors for the five key constituents predicted by the Johnson and Ettinger model are less than one order of magnitude different. However, the apparent attenuation factors among these chemicals differ by two to three orders of magnitude in the same house (Table 6.6) and do not agree with the model prediction. Furthermore, many of the apparent attenuation factors are significantly higher than typical attenuation factors reported for shallow soil vapor intrusion (0.03 or less). In fact, some values are close to or even exceeding 1, which is the theoretical upper bound for vapor intrusion (i.e., the indoor air concentration is approaching the soil vapor concentration). Therefore, it is

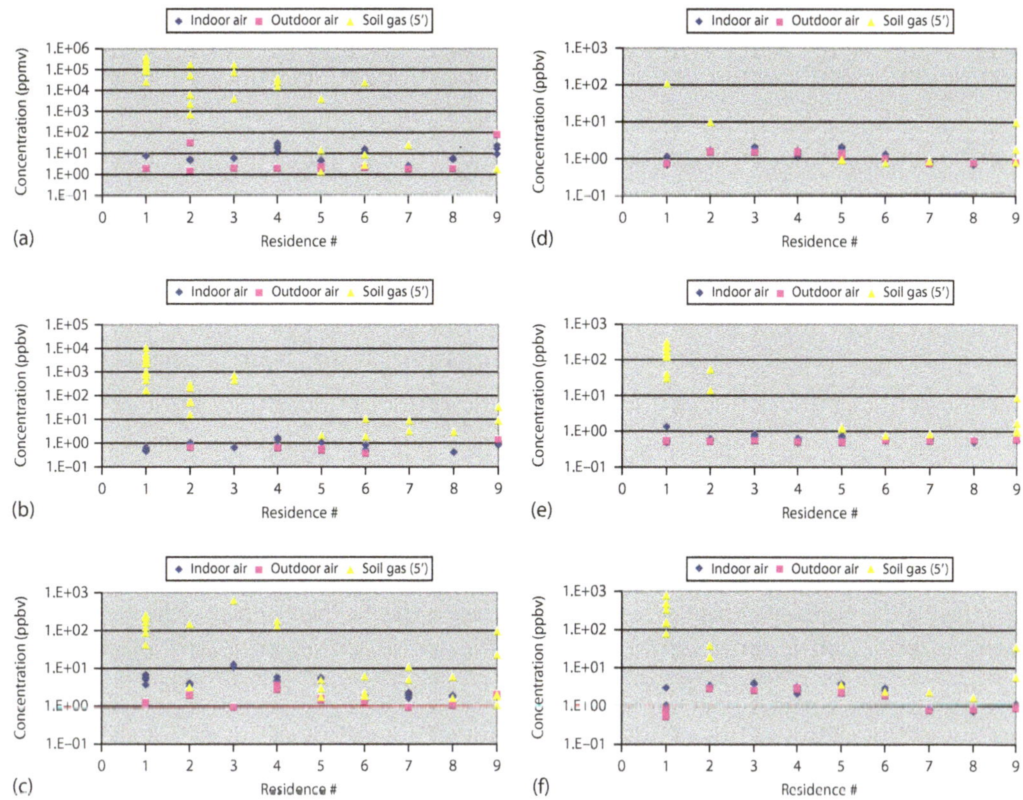

FIGURE 6.6 Plots of indoor air, outdoor air, and soil gas data by residence: (a) methane, (b) benzene, (c) toluene, (d) ethylbenzene, (e) o-xylene, and (f) m/p-xylene.

TABLE 6.6
Apparent Attenuation Factors for Select Residences

Residence #	Methane	Benzene	Toluene	Ethylbenzene	m/p-Xylenes
1	2.0×10^{-5}	2.7×10^{-5}	1.8×10^{-2}	7.8×10^{-3}	1.7×10^{-3}
3	3.8×10^{-5}	9.5×10^{-4}	1.9×10^{-2}	5.3×10^{-3}	5.1×10^{-3}
5	1.2×10^{-3}	5.0×10^{-1}	1.1×10^{0}	1.9×10^{0}	9.7×10^{-1}
7	9.2×10^{-2}	1.8×10^{-3}	1.9×10^{-1}	9.1×10^{-1}	3.5×10^{-1}
9	9.5×10^{0}	2.9×10^{-2}	2.0×10^{-2}	7.8×10^{-2}	3.2×10^{-2}

(Average of apparent attenuation factors in each residence.)

unlikely that vapor intrusion is occurring at a significant level at these homes, and the measured indoor air concentrations are likely attributed to sources not related to the soil gas plume (e.g., ambient air and/or indoor sources).

In summary, a step-wise approach was used to assess the potential significance of subsurface petroleum vapor migration into residential and commercial buildings near an oil refinery. The data sets collected from these buildings were analyzed, and several techniques were used to assess the significance of subsurface vapor intrusion. It was concluded that subsurface vapor intrusion does not have a significant effect on indoor air quality in these buildings based on the MLE from these analyses.

6.4.2 Risk Evaluation and Management

Exposure to contaminants due to vapor intrusion does not necessarily mean that adverse health effects will occur. Relevant site data collected during the site investigation should be properly evaluated, and if deemed appropriate, they may be used to estimate the cancer risk and noncancer hazard posed by vapor intrusion. Chapter 3 provides an overview of the human health risk assessment in site remediation and discusses the factors that are important in determining whether individuals may experience health effects. They include the frequency and duration of exposure, the concentration and toxicity of the contaminant, and the individual's sensitivity to the chemical. For most VOCs, the main concern is an increase in the cancer risk and chronic health effects as a result of long-term exposure (e.g., a minimum of 1–7 years), although some chemicals are also known to cause physiological effects or lead to adverse health effects (e.g., TCE) with short-term exposure to elevated concentrations. In those cases, an urgent response may be needed to mitigate the immediate hazard or acute exposure condition (see Section 6.4.3).

The inhalation route inside buildings is the primary means of human exposure for the vapor intrusion pathway. Typically, exposure concentrations in existing buildings can be estimated using time-integrated indoor air measurements (e.g., Summa canisters and passive samplers), while modeling is used to estimate exposure concentrations for future buildings. The risks from all site-related volatile chemicals should be added to obtain the total potential cancer risk and noncancer hazard for the indoor air exposure pathway. Because vapor intrusion may be one of many pathways that need to be evaluated for a contaminated site, the risk and hazard from the vapor intrusion pathway should be included with the other complete exposure pathways at the site to calculate the cumulative risk and hazard. Chapter 3 discussed the methodology in regulatory guidance documents for conducting a complete human health risk assessment.

Characterization of the health risk associated with the vapor intrusion pathway is a critical step for determining whether response actions may be warranted. Table 6.7 shows an example of the risk management decision matrix, based on the estimated cancer risk and noncancer hazard, for vapor intrusion sites and a list of corresponding response action options, which are discussed further in the next section.

Other factors that may affect the risk management decision and response action selection include background concentrations from indoor and outdoor (ambient) sources, building types (see Appendix C, ITRC, 2007), existing ventilation/HVAC systems, and land and building use restriction as institutional control. For example, the presence of VOC sources from household items such as cleaning reagents,

TABLE 6.7
Example Risk Management Decision Matrix for Vapor Intrusion

Cancer Risk/Noncancer Hazard Index (HI)	Risk Management Decision	Response Action
Risk $<1\times10^{-6}$ HI $\leq$ 1	No further action	None
$1\times10^{-6}<$ Risk $<1\times10^{-4}$ HI > 1	Evaluate need for action	Possible actions: • Additional data collection • Monitoring • Additional risk characterization • Mitigation • Source remediation
Risk $>1\times10^{-4}$	Response action needed	Vapor intrusion mitigation Source remediation

Source: California Department of Toxic Substances Control (DTSC), *Guidance for the Evaluation and Mitigation of Subsurface Vapor Intrusion to Indoor Air*, October, 2011.

cosmetics, dry-cleaned clothes, and paints may confound the interpretation of indoor air sample results, and these should be identified (and removed if possible) prior to the sampling and considered in the risk evaluation and decision-making. Thus, ambient air data are often collected during indoor air sampling to help guide risk management decisions, as the selected response action is not expected to reduce chemical concentration in indoor air below the ambient levels. Similarly, the response action targeting the subsurface contamination source would have no or little effect on chemical levels that are attributed to indoor or ambient sources (e.g., emissions from vehicles).

6.4.3 Response Action Options

If response actions are required to address vapor intrusion concerns as a result of the risk management decision, as discussed in Section 6.4.2, they typically are grouped into the following three categories based on time frame, hazard type, and implementability:

An Immediate/Urgent Response is needed when a potential acute hazard or explosive condition (see Section 6.4.2) is present in a building. For example, exposure to elevated levels of certain VOCs may cause physiological effects such as headaches, nausea, irritation, vomiting, and confusion. In these situations, an immediate response should be considered and may include one or more of the following activities:

- Notifying the local fire department about potential explosion hazards
- Relocating occupants temporarily until the condition improves
- Increasing building ventilation/pressurization
- Treating indoor air (e.g., carbon filtration air purifiers)
- Sealing potential conduits/cracks where vapors are entering the building

Interim Mitigation Measures are effective methods for reducing or eliminating subsurface vapors entering into buildings when there is no immediate acute or safety hazard. Interim mitigation measures for vapor intrusion are typically not considered a final remedy, as the main goal of mitigation systems is to reduce or eliminate current indoor air exposure until the contamination source in the subsurface is remediated. Typical mitigation measures that can reduce subsurface vapor entry into a building or minimize current indoor air exposure include

- Passive vapor barrier
- Sub-slab depressurization (SSD)
- Sub-slab venting (SSV)
- Sub-membrane depressurization (SMD)
- Building pressure/HVAC control
- Indoor air treatment
- Building design (e.g., podium-style ground floor)
- Institutional control (e.g., restriction of land/building use)

Section 6.5 provides additional discussion of common mitigation methods for vapor intrusion, system selection and design considerations, and operation and maintenance (O&M) and monitoring requirements.

Source Remediation is required to remove or reduce contaminant concentrations in the source areas. As discussed earlier, mitigation and source remediation can be used as complementary components of an overall response action for addressing vapor intrusion. Most regulatory agencies require the remediation of vapor sources that pose high vapor intrusion risks to existing buildings and at properties where new buildings will be constructed with no institutional or engineering controls. Chapters 7 through 10 present conventional and innovative remedial technologies for treating volatile contaminants in the subsurface and are not repeated here.

6.5 VAPOR INTRUSION MITIGATION OPTIONS

This section presents detailed discussion on various mitigation measure options that are used to reduce subsurface vapor entry into a building. The topics include evaluating and selecting mitigation measures for vapor intrusion, design considerations for common mitigation methods, and O&M requirements for mitigation systems.

6.5.1 Vapor Intrusion Mitigation Methods

Vapor intrusion mitigation methods all rely on the conceptual model of a vapor migration mechanism. Vapor flows from a high-pressure regime to a low-pressure regime when there is a gradient. For buildings that are not pressurized (e.g., through HVAC operation), the indoor air pressure is typically lower than the outside pressure surrounding the building due to the temperature differences between indoor air and surrounding soils, wind and barometric changes, and stack effects. Vapor may travel from the subsurface into buildings under this pressure gradient. Vapor intrusion mitigation measures focus on interruption of this migration pathway. Figure 6.7 lists the commonly available vapor intrusion mitigation technologies and their applicability to different types of buildings.

The two most commonly used vapor intrusion mitigation technologies are SSV and SSD. Both SSV and SSD focus on adjusting the sub-slab condition to minimize or eliminate the vapor intrusion pathway. The difference between these two lies in that SSV is intended to reduce or dilute VOC concentrations in the sub-slab by venting it out and drawing in fresh air from outside, while SSD is designed to eliminate vapor intrusion by constantly maintaining a sub-slab pressure lower than the indoor air pressure. Sections 6.5.1.1 and 6.5.1.2 discuss SSV and SSD, respectively. Section 6.5.1.3 discusses additions or alterations to these two technologies, including sub-slab liners, SMD, building pressurization, indoor air treatment, and so on.

6.5.1.1 Sub-Slab Venting (SSV)

An SSV system functions by creating a pathway to vent sub-slab vapors outside through pipes and stacks and/or drawing in air from outside to dilute the undesired volatile compounds in sub-slab air. It consists of a layer of permeable material (sand or pea gravel) and an impermeable sub-slab liner (barrier) between the permeable layer and the floor slab. This layout allows soil vapor to

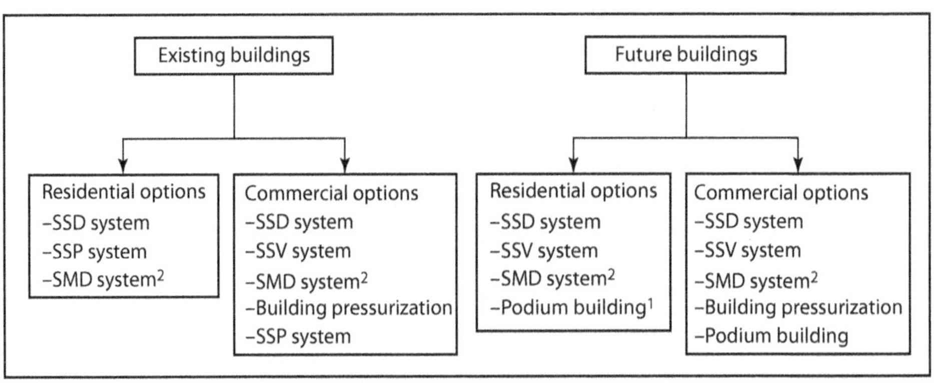

FIGURE 6.7 Vapor intrusion mitigation technologies. (Adapted from California Department of Toxic Substances Control (DTSC), *Vapor Intrusion Mitigation Advisory*, Final Revision 1. October, 2011.)

[1] In general, appropriate for multi-family dwellings only.

[2] For buildings with crawl space. SMD: sub-membrane depressurization; SSD: sub-slab depressurization; SSP: sub-slab pressurization; SSV: sub-slab venting.

move laterally under the slab and eventually discharge to the outside rather than move vertically upward to the inside of the building. SSV systems use perforated collection pipes placed to the periphery of the building footprint, which are connected into a central collection manifold. The central collection point is then connected to a vertical pipe that runs up through the roof for gas discharge into the atmosphere. An SSV system can also include a vertical inlet system within or next to the building, which allows fresh air to enter the permeable layer to dilute volatile chemical concentrations.

SSV systems can be either passive, relying solely on the natural pressure gradient from thermal or wind effects, or active, by installing fans to extract sub-slab vapor and blow in fresh air (refer to sub-slab pressurization, discussed later). A passive SSV can be easily converted to an active SSV with minimal modification.

Considering the emplacement of various layers in building foundations, SSV systems are more applicable to new constructions than existing buildings. Because the venting layer will lose its function when saturated with water, SSV systems are not recommended for areas where the groundwater table is shallow or there is surface drainage nearby.

6.5.1.2 Sub-Slab Depressurization (SSD)

An SSD system functions by maintaining a lower sub-slab air pressure relative to indoor air pressure, therefore preventing soil gas migration into the buildings. The negative pressure gradient (depressurization) is achieved by a motorized blower that draws sub-slab vapor and discharges it to the atmosphere. Collection manifolds and discharge pipes are used to channel the air from the sub-slab to the discharge point on the roof.

Sub-slab liners are commonly used in SSD systems as a redundant feature. If the blower of the SSD system fails, the sub-slab liner will function as a barrier to prevent soil vapor from entering the building. In addition, the liner can increase the blower efficiency of the SSD system by preventing air entry from the building interior into the sub-slab, such that a smaller blower is required.

From the preceding discussion, one can tell that the mechanisms of SSD and SSV are very similar. Both systems reduce vapor concentrations in the sub-slab through venting. The difference of these two systems lies in the design objectives. The SSD system is designed to maintain a negative pressure gradient from the building interior to the sub-slab; therefore, the magnitude of vapor concentration reductions in SSD systems is less critical than that in SSV systems. SSV systems are normally less aggressive than SSD systems in term of depressurization extent and lower air flow rates. Therefore, SSD is preferred over SSV in areas with high vapor concentrations.

The different design objectives of SSD and SSV lead to different monitoring requirements. The effectiveness of SSD can be monitored by pressure monitoring (once its protectiveness is verified with indoor air monitoring), while SSV systems are monitored by measuring either sub-slab vapor concentrations or indoor air concentrations. The monitoring requirements of SSD and SSV will be further discussed in Section 6.5.3.

The installation of SSD systems involves drilling or cutting holes in the existing slab, removing soil from beneath the slab, and placing vertical suction pipes into the openings. The suction pipes are connected into a manifold together for gas discharge outside. The installation requirement of an SSD system makes it suitable for both existing buildings and new constructions.

6.5.1.3 Additions and Alterations to SSV and SSD

Other mitigation measures as additions or alterations to SSD and SSV systems are available to address site-specific conditions.

Sub-membrane depressurization: SSD and SSV systems are applicable for slab-on-grade building construction. For buildings with crawl spaces, a sub-membrane depressurization (SMD) system is more appropriate. An SMD system uses membrane (liner) as a surrogate for a slab to allow depressurization. A blower is used to draw air from the space between the exposed dirt surface and the membrane, therefore inhibiting vapor migration into the overlying crawl space. Because the membrane can be easily damaged, periodic inspection is required to ensure its integrity and proper seal. SMD can be installed by retrofitting for existing buildings with crawl spaces.

Building pressurization (HVAC optimization): Building pressurization prevents vapor intrusion by maintaining a higher indoor air pressure relative to the sub-slab area by manipulation of the building HVAC systems or installation of a new HVAC system. Building pressurization is efficient only when the building is relatively tight, with few doors or other openings. Positive pressurization will require higher energy costs, particularly if significant heating and cooling are required. Regular maintenance and air balancing of the system are required to maintain positive pressurization, which renders this technology more suitable for commercial buildings than residential buildings.

Sub-slab pressurization: Sub-slab pressurization (SSP) systems work in a similar mechanism to SSD systems, except that fans are used to push air into the soil or venting layer below the slab instead of drawing it out. The increased sub-slab air pressure will force soil gas to migrate laterally to the sides of the building, provided that the entry routes from the sub-slab to the building are well blocked (i.e., cracks and openings are well sealed; see later). This technology requires highly permeable soils to induce effective lateral air movement. SSP systems can be applied to either existing buildings or new constructions.

Indoor air treatment: This method collects indoor air to central treatment equipment to remove contaminants. Available air treatment devices include both in-duct models and portable air cleaners. The contaminant removal mechanisms include adsorption using zeolite or granular activated carbon, chemical oxidation using ozone, and photocatalytic oxidation. Because indoor air treatment involves collection of contaminant vapor at the exposure point (building interior), and its protectiveness can be compromised by interrupted operation of the treatment system, this technology is normally not considered a proven stand-alone technology. However, it can be combined with other technologies to control vapor concentrations in problem rooms. It is normally used as the last resort when other engineering controls such as SSV, SSD, or SMD are not feasible (e.g., shallow groundwater or wet soils). Indoor air treatment is generally only considered applicable in existing buildings, since new constructions can install more cost-effective systems.

Sealing cracks and openings: Volatile chemicals can enter a building through cracks or perforations in slabs or basement floors and walls, openings around sump pumps, elevator shafts, or where pipes and electrical wires go through the foundation. As a preliminary first step (rather than a stand-alone option) in preventing vapor intrusion, it is intuitive to seal these cracks and openings.

Sub-slab liners (passive barriers): Sub-slab liners are impermeable materials or structures installed below a building to block vertical upward vapor migration into the buildings. Due to the potential for punctures, perforations, tears, and incomplete seals, sub-slab liners cannot eliminate vapor intrusion completely by themselves and are not generally used as a stand-alone technology. As discussed in Sections 6.5.1.1 through 6.5.1.3, sub-slab liners are used in combination with SSV, SSD, or SMD systems.

Podium-style buildings: Podium-style building design designates the first floor of the building as some structure with sufficient ventilation, such as parking garages, to eliminate the vapor intrusion exposure route. All potential vapor entry routes to upper floors, including utility lines, elevator shafts, and ventilation systems, must be engineered and properly sealed in to reduce the risk of vapor intrusion to occupants in the upper floors. Specific measures may include strategic layout of the elevator (on an exterior wall instead of at an interior, central entrance), sealing the base of the elevator, possible venting, and increased ventilation of the elevator. Podium-style building can obviously only be considered for new constructions and is generally not recommended for residential buildings due to potential alteration of the first floor into livable space.

Institutional controls (ICs): ICs are not engineered instruments but rather, administrative or legal controls to prevent human exposure to vapor intrusion risks. ICs typically operate by restricting land uses for certain purposes (e.g., residential buildings, hospitals, schools and day care facilities, etc.). In some situations, ICs place restrictions on access to a property to ensure the uninterrupted implementation of response activities, such as the installation or maintenance of vapor intrusion mitigation systems, or to ensure the integrity of vapor mitigation systems. They may also include an informational device to notify potentially involved parties regarding subsurface contamination. ICs can be used either as a short-term interim measure until site cleanup goals are reached or as part of a long-term response where hazardous vapor remains in soil gas.

6.5.2 Selection and Design of Mitigation System

Available vapor intrusion mitigation measures for a building should be evaluated based on the criteria normally used in feasibility studies as discussed in Chapter 1, Section 1.3.1, of this book. The following sections discuss technology-based factors, including building type and use, foundation type, subsurface conditions, and the type of contaminants, which should be considered in the selection and design of vapor intrusion mitigation measures. More in-depth discussion can be found in the guidance documents from the Interstate Technology Regulatory Council (ITRC, 2007) and Department of Toxic Substances Control (DTSC, 2011b).

New versus existing buildings: Section 6.5.1 mentioned the applicability of various vapor intrusion mitigation options to new buildings and existing buildings. Typically, new buildings have more choices than existing buildings. For example, passive barriers and venting layers can be placed below the foundation of new buildings easily but are hard to "insert" under existing buildings. Other vapor intrusion mitigation features, such as collection pipes, suction pits, and vent risers, if integrated into the design of new buildings, can be installed with less effort.

Building uses/sizes: The differences in building sizes are normally associated with different building uses. Large buildings are normally for commercial or industrial uses, such as warehouses. Vapor intrusion mitigation technology selection should consider current and future building use. Building pressurization (HVAC optimization) is more suitable for commercial and industrial buildings, because their relatively large HVAC systems are more readily adapted for pressurization, and they are more likely to receive routine maintenance and monitoring. The layout of vapor intrusion mitigation pipes needs to consider building use. For example, warehouse areas should not have vent risers placed in the middle, to avoid damage by traffic.

The building size may have some effect on technology selection and system design. Based on the concept borrowed from radon mitigation experience (USEPA, 1993), a general rule of thumb is that an SSD system for a building up to 50,000 ft^2 can use one suction point with permeable sub-slab material and proper seal of the suction field. In practice, conservative designs will normally use multiple suction points for such large buildings.

In some cases, larger buildings simply require sized-up systems; in other cases, different building sizes may call for different mitigation measures. Because of their large indoor air volume, higher air exchange rates, more openings (e.g., open bay doors), and thick floor slabs, large buildings are less likely to have complete exposure pathways to vapor intrusion risks. ICs, either in combination with engineering control measures or as the mitigation itself (such as building occupancy or use restrictions), may provide sufficient protection of human health for occupants/workers in these large buildings. But portions of these large buildings, such as office areas with lower ceilings and partitions blocking air flow, may still require engineering vapor intrusion mitigation measures.

Foundation types and conditions: The foundation types (basements, crawl spaces, slabs on grade, and any number of combinations) should be considered in the selection of vapor intrusion mitigation measures. SSD is applicable to slab over soil, whereas SMD is applicable to crawl spaces or anywhere that large areas of soil are exposed without potential future damage to the membrane. Some basements may have exposed earthen floors or walls, which need to be covered by membranes or concrete slab to enable effective depressurization.

Vapor intrusion mitigation selection and design need to consider the materials used in foundation construction. Foundation walls with fieldstone may be too permeable for depressurization to be effective. Cinder block foundation walls may provide vapor entry points through the interconnected voids in the blocks. Solutions to these situations include sealing the cracks and the openings at the top of the wall, the installation of additional suction points, and in extreme cases, considering alternative mitigation technology, such as building pressurization or slab replacement.

Soil conditions: Foundation soil properties, particularly air permeability, affect the performance of vapor intrusion mitigation systems, including SSD and SMD. Tight soils will have limited radii of influence of depressurization beyond the suction point. Solutions to conquer the low air permeability include increasing the number of suction points, installing a permeable venting layer below the slab, or using high-vacuum/low flow blowers (Folkes and Kurz, 2002). Permeable soils in foundations generally render very effective depressurization; however, in very permeable soils, depressurization might become difficult, and SSP might be a better option to pressurize the sub-slab and divert soil gas around the structure (see Section 6.5.1.3).

Soil saturation would impact the effectiveness of vapor intrusion mitigation or depressurization systems. When soil pores are filled by water, they become inaccessible for air migration. This may be desired or undesired, depending on the sources of pore water. Soil saturation might be a result of high groundwater table, surface drainage, or infiltration from precipitation. If the wet conditions are due to a high groundwater table containing VOCs, it becomes more important to address the groundwater contamination before addressing the vapor intrusion issue (see Section 6.5.1.1). If soil is saturated due to surface drainage or precipitation that brings in "clean" water, the blocked pore space below the floor slab may actually function as a barrier that prevents or minimizes vapor intrusion into the building. In this case, the SSD can either still function by drawing air into the headspace of the drain pipes or be operated in a cyclic mode and run in low–water table periods (e.g., dry seasons), when depressurization would be feasible.

Chemicals of concern (COC): Three factors associated with the COCs need to be considered in mitigation system selection and design: (1) flammability (potentially combustible COCs [e.g., methane] require intrinsically safe blowers and monitoring or alarm systems); (2) compatibility with vapor intrusion system materials (certain vapors may cause degradation of membranes, pipes, or the solvents used to join pipes); and (3) COC action levels, which will determine the target reduction. Reductions of 80% or more will need a more aggressive system such as SSD or SMD rather than SSV.

6.5.3 Operation and Maintenance (O&M) and Monitoring

Periodic O&M and monitoring are needed to ensure the continuous operation and protectiveness of vapor intrusion mitigation systems. O&M activities and monitoring activities are often performed concurrently, and they are discussed together in the following subsections.

6.5.3.1 Frequency of O&M and Monitoring

An O&M and monitoring program should generally include two phases: (1) an initial post-construction phase, which is generally more intensive; and (2) a long-term maintenance phase at reduced frequencies. The frequency and duration of each phase can vary from case to case.

The initial phase normally consists of quarterly or semi-annual O&M and sampling for the first 1–3 years. Situations that warrant frequent O&M activities include

- Old buildings
- High COC concentrations in soil vapor
- Passive mitigation systems, such as passive SSV, that are less predictable or less efficient
- Change of hydrogeology conditions, such as migration of groundwater plume, or implementation of remediation measures

Once an adequate demonstration of effectiveness has been made for the vapor intrusion mitigation system, O&M and monitoring activities can be conducted at a reduced frequency (e.g., once per year or biennially), and potentially with fewer types of tests, to verify that this performance is sustained.

Examples of various monitoring scenarios for these two phases can be found in various guidance documents (DTSC, 2011b; NJDEP, 2012; MADEP, 2011).

Additional O&M activities can be triggered by alarms of system failures, system modification, or the onset of remediation activities that may alter soil vapor conditions.

The scheduling of O&M, and particularly sampling activities, should consider weather conditions. Temperature and precipitation can affect the performance of a vapor intrusion mitigation system. For example, cold temperatures may increase the building depressurization created by the thermal stack effect and thus, increase the driving force for soil gas entry. Consequently, active depressurization systems may need to generate higher-vacuum conditions than originally considered when they were designed to maintain the desired pressure differential. Precipitation may also cause saturation in the fill under the slab, which may affect the performance of the system and limit the migration of VOCs in soil gas. Also, VOCs can partition into the clean infiltrating rainwater. Therefore, soil gas sampling should not be conducted during or immediately after heavy rainfall events (CalEPA, 2015).

6.5.3.2 System Startup and Initial Testing

After its installation, the vapor intrusion mitigation system should be tested for proper functioning, liner integrity, and achievement of performance goals. This could include the following activities:

Tracer testing: Openings within the building or leaks in the mitigation system can affect system performance. Tracers can be used to verify integrity of barriers, building materials, and piping system. Smoke testing is a widely used qualitative method to detect leaks through floors above sealed crawl space systems or any other possible cracks and openings, or to test airflow patterns. Quantitative tracer testing methods are also available with instrumentation for quantitative results.

Proper function: For active vapor intrusion mitigation systems, the system components, including blowers, gauges, and alarms, should be tested for proper functioning at system startup. Alarms are integrated into the system to ensure that any malfunction or system failure is detected and addressed in a timely manner. At system startup, artificial failure signals should be used to activate the alarm and to verify proper response actions. For example, methane gases are normally used to trigger high methane alarms, and corresponding response actions, such as increased air flow in the HVAC system, should be verified.

Achievement of performance goals: Performance goals might vary for different vapor intrusion mitigation systems. For pressure manipulation systems such as SSD, SSP, and building pressurization, the pressure differential needs to be verified by measurement in sub-slab probes and the building interior for SSD, SSP, and building pressurization systems. A pressure differential of −4 to −10 Pascal beneath the sub-slab liner is generally considered adequate for an SSD system (EPA, 2008; DTSC, 2011b).

Normally, indoor air quality will be acceptable if an adequate pressure difference is maintained. However, indoor air sampling is generally recommended at least for initial verification, since it provides more direct confirmation of system effectiveness.

6.5.3.3 O&M Activities and Monitoring Parameters

Typical O&M and monitoring activities include the following:

- Inspection of the area of concern, including all visible components of the mitigation system and collection points, for any significant changes in site condition and any sign of degradation of the mitigation system. For example, noisy fans typically indicate problems with ball bearings and should be replaced. Some changes made to the building (e.g., remodeled basement, new floor penetrations, new furnace) might impact the performance of the mitigation system.
- Measurement of flow rates and pressures generated by the fan at the vent risers to verify that the system is working and moisture is draining correctly for SSD and SMD systems.

- Routine maintenance, calibration, and testing of functioning components of the VOC venting systems according to manufacturers' specifications.
- Monitoring of VOCs in indoor air for protectiveness. Indoor air sampling is normally required for passive systems such as SSV to evaluate long-term performance. For active systems such as SSD, it can be either dropped or reduced to a portion of monitoring locations during the second phase after verification in the first phase.
- Monitoring of VOC concentrations in sub-slab probes or vent risers for any significant change in subsurface gas concentrations. Pressures at the probes, and also ambient pressures, should also be measured at the time of sampling.
- Periodic measurement of pressure differentials across the slab. This is more critical for systems of pressure adjustment (SSD, SMD, SSP, and building pressurization) than for passive venting systems (SSV).

Specific O&M and monitoring requirements for a mitigation system depend on its objectives and design details and may be modified during the operation based on performance data to support continuing operation or shutdown.

PRACTICE AND EXERCISES:

General Instruction: Physical properties and toxicity information of the chemicals in the problem sets can be found in the USEPA RSL Generic Tables (www.epa.gov/risk/regional-screening-levels-rsls).

Question 1 (Section 6.2.2): (a) At a former underground storage tank (UST) site, the average concentration of benzene measured in the soil samples collected above the water table is 10 mg/kg on a wet-weight basis (C_t). Use the following equations to calculate the equilibrium concentrations in soil pore air (C_a, mg/L), in soil pore water (C_w, mg/L), and onto the soil surface (C_s, mg/kg):

$$C_t = \frac{C_s \rho_b + C_w \theta_w + C_a \theta_a}{\rho_b}$$

$$H' = C_a / C_w$$

$$K_d = K_{oc} \times f_{oc} = C_s / C_w$$

where:

H' is the Henry's law constant (unitless)
K_d is the soil-water partition coefficient (L/kg)
K_{oc} is the organic carbon partition coefficient (L/kg)
f_{oc} is the fraction of organic carbon in soil (unitless, 0.002)
ρ_b is the dry soil bulk density (1.5 kg/L)
θ_w is the water-filled porosity (0.15, unitless)
θ_a is the air-filled porosity (0.20, unitless)

(b) Calculate the mass fraction in each of the three phases (pore air, pore water, and sorbed to solid) in the soil matrix (hint: assuming a fixed volume of soil matrix).

Solution 1:

(a) For benzene, $H' = 0.23$ and $K_{oc} = 150$ L/kg from the USEPA RSL table.

Substituting the C_s and C_a terms into the first (three-phase equilibrium) equation and solving C_w first,

$$C_t = \frac{(K_{oc} \times f_{oc} \times C_w)\rho_b + C_w \theta_w + (H' \times C_w)\theta_a}{\rho_b}$$

$$10 \text{ mg/kg} = \frac{(150 \times 0.002 \times 1.5) \times C_w + (0.15 \times C_w) + (0.23 \times 0.20) \times C_w}{1.5}$$

$$C_w = 23.2 \text{ mg/L}$$

$$C_a = H' \times C_w = 0.23 \times 23.2 = 5.34 \text{ mg/L}$$

$$C_s = K_{oc} \times f_{oc} \times C_w = 150 \times 0.002 \times 23.2 = 6.96 \text{ mg/kg}$$

(b) Assuming a soil volume of 1 liter ($V_s = 1$ L),

The contaminant mass sorbed to solid, $M_s = C_s \times \rho_b \times V_s = 6.96 \times 1.5 \times 1 = 10.4$ mg
The contaminant mass in pore water, $M_w = C_w \times \theta_w \times V_s = 23.2 \times 0.15 \times 1 = 3.48$ mg
The contaminant mass in pore air, $M_a = C_a \times \theta_a \times V_s = 5.34 \times 0.2 \times 1 = 1.07$ mg
Thus, the mass fractions of benzene in solid, pore water, and pore air are approximately 70%, 23%, and 7%, respectively.

Question 2 (Section 6.2.2): (a) Repeat Question 1 for the same average concentration (10 mg/kg on a wet-weight basis) of naphthalene measured in the soil samples. (b) Explain why the distributions of contaminant concentration and mass in the soil matrix are different for benzene and naphthalene. (c) Which contaminant is more likely to be found in indoor air if a building is to be constructed without mitigation or remediation at the site?

Solution 2:

(a) For naphthalene, $H' = 0.018$ and $K_{oc} = 1500$ L/kg from the USEPA RSL Generic Table; using the same equation to solve C_w first,

$$C_t = \frac{(K_{oc} \times f_{oc} \times C_w)\rho_b + C_w \theta_w + (H' \times C_w)\theta_a}{\rho_b}$$

$$10 \text{ mg/kg} = \frac{(1500 \times 0.002 \times 1.5) \times C_w + (0.15 \times C_w) + (0.018 \times 0.20) \times C_w}{1.5}$$

$$C_w = 3.22 \text{ mg/L}$$

$$C_a = H' \times C_w = 0.018 \times 3.22 = 0.058 \text{ mg/L}$$

$$C_s = K_{oc} \times f_{oc} \times C_w = 1500 \times 0.002 \times 3.22 = 9.66 \text{ mg/kg}$$

Assuming a soil volume of 1 liter ($V_s = 1$ L),
The contaminant mass sorbed to solid, $M_s = C_s \times \rho_b \times V_s = 9.66 \times 1.5 \times 1 = 14.5$ mg
The contaminant mass in pore water, $M_w = C_w \times \theta_w \times V_s = 3.22 \times 0.15 \times 1 = 0.48$ mg
The contaminant mass in pore air, $M_a = C_a \times \theta_a \times V_s = 0.058 \times 0.2 \times 1 = 0.012$ mg
The mass fractions of naphthalene in solid, pore water, and pore air are 96.7%, 3.2%, and 0.1%, respectively.

(b) Benzene is more volatile (higher H') and has less affinity for organic matter (lower K_{oc}) than naphthalene, so the concentrations and mass fractions of benzene in pore air and pore water are higher than those of naphthalene.
(c) Benzene is more likely to be found in indoor air due to the much higher soil vapor concentration predicted from the soil data.

Question 3 (Section 6.2.3): According to Fick's law of diffusion, the one-dimensional rate of diffusion flow across a soil layer can be approximated by the expression

$$E = \frac{(C_{\text{source}} - C_0) \times D_{\text{eff}}}{L_s}$$

where:

E	is the rate of diffusion flux (mg/m²-s)
C_{source}	is the vapor concentration at the contamination zone (mg/m³)
C_0	is the vapor concentration at the top of the soil layer (mg/m³)
D_{eff}	is the effective diffusion coefficient across the soil layer (m²/s)
L	is the thickness of soil layer (m)

The effective diffusion coefficient depends on the soil properties and can be estimated using the Millington and Quirk equation:

$$D_{\text{eff}} = D_a \left(\theta_a^{3.33} \Big/ n^2 \right) + \frac{D_w}{H'} \left(\theta_w^{3.33} \Big/ n^2 \right)$$

where:

D_a	is the diffusivity in air (m²/s)
D_w	is the diffusivity in water (m²/s)
θ_a	is the soil air-filled porosity (unitless)
θ_w	is the soil water-filled porosity (unitless)
n	is the soil total porosity ($=\theta_a + \theta_w$, unitless)
H'	is the Henry's Law constant (unitless)

At a dry cleaner's site with releases of tetrachloroethylene (PCE), the measured soil vapor concentration was 200 mg/m³ at 3 m below the surface. Assuming no PCE was detected at the surface, calculate the diffusion flow rate for (a) sandy soil ($n = 0.375$ and $\theta_w = 0.054$) and (b) silty clay soil ($n = 0.481$ and $\theta_w = 0.216$). (c) Which soil type would have greater potential for vapor intrusion into the dry cleaner's building, and why?

Solution 3:

PCE properties: $D_a = 5.0 \times 10^{-6}$ m²/s, $D_w = 9.5 \times 10^{10}$ m²/s, $H' = 0.72$

(a) For sandy soil,

$$D_{\text{eff}} = 5.0\times10^{-6}\left(0.321^{3.33}\Big/0.375^2\right)+\frac{9.5\times10^{-10}}{0.72}\left(0.054^{3.33}\Big/0.375^2\right)$$

$$= 8.1\times10^{-7}\ \text{m}^2/\text{s}$$

$$E = \frac{(200-0)\times8.1\times10^{-7}}{3} = 5.4\times10^{-5}\ \frac{\text{mg}}{\text{m}^2\times\text{s}}$$

(b) For silty clay soil ($n = 0.481$ and $\theta_w = 0.216$),

$$D_{\text{eff}} = 5.0\times10^{-6}\left(0.265^{3.33}\Big/0.481^2\right)+\frac{9.5\times10^{-10}}{0.72}\left(0.216^{3.33}\Big/0.481^2\right) = 2.6\times10^{-7}\ \text{m}^2/\text{s}$$

$$E = \frac{(200-0)\times2.6\times10^{-7}}{3} = 1.7\times10^{-5}\ \frac{\text{mg}}{\text{m}^2\times\text{s}}$$

(c) The site with sandy soil, as the contaminant vapor flux is approximately three times higher through sand than silty clay due to higher air-filled porosity (85% vs. 55%).

Question 4 (Section 6.2.4): In the Johnson and Ettinger model (1991), the building ventilation rate, Q_{bldg}, may be calculated as

$$Q_{\text{bldg}} = \frac{L_B \times W_B \times H_B \times AER}{3600\ \text{s/h}}$$

where:

- Q_{bldg} is the building ventilation rate (m³/s)
- L_B is the length of building (m)
- W_B is the width of building (m)
- H_B is the height of building (m)
- AER is the air exchange rate (1/h)

(a) Calculate the building ventilation rate for a one-story residence 10 m in length, 10 m in width, and 2.5 m in height, and with an average air exchange rate of 0.5 per hour. (b) The measured concentration of trichloroethylene (TCE) beneath the building slab was 500 μg/m³, and the volumetric flow rate of soil gas entering the building through cracks (Q_{soil}) is estimated at 5 L/min. Assuming that the total air volume within the structure is well mixed and that any vapor contaminant entering the structure is homogeneously distributed, what is the TCE concentration in indoor air due to subsurface vapor intrusion?

Solution 4:

$$Q_{\text{bldg}} = (10 \times 10 \times 2.5 \times 0.5)/3600 = 0.035 \text{ m}^3/\text{s}$$

$$Q_{\text{soil}} = 5 \text{ L/min} \times (1 \text{ m}^3/1000 \text{ L}) \times (1 \text{ min}/60 \text{ s}) = 8.3 \times 10^{-5} \text{ m}^3/\text{s}$$

$$C_{\text{bldg}} = 500 \times 8.3 \times 10^{-5}/(0.035 + 8.3 \times 10^{-5}) = 1.2 \text{ µg/m}^3$$

Question 5 (Section 6.3.2): (a) At the same UST site discussed in Questions 1 and 2, the measured concentrations (C_{gw}) of benzene and naphthalene in a grab groundwater sample are 5 and 2 mg/L, respectively. Calculate their concentrations in soil vapor (C_{sg}) above the water table. (b) Are the soil and groundwater sampling results consistent with each other, and what may be the reasons for discrepancy in these lines of evidence, if any? (c) A health risk assessment was conducted to establish the soil gas screening levels for off-site residents of 2 and 0.1 mg/L for benzene and naphthalene, respectively. What would you recommend doing next for residences adjacent to the UST site?

Solution 5:

(a) Using the following equation,

$$C_{sg} = H' \times C_{gw}$$

where:

H'	is the Henry's Law constant (0.23 and 0.018 for benzene and naphthalene, respectively)
For benzene, C_{sg}	is $0.23 \times 5 = 1.2$ mg/L
For naphthalene, C_{sg}	is $0.018 \times 2 = 0.036$ mg/L

(b) The calculated concentrations of benzene and naphthalene in soil pore air are 5.34 and 0.058 mg/L in Questions 1 and 2, respectively. While naphthalene results agree with each other in general, the predicted benzene concentrations differ four-fold.
There are several possible reasons for the observed discrepancy in benzene: (1) soil samples were less homogeneous compared with groundwater samples, and thus, the soil detections for benzene might be biased high in this case; (2) benzene was more likely biodegraded in groundwater than naphthalene; or (3) there may be different contamination sources for soil and groundwater (e.g., surface release for soil and off-site plume for groundwater).

(c) For naphthalene, the calculated soil gas concentrations are below the screening level of 0.1 mg/L, so no further action is needed. For benzene, the calculated soil gas concentration based on soil data is higher than the screening level, while that based on groundwater data is below it; thus, additional lines of evidence are needed (e.g., installing groundwater wells and collecting soil gas samples near residences) before deciding whether indoor air sampling should be considered.

Question 6 (Section 6.3.3): For the same one-story residence as in Question 4: (a) Calculate the predicted indoor air concentration using the California default sub-slab vapor attenuation factor in Table 6.2. (b) Calculate the predicted indoor air concentration, based on the soil vapor sampling conducted around the residence with the average TCE concentration detected at 5 ft below surface of 900 μg/m³. (c) How do these results compare with the answer to Question 4, and why are they different?

Solution 6:

(a) $C_{bldg} = 500 \times 0.05 = 25$ μg/m³
(b) $C_{bldg} = 900 \times 0.002 = 1.8$ μg/m³
(c) The default attenuation factors are conservative, likely to generate "high-end" values, and intended for preliminary screening. The calculations in Question 4 use building-specific information and may be more appropriate for site-specific evaluation.

Question 7 (Section 6.4.2): Continuing the information in Question 6, the owner decided to conduct indoor air sampling to address the uncertainty associated with the available lines of evidence based on subsurface measurements. The sampling measured an average TCE concentration of 1.5 μg/m³ in indoor air. (a) Calculate the inhalation cancer risk for the residents. (Hint: use Equations 3.3 and 3.5a and Table 3.1 in Chapter 3.) (b) For TCE, the USEPA short-term (noncancer) accelerated response action level for residents is 2 μg/m³. Based on Table 6.5, what response action may be appropriate in this case?

Solution 7:

(a) Using Equation 3.3 and the exposure parameters for residents in Table 3.1,

$$EC(\text{cancer}) = 1.5\ \mu\text{g/m}^3 \times 24\ \text{h/d} \times 350\ \text{d/yr} \times 26\ \text{yr} \big/ \left(70\ \text{yr} \times 365\ \text{d/yr} \times 24\ \text{h/d}\right)$$

$$= 0.53\ \mu\text{g/m}^3$$

Using Equation 3.5a and the inhalation unit risk of 4.1×10^{-6} $(\mu g/m^3)^{-1}$ from the USEPA RSL Generic Table,

$$\text{Cancer risk} = 0.53\ \mu\text{g/m}^3 \times 4.1 \times 10^{-6} \left(\mu\text{g/m}^3\right)^{-1} = 2 \times 10^{-6}$$

(b) Because the measured indoor air concentration is less than the short-term response action level, no immediate action is needed. However, the estimated cancer risk is between 1×10^{-6} and 1×10^{-4}, so one or more of the possible actions listed in Table 6.5 should be considered if the residence will remain occupied.

Question 8 (Section 6.4.2): Based on the evaluation in Question 7, the owner of the residence plans to demolish the residence and build a new office building. (a) Estimate the indoor air concentration in the office building (hint: use the information in Question 6). (b) Is this a good plan to mitigate potential vapor intrusion concern?

Solution 8:

(a) If the residence is demolished, the sub-slab vapor and indoor air data will no longer be a good indicator for the new building. Thus, using the soil vapor sampling results and the USEPA default attenuation factor of 0.03 (the most conservative value in Table 6.4) for a future commercial building,

$$C_{bldg} = 900 \times 0.03 = 2.7\ \mu g/m^3$$

(b) To calculate the cancer risk for an indoor worker,

$$\text{EC(cancer)} = 2.7\ \mu g/m^3 \times 8\ h/d \times 250\ d/yr \times 25\ yr / (70\ yr \times 365\ d/yr \times 24\ h/d)$$

$$= 0.22\ \mu g/m^3$$

$$\text{Cancer risk} = 0.22\ \mu g/m^3 \times 4.1 \times 10^{-6} \left(\mu g/m^3\right)^{-1} = 9 \times 10^{-7}$$

The estimated cancer risk is less than 1×10^{-6}, so this new land use plan would justify no further action according to Table 6.7.

REFERENCES

Abreu, L.D.V. and P.C. Johnson. 2005. Effect of vapor source-building separation and building construction on soil vapor intrusion as studied with a three-dimensional numerical model. *Environmental Science and Technology* 39 (12): 4550–4561.

Abreu, L.D.V. and P.C. Johnson. 2006. Simulating the effect of aerobic biodegradation on soil vapor intrusion into buildings: Influence on degradation rate, source concentration, and depth. *Environmental Science and Technology* 40 (7): 2304–2315.

American Petroleum Institute (API). 2012. *BIOVAPOR – A 1-D Vapor Intrusion Model with Oxygen-Limited Biodegradation* (Version 2.1). Washington, DC. November 2012.

California Department of Toxic Substances Control (DTSC). 2011a. *Guidance for the Evaluation and Mitigation of Subsurface Vapor Intrusion to Indoor Air.* October 2011.

California Department of Toxic Substances Control (DTSC). 2011b. *Vapor Intrusion Mitigation Advisory.* Final Revision 1. October 2011.

California Environmental Protection Agency (Cal EPA). 2015. *Advisory – Active Soil Gas Investigation* (www.dtsc.ca.gov/SiteCleanup/upload/VI_ActiveSoilGasAdvisory_FINAL.pdf).

Folkes, D.J. and D.W. Kurz. 2002. Efficacy of sub-slab depressurization for mitigation of vapor intrusion of chlorinated compounds. Presented at *Indoor Air 2002, the 9th International Conference on Indoor Air Quality and Climate, International Academy of Indoor Air Sciences*, Monterey, CA.

Interstate Technology and Regulatory Council (ITRC). 2007. *Vapor Intrusion Pathway: A Practical Guideline* (www.itrcweb.org/Documents/VI-1.pdf). January 2007.

Jeng, C.Y., S. Hariri, and R. Chou. 2008. A step-wise approach to assess the significance of subsurface petroleum vapor intrusion to indoor air. *Overseas Chinese Environmental Engineers & Scientists Association (OCEESA) Journal* 25: 2.

Johnson, P.C. and R.A. Ettinger. 1991. Heuristic model for predicting the intrusion rate of contaminant vapors into buildings. *Environmental Science and Technology* 25: 1445–1452.

Little, J.C., J.M. Daisey, and W.W. Nazaroff. 1992. Transport of subsurface contaminants into buildings. *Environmental Science and Technology* 26: 2058–2066.

Loureiro, C.O. 1990. Three-dimensional simulation of radon transport into houses with basements under constant negative pressure. *Environmental Science and Technology* 24: 1338–1348.

Massachusetts Department of Environmental Protection (MADEP). 2011. Interim Final Vapor Intrusion Guidance. WSC#-11-435. December.

Nazaroff, W.W., H. Feustel, A.V. Nero, K.L. Revzan, and D.T. Grimsrund. 1985. Radon transport into a detached one-story house with a basement. *Atmospheric Environment* 19(1): 31–46.

Nazaroff, W.W., S.R. Lewis, S.M. Doyle, B.A. Moed, and A.V. Nero. 1987. Experiments on pollutant transport from soil into residential basements by pressure-driven air flow. *Environmental Science and Technology* 21(5): 459–466.

New Jersey Department of Environmental Protection (NJDEP). 2012. Vapor Intrusion Technical Guidance, Version 2.0. Site Remediation Program. Currently available online at: www.state.nj.us/dep/srp/guidance/vaporintrusion/vig_main.pdf

United States Environmental Protection Agency (USEPA). 1993. *Radon Reduction Techniques for Existing Detached Houses, Technical Guidance for Active Soil Depressurization Systems.* EPA/625/R-93/011.

USEPA. 2008. *Brownfields Technology Primer: Vapor Intrusion Considerations for Redevelopment.* EPA 542-R-08-001. March (www.brownfieldstsc.org).

USEPA. 2012a. *Conceptual Model Scenarios for the Vapor Intrusion Pathway.* EPA 530-R-10-003. February 2012.

USEPA. 2012b. *Petroleum Hydrocarbons and Chlorinated Solvents Differ in Their Potential for Vapor Intrusion.* Office of Underground Storage Tanks, Washington, DC, March 2012.

USEPA. 2012c. *Vapor Intrusion Database: Evaluation and Characterization of Attenuation Factors for Chlorinated Volatile Organic Compounds and Residential Buildings.* EPA 530-R-10-002. March 16, 2012.

USEPA. 2012d. *Fluctuation of Indoor Radon and VOC Concentrations Due to Seasonal Variations.* EPA/600/R/12/673. September 2012.

USEPA. 2014. *Vapor Intrusion Screening Level (VISL) Calculator User's Guide* (www.epa.gov/vaporintrusion/vapor-intrusion-screening-levels-visls). May 2014.

USEPA. 2015. *Technical Guide for Assessing and Mitigating the Vapor Intrusion Pathway from Subsurface Vapor Sources to Indoor Air.* OSWER Publication 9200.2-154. June 2015.

USEPA. 2017. *Johnson and Ettinger Model Spreadsheet Tool*, Version 6.0. (www.epa.gov/vaporintrusion/epa-spreadsheet-modeling-subsurface-vapor-intrusion)

Wisconsin Department of Natural Resources (WDNR). 2012. *Addressing Vapor Intrusion at Remediation & Redevelopment Sites in Wisconsin (RR-800) Update.* July 2012.

7 Conventional Site Remediation Methods

Yue Rong and Xihong Zhai

CONTENTS

7.1 INTRODUCTION

After finishing the site assessment and characterization, we would like to know the types of pollutants, the extent of the pollution, and the magnitude of the problem. The next step would be to evaluate remediation options at the impacted site. Remediation means to reduce the contamination to an acceptable level. To start the remediation, the pollution sources must be eliminated. For example, if the pollution originates from the leakage of an underground storage tank (UST) (shown in Figure 7.1), the UST as a leaking source should be removed before the start of remediation of impacted soil and/or groundwater. The types of contaminants should also be factored into remedial selections. Typical contaminants are generally classified into two broad categories: organic and inorganic. Organic contaminants can be volatile (i.e., volatile organic compounds [VOCs]) such as benzene, or semi-volatile (SVOCs) such as pesticides and polychlorinated biphenyls (PCBs). Inorganic contaminants include metals such as hexavalent chromium and anions such as perchlorate. In addition, there are "emerging contaminants" such as perchlorate, methyl tertiary butyl ether (MTBE), and N-nitrosodimethylamine (NDMA) (see Chapter 2). Different remedial methods are suitable for different types of contaminants. The discussion of each treatment method in the following sections, and in subsequent chapters, reflects its applicability to different types of contaminants. Soil and groundwater cleanup is site specific. The choice of cleanup technology depends on site-specific conditions, such as type of pollutants, concentration distribution, characteristics of underlying soil, and site geology and hydrogeology. Other factors, such as remediation time frame, method efficiency, land use change, sensitive receptors around the location, risk assessment and

FIGURE 7.1 (a) Leaking underground storage tank removal and (b) the corroded underground storage tank as a soil and groundwater pollution source.

management, economical or physical constraints, and so on, also play important roles in the selection of remediation methods.

Remediation methods can be either conventional or innovative. The boundary between conventional and innovative remediation methods is not always clearly defined, and it evolves with time. A method that was considered innovative 20 years ago might now have become conventional. Conventionally, site remediation methods usually include physical, chemical, and biological means. The remediation method type can be classified as active and passive. Active methods are methods that actively perform to remediate the pollution situation. Passive methods are methods that do not actively perform to remediate the pollution; rather, they allow nature to reduce the pollution. A general review of all current remedial methods is well documented and can be seen at https://clu-in.org/remediation/.

7.2 PHYSICAL METHODS

Basic physical remedial methods include excavation, soil vapor extraction, groundwater pump and treat, dual-phase extraction, multi-phase extraction, high vacuum pump, air sparging, thermal, and granular activated carbon sorption (GAC).

7.2.1 Soil Excavation

Excavation is usually used to remove contaminant sources, for example, impacted soil, and transport off-site for disposal or treatment.

The excavated materials normally are characterized for their contaminant concentrations (i.e., profiling) and transported off-site for treatment or disposal in appropriate landfills based on the contaminant profile. Excavation needs to be carefully designed with protective systems to avoid potential hazards such as cave-in or falling loads. Protective systems include sloping, shoring and boarding, and so on. Sloping involves cutting back the excavation wall at an angle inclined away from the excavation. Shoring requires installing some supports to prevent soil movement and cave-ins (refer to Figure 7.2). The design of a protective system considers many factors, including soil type, excavation depth, and water content of soil.

Soil excavation is good for a quick solution, particularly for sites soon to be re-developed (Figure 7.3). The advantages of soil excavation also include complete removal of a pollution source area, and it is good for fine-grained materials where soil vapor extraction or *in situ* injection does not work effectively. The disadvantages of soil excavation include that it is costly, disruptive to the site business, limited to the open area for operation, and usually not used to dig very deep to remove deep soil contamination. Ultimately, excavation does not destroy the contaminants; in some cases without treatment, it simply transfers them from one place to another.

FIGURE 7.2 (a) Open pit soil removal and (b) deep soil digging with shoring and boarding to prevent soil cave-in.

FIGURE 7.3 Large-diameter drilling rig removing deep impacted soil.

7.2.2 Non-Aqueous Phase Liquid (NAPL) Removal

Some contaminants are in the form of non-aqueous phase liquid (NAPL), also referred to as *free product*, in the subsurface. NAPLs, which are not soluble in water, exist because the contaminant exceeds its solubility in water, and therefore, they are in a liquid form separated from groundwater. Some NAPLs with density greater than water, such as chlorinated solvent compounds, including trichloroethylene (TCE) and tetrachloroethylene (PCE), are often referred to as *dense non-aqueous phase liquids* (DNAPLs). DNAPLs will sink through the groundwater aquifer until they encounter a low-permeability layer (aquitard). The removal of DNAPL is very difficult due to the challenges associated with depth, aquifer lithology, and complicated interaction with groundwater.

Compared with DNAPLs, a light non-aqueous phase liquid (LNAPL) has lower density than water, such as gasoline; therefore, it will float on top of the groundwater table. Because it floats on the groundwater table, fluctuation of the groundwater table can cause upward or downward movement of LNAPLs, creating a "smear zone" where soil pores might have residual LNAPLs. LNAPLs can be removed in the smear zone or the current groundwater table by using bailer or pump via recovery wells. The LNAPL mass recovery rate can be limited by the equilibrium process between the surrounding aquifer and the recovery wells. When there is not enough LNAPL in the aquifer, the rate of LNAPL accumulation in the recovery wells may be too slow to support continuous pumping, and periodic bailing or sorbent socks can be more effective ways to recover LNAPL (Figure 7.4).

The advantages of free product bailing/pumping are that it is direct, good for a small quantity of free product, and has limited disruptive impact on the on-site business. The disadvantages of free product bailing/pumping are that it may be very slow, it may not be able to operate continuously, it may not be effective for slow recovered groundwater formation, and it is hard to completely remove

FIGURE 7.4 Bailing LNAPL from a well.

the residual free product from the aquifer. The removal of residual free product may need some more innovative technology, such as *in situ* soil flushing, which is discussed further in Chapter 8.

7.2.3 Groundwater Pump and Treat

Groundwater pump and treat (P&T) is one method to treat contaminated groundwater by pumping the water from the aquifer and treating or removing pollutant from the water using physical, mechanical, or other means such as chemical or biological methods on the surface. Since the method involves above-ground treatment, this type of method is also called an *ex situ* method, in contrast to *in situ* methods, which treat the contaminated soil and groundwater in the subsurface. Chapter 8 will discuss more *in situ* methods in detail.

Groundwater is pumped from the underlying aquifer. The pumping will cause a cone of depression at the water table, as demonstrated in Figure 7.5. The radius of the depression in plan view is called the *zone of influence*, which is usually associated with the volume of water pumped out. The natural groundwater elevation is therefore altered in the zone of influence, which leads to a change of the groundwater flow field. The hydraulic containment goal of P&T is achieved by drawing groundwater flow toward the cone of depression around the recovery wells, normally located in the plume core, which then prevents expansion of the contaminant plume. The design of an effective P&T requires an estimate of the capture zone, that is, the portion of the aquifer that will ultimately discharge to the pumping wells.

Groundwater P&T is more effective to remove soluble contaminants, such as hexavalent chromium and perchlorate. As existing dissolved contaminants are removed by groundwater pumping,

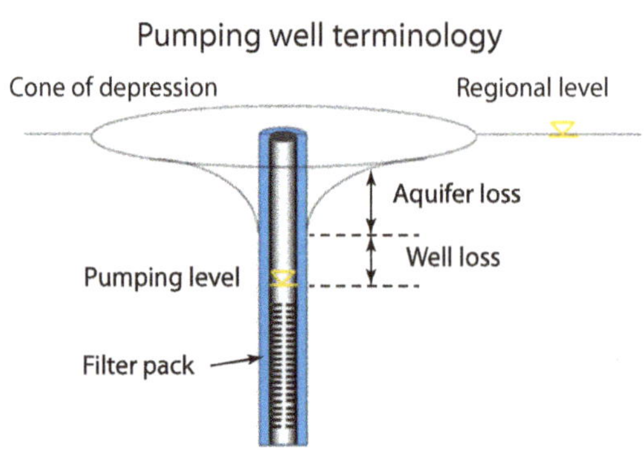

FIGURE 7.5 Conceptual model of zone of influence (radius of influence).

more contaminant will enter the groundwater by desorbing from porous media, dissolving from precipitates or NAPL, and/or diffusing from low-permeability zones. As the remediation progresses, contaminant concentration gets lower and lower, and the mass recovery reaches an asymptotic level. This phenomenon is often referred to as *tailing*. Contaminant concentration can also bounce back if P&T operation stops, referred to as *rebound*. If the treatment goal is not achieved before this asymptotic level, this treatment method becomes less cost effective.

After groundwater is pumped to the surface, many methods can be used to treat and remove pollutant. These methods rely on physical, chemical, and/or biological processes to reduce contaminant concentrations to acceptable levels. The selection of an above-ground treatment method considers many factors, including the physical and chemical properties of contaminants, technical feasibility, and costs. For organic contaminants, the commonly used treatment technologies include air stripping, GAC sorption, chemical/ultraviolet (UV) oxidation, and aerobic biological reactors. For inorganic constituents, the common treatment methods include chemical precipitation, ion exchange/adsorption, reverse osmosis, and electrochemical methods (US EPA 1997). The post-treated water is generally either discharged to the sewer system or re-injected into the aquifer to conserve water resources. Re-injection points are usually placed outside of the pollution source areas.

One of the common physical methods to treat groundwater after it is pumped up is to apply either GAC or the reverse osmosis (RO) method. GAC is a sorption process to remove the dissolved phase of contaminants. RO is a membrane method to screen out the particles and pollutants dissolved in water. The pores of the specific membrane are so minute that only tiny molecules, such as water molecules, can pass through it. Other "bigger" molecules, such as organic pollutants, are retained by the membrane. Therefore, the keys to RO are to have a good-quality membrane and to apply pressure to get the water to be treated through the membrane (Figure 7.6).

The advantages of the RO method are that it conserves water, if water is re-injected into the aquifer, and it can be used in combination with well-head treatment. Another advantage of the RO method is that it is effective for specific compounds, since different types of pollutants can be treated by different matching membranes. The disadvantages of the RO method are that it is costly and slow, and that it wastes water resources if it discharges to surface water. Sometimes, well-head treatment is an important and better option for situations where the pollutant plume is widespread with relatively low concentrations, but not meeting drinking water standards, and there is no hot spot pollution source within the plume. In that case, pumping from the plume to treat the water may not be cost effective. Therefore, well-head treatment may be a good option to treat the water for a drinking water supply.

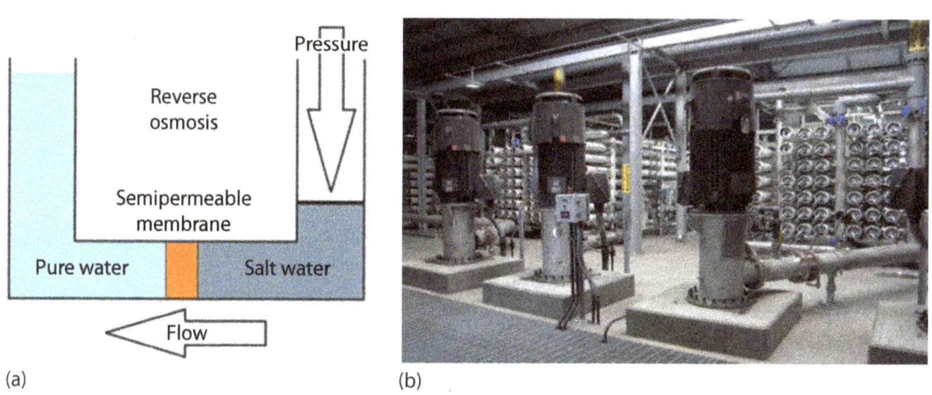

FIGURE 7.6 (a) Schematic of how RO works. (b) RO operating room.

7.2.4 Soil Vapor Extraction

Soil vapor extraction (SVE) uses vacuum pressure to extract volatile compounds from porous soil. It is a physical process in the vadose zone, which requires vacuum pressure, porous soil medium, and volatile compounds. There is similarity between groundwater P&T and SVE, as both involve removal of contaminant media (vapor or groundwater) and subsequent treatment. The SVE will cause a zone of influence, or radius of influence in the vadose zone, similar to groundwater pumping (Figure 7.7).

SVE wells have casings and screens just like groundwater monitoring/recovery wells. The extraction is more effective in coarse soils, since there are porous spaces in soils for the vapor to travel and move. The well screens should be installed across the soil zones where high contaminant concentrations and high-permeability layers of soils are located.

SVE is effective for coarse media soil and volatile contaminants and can be operated continuously over time until it is complete. Pilot tests are normally performed during the remedial investigation stage to determine the radius of influence and aid full-scale SVE design. Numerous references are available for design guidelines (USEPA 1992, 1995; USACE 1995). Similarly to groundwater P&T, the contaminant mass removal rate of SVE diminishes as treatment progresses, and the vapor concentration will also show tailing and rebound. The common practice is to switch to a pulsing mode to allow contaminants to re-equilibrate between soil and vapor phases (i.e., rebound) between the two operation cycles. Eventually, the pulsed SVE operation will also reach an asymptotic level in mass recovery.

A side bonus of SVE is that the accelerated air flow can bring in abundant oxygen, which can enhance aerobic biodegradation processes of certain organic compounds (e.g., petroleum-related contaminants such as benzene). A variation of SVE, bioventing, uses this process by aerating the vadose zone at a low flow rate compared with SVE (see details in Chapter 8).

Extracted soil vapor needs to be treated to remove contaminants before discharge to atmosphere. Unlike groundwater P&T, treatment options for SVE process are limited to those for volatile organic constituents, since they are more amenable to SVE. Commonly used vapor treatment options include GAC, catalytic oxidation, and thermal oxidation. The discharge of treated vapor to atmosphere requires extensive air quality permitting process, required by regulatory agencies for soil vapor extraction the method can be very extensive and long, which can be a disadvantage of SVE technology. In addition, SVE may be ineffective in fine-grained material soils.

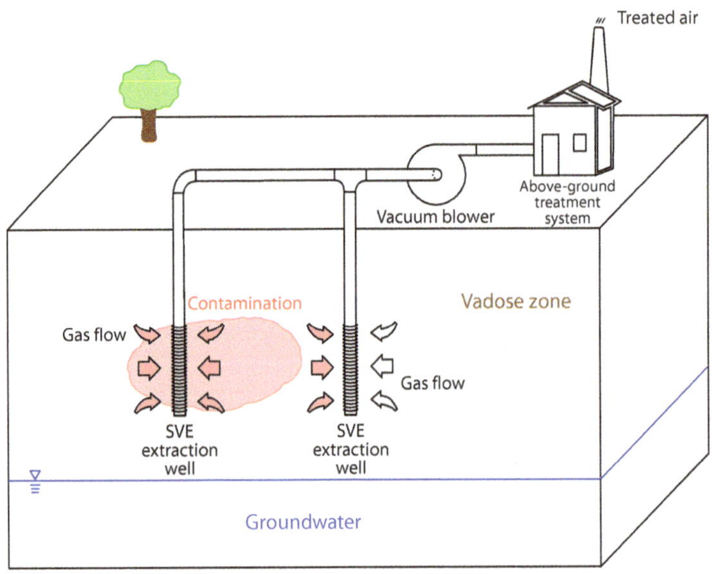

FIGURE 7.7 Schematic of vapor extraction system.

7.2.5 Thermal Enhancement by Electrical and Natural Gas Heating

Since SVE is targeting volatile compounds, it is assumed that the method may be more effective at encouraging volatilization if the soil is heated up. Soil heating can be achieved electrically or using natural gas. The electrical heating method is to apply electricity as an energy source to heat up impacted soil and use a vapor extraction method to remove volatile compounds that are volatilized more by increasing temperature. Electricity may be expensive in terms of consuming energy, rendering the cost of this technology prohibitive. Natural gas can also be used as an energy source to heat the subsurface as a suitable alternative to electrical heating (Figures 7.8 and 7.9).

The advantages of thermally enhanced SVE are that it is an effective heating method for tight soils, where SVE does not work effectively, and it is also effective for semi-volatile contaminants, which may be difficult for SVE to extract effectively under ambient temperature conditions. Another advantage of thermal treatment is the quick results within a relatively short time frame. The disadvantage of heating methods is that they are energy consuming and therefore costly. Thermal treatment is normally the preferred method for sites with expedited re-development plans and ample financial resources. Due to heat transport in the soil media, thermal enhancement may have only a limited impact area.

In addition to aiding SVE treatment in the vadose zone, thermal enhancement can also be applied in saturated zone contamination. A more detailed discussion on soil and groundwater thermal treatment is presented in Chapter 8.

FIGURE 7.8 (a) Installed electrical pole in soil pile where the temperature in soil is heated up as seen, and (b) the size of one electrical pole, which needs to be carried by two people.

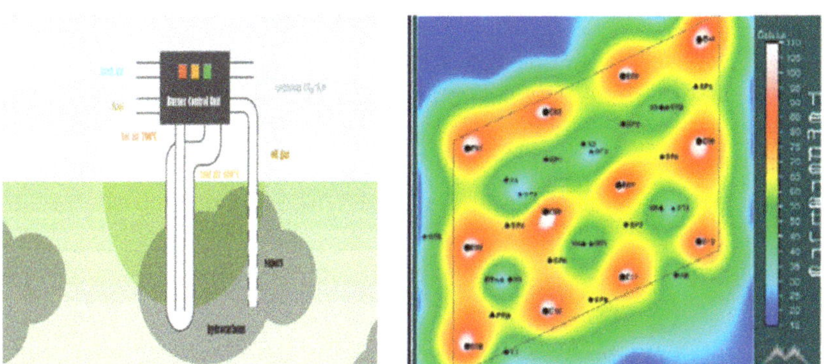

FIGURE 7.9 (a) Gaseous heating system schematic and (b) a heating monitoring map showing the heat effect in an aerial perspective and temperature distribution.

7.2.6 Dual-Phase or Multi-Phase Extraction and Air Sparging

Dual-phase usually means in the vapor phase and the liquid phase, and the extraction principle applies to both phases. The key consideration is to install the extraction points at the right locations in the soil profile. The soil extraction and groundwater extraction may still work separately in a dual-phase system. Particularly, this method may work better at the capillary fringe zone, where a high concentration of dissolved pollutants or even free product exists. Sometimes, the extraction well screen may be across three phases of pollutant (vapor, dissolved, and LNAPL). In that case, we call it *multi-phase extraction*. To accelerate the extraction, air is injected into the polluted water body to encourage volatilization from the dissolved phase. This procedure is called *air sparging*. Similarly to the SVE process in the vadose zone, air sparging brings in oxygen to the saturated zone and enhances the aerobic biological activities. The analog of bioventing for the saturated zone is called *biosparging*, which aerates the saturated zone to aid the biodegradation of certain contaminants (such as benzene) susceptible to aerobic processes (Figures 7.10 and 7.11).

The advantage of multi-phase extraction is that it can clean up multiple phases of pollutants at one location. It works well for pollutants distributed in the capillary fringe. Like SVE, it can run continuously over time until it is complete. Its disadvantage is that it needs a complicated system of piping and engineering units and that the operating units take up a lot of surface space. Sometimes, it may not work well without a clear understanding of the different phases of pollutant distribution in the subsurface.

7.2.7 Vapor Barrier Installation

For volatile pollutants, one of the potential impacts to human health is by transport through the building floor into people's living space. This is called *vapor intrusion* (see Chapter 6 for details). In addition to the SVE method, one method to remediate the vapor intrusion situation is to install

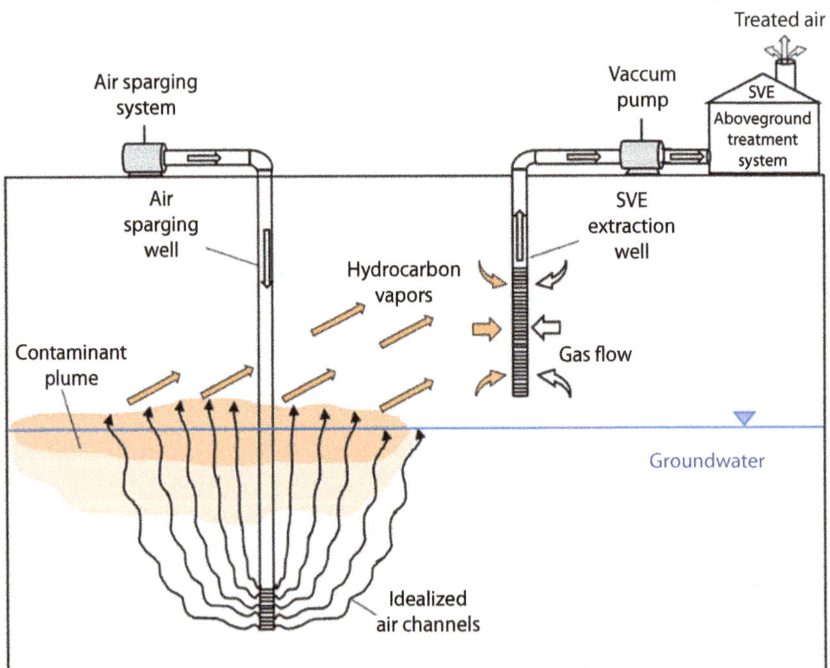

FIGURE 7.10 Schematic of air sparging combined with soil vapor extraction system. Sometimes, the vadose zone extraction well has a screen across the capillary fringe to extract dual-phase pollutants.

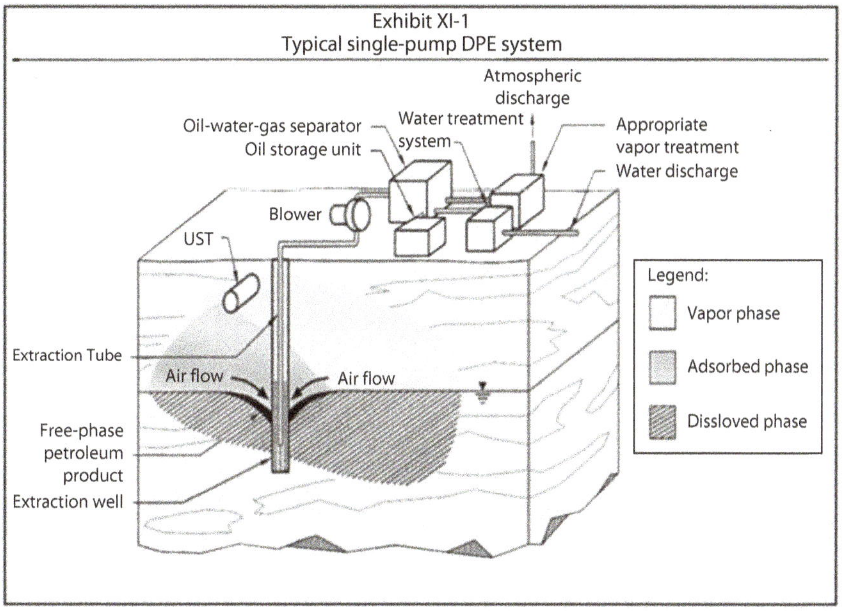

FIGURE 7.11 System of multi-phase extraction, including vapor phase, free product phase, and dissolved phase.

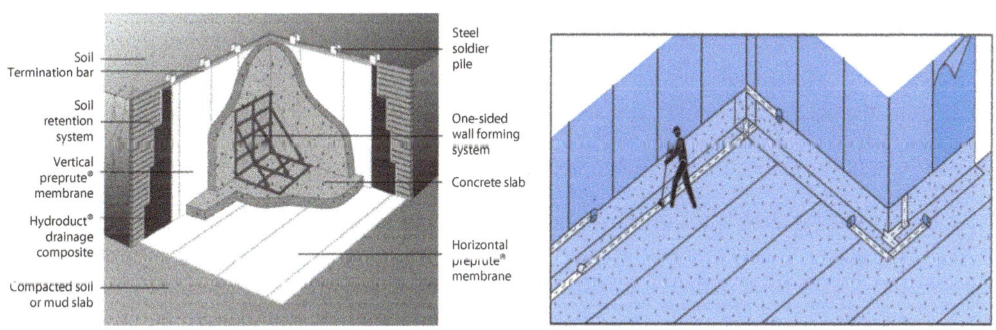

FIGURE 7.12 How the vapor barrier is installed.

a vapor barrier between the contaminated soil or groundwater and the floor of the building. This method is not a treatment method but rather, a remedial method to prevent pollutant impact on humans on the surface. The vapor barrier usually contains a solid surface and a porous sandwich layer of sorption material such as activated carbon (Figure 7.12).

7.3 CHEMICAL METHODS

The fundamental concept for this method is to have chemical reactions to transform, destroy, or precipitate pollutants. Basic chemical remedial methods include in-situ injection and ex-situ treatment using chemical oxidation agents such as hydrogen peroxide (H_2O_2), ozone (O_3), potassium permanganate ($KMnO_4$), sodium permanganate ($NaMnO_4$), and chemical reduction agents such as ferrous chloride and zero-valent iron. An example of such a list of materials for in-situ injection can be seen at http://www.waterboards.ca.gov/losangeles/board_decisions/adopted_orders/docs/AttachmentA.pdf.

Chemical methods apply chemicals for chemical reactions. For example, *Fenton reagent* is a method to create an oxidation scenario by using H_2O_2, elemental iron, and acidic conditions (lower pH). Fenton reagent can be used in *ex situ* treatment on the surface or *in situ* injection to the subsurface. An example of the Fenton reagent chemical reaction to transform chemical pollutants by hydroxylation is shown in Equation 7.1.

$$Fe^{2+} + H_2O_2 \rightarrow Fe^{3+} + OH^- + OH \tag{7.1}$$

The second type of chemical method is to use chemical-generating compounds that will generate oxygen in the presence of water. This oxygen can trigger an oxidation process or enhance bioremediation by the oxygen-using microbial population. Equation 7.2 shows how the oxygen-generating compound magnesium peroxide produces O_2 given water:

$$MgO_2 + H_2O \rightarrow \tfrac{1}{2}O_2 + Mg(OH)_2 \tag{7.2}$$

The third type of chemical method is to chemically transform or degrade a contaminant. Equations 7.3 and 7.4 show the chemical transformation of benzene (C_6H_6) oxidized by oxygen (O_2) (7.3) and hydrogen peroxide (H_2O_2) (7.4), respectively.

$$7.5\,O_2 + C_6H_6 \rightarrow 6CO_2 + 3H_2O \tag{7.3}$$

$$15(H_2O_2) + C_6H_6 \rightarrow 6CO_2 + 18H_2O \tag{7.4}$$

The goal of a chemical method is to use a limited quantity of chemicals that are benign and will not exacerbate the pollution problems. There is concern that adding chemicals into the groundwater may cause new groundwater pollution. To address this concern, proper procedures are put in place to ensure that the chemicals applied will work as planned. Therefore, any chemicals to be considered for reaction with pollutants should be evaluated and tested in the laboratory. After it has been demonstrated in the laboratory that the reaction might work to remediate a particular pollutant, which is called a *bench-scale* demonstration, this chemical can be used at the pollution site on a small scale, which is called a *pilot test*. After the pilot project has sufficiently demonstrated that this chemical works in the field conditions at the site, a full-scale application can proceed.

In situ treatment methods use chemical (oxidation or reduction) and biological processes to destroy or transform the contaminants without removing the contaminated media. The chemical or biological amendment(s) are delivered to the contaminated zone. Contact between the contaminants and the delivered amendments is the key to effective *in situ* treatment.

In situ chemical methods can proceed with direct injection. Therefore, the subsurface physical composition and hydrogeology may become key factors in the chemical reaction. The key point is to have the injected chemicals react with the pollutants as much as possible. Currently, the most commonly used method to facilitate *in situ* injection is the direct-push method. A power tool is used to push a hollow rod to a certain depth, and then the chemical injection is performed with pressure through the rod. At the tip of the rod, there is designed an open segment with a screen to connect with the subsurface (see Figure 7.13). How far the chemical can be dispersed in the aquifer is proportional to the pressure applied for the injection and the subsurface conditions of porousness. One measurement is to determine the distance the reagent can be distributed in the subsurface, often referred to as the *zone of influence* of the chemical injected. More porous media, such as a sandy aquifer, likely allow greater impact from the injection. Less porous media, such as a silt or clayey type of aquifer, produce less impact from the injection.

In situ treatment greatly depends on the subsurface conditions. It is very possible that the *in situ* treatment area is very limited, to just around the injection points. Therefore, the grid of injection points is crucial to this method. Another consideration is to regulate the injection pressure. Incidents have happened where the injected chemical materials were forced to come to the ground

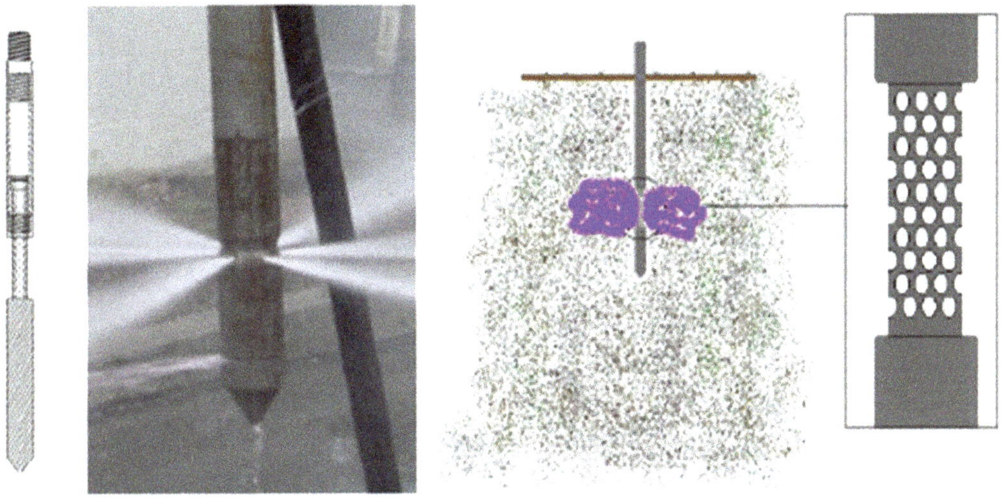

FIGURE 7.13 Tip and screen of the rod by direct-push method to disperse liquid chemicals.

surface (called *re-surfacing*) as the pressure was too high. Usually, the injection wells are not used as monitoring points. In field monitoring, people usually monitor only the concentration changes after the injection; they rarely monitor how the chemical reactions actually take place. However, the concentration changes might be attributed to other reasons, such as pollution source removal, or hydrological dispersion and diffusion.

The advantage of *in situ* injection is that it conserves water resources. It is not necessary to pump the water out of the aquifer and then put it back in. Another advantage of *in situ* injection is low cost. The disadvantage of *in situ* injection is that it is difficult to monitor its effect; it may only have very limited impact around the injection points.

The *in situ* chemical injection method may work very effectively to reduce the residual concentration as a polishing step after a major physical removal method. However, this method may not be effective to treat very high-concentration pollution plumes, such as those in LNAPL sites. A more comprehensive discussion of *in situ* treatment methods is presented in Chapter 8.

The chemical method can also treat contaminants *ex situ*; that is, groundwater can be pumped out to the surface and chemical methods can then be used to destroy or transform the contaminants. Compared with *in situ* methods, *ex situ* chemical methods can usually engage pollutants completely and react effectively as planned. However, chemical reactions may happen slowly, and an above-ground facility may take up a lot of space. In addition, water needs to be re-injected into the aquifer after treatment to conserve water resources. Therefore, there is additional engineering design and well drilling, which may increase the total cost of the remediation.

7.4 BIOLOGICAL METHODS

Biological methods are very similar to the chemical methods discussed in the previous section in terms of their application, operation, conceptual model, injection method, monitoring, verification, and advantages and disadvantages. The only difference is the materials used. Chemical methods apply chemical materials. Biological methods apply a bio-agent such as bacteria. Bacteria are very complex in their characteristics and have many species. The method to be used for site cleanup and remediation can be very complicated due to biological processes. The subject of bacteria will be discussed further in Chapters 8 and 10.

The fundamental concept of a biological method for site remediation is to encourage bacteria and other microbes to destroy pollutants, or to let the living things "consume" the organic pollutant, which they use as their "food" or energy source. Biological methods inject or introduce something to reduce the contaminant concentration. The injected material may be oxygen (O_2), which is used to feed native bacteria and let them grow, or oxygen-generating compounds to produce O_2 in water, and can also be bacterial nutrients such as sugar or other organic chemicals to encourage them to grow. Another injected material is bacteria themselves. Bacteria are cultivated in the laboratory and developed to destroy certain types of pollutant. For example, the bacteria named Dehalococcoides can reduce halogenated organic pollutants such as tetrachloroethylene (PCE).

The biggest concern with the biological method is the addition of bacteria or other microbial species to the water resources. Bacterial injection is not like chemical injection. Chemicals will eventually be diluted in the environment after the injection finishes. However, bacteria can grow exponentially even after the injection finishes. It is a challenge to control unwanted growth and just have the bacteria consume and destroy pollutants. The hope is to have the right species, which is not harmful to human beings but can be effective in remediating the contamination. In the meantime, the scale and quantity of the biological materials to be injected can be controlled to minimize the potential impact on humans.

7.5 PASSIVE METHODS

Basic passive remedial methods include monitoring natural attenuation (i.e., monitoring soil and groundwater concentrations while not applying any active remedial methods) and phytoremediation (i.e., using plants to remove contaminants).

7.5.1 Monitoring Natural Attenuation

The U.S. Environmental Protection Agency (USEPA) acknowledges that natural attenuation is an acceptable method for site remediation. Natural attenuation is a method to monitor pollutant concentration changes due to natural means and the subsurface fate and transport processes, such as advection, dispersion, diffusion, volatilization, sorption, biodegradation, or simply dilution. Natural attenuation is called a *passive method* because it uses natural means, not active human-induced means.

Besides monitoring pollutant concentration changes, how can we quantify the so-called *natural attenuation*? There are three levels of lines of evidence to monitor natural attenuation: (1) the primary line of evidence is to monitor concentration and mass; (2) the secondary line of evidence is to monitor geochemistry parameters (e.g., O_2, NO_3^-, Fe^{3+}); and (3) the tertiary line of evidence is to use environmental molecular diagnostics (EMD) tools (e.g., isotope and DNA technologies) to monitor microbial species. With the pre-condition that the primary line of evidence (concentration and mass data) has been produced and adequately characterized at a site, this section only discusses the secondary line of evidence. The tertiary line of evidence is discussed in Chapters 5 and 10.

Monitoring natural attenuation can include testing the water chemistry and other indicative parameters that can show whether an oxidation or reduction process is ongoing or not. These indicative parameters can be oxygen, nitrate, iron, sulfate, and methane, generally referred to as *geochemical parameters*. Table 7.1 shows the typical oxidation or reduction reactions for commonly observed geochemical parameters. Under oxidative conditions, reduced forms of nitrogen and sulfur (e.g., nitrite [NO_2^-] and sulfite [SO_3^{2-}] or sulfide [S^{2-}]) lose electrons and become nitrate (NO_3^-) and sulfate (SO_4^{2-}), and increased nitrate and sulfate concentrations will be observed in groundwater (not likely to occur under natural conditions unless oxidant is introduced). Under reductive conditions, dissolved oxygen is probably depleted, and groundwater concentrations of nitrate and sulfate are likely to decrease as they gain electrons and transform into reduced forms as nitrite and sulfite. The theory of natural attenuation monitoring is that the natural attenuation processes can result in the change (spatially and temporally) of these geochemical parameters, typically characterized as oxygen depletion and/or nitrate and sulfate decrease when they are being used as oxidants (also termed *terminal electron acceptors*; see Chapter 8), or methane increase.

TABLE 7.1
Typical Oxidation and Reduction Reactions for Geochemical Parameters

Oxidation (Lose Electrons)		Reduction (Gain Electrons)	
$O_2 \rightarrow 2O^{-2}$	$[O_2]\downarrow$		
	ORP $\downarrow$		
(OPR = Oxidation Reduction Potential [mV])			
$NO_3^- \leftarrow NO_2^-$	$[NO_3^-]\uparrow$	$NO_3^- \rightarrow NO_2^-$	$[NO_3^-]\downarrow$
$SO_4^{2-} \leftarrow SO_3^{2-}$	$[SO_4^{2-}]\uparrow$	$SO_4^{2-} \rightarrow SO_3^{2-}$	$[SO_4^{2-}]\downarrow$
$Fe^{+3} \leftarrow Fe^{+2}$	$[Fe^{3+}]\uparrow[Fe^{2+}]\downarrow$	$Fe^{+3} \rightarrow Fe^{+2}$	$[Fe^{3+}]\downarrow[Fe^{2+}]\uparrow$
$Mn^{4+} \leftarrow Mn^{2+}$	$[Mn^{4+}]\uparrow$	$Mn^{4+} \rightarrow Mn^{2+}$	$[Mn^{4+}]\downarrow$
$CO_2 \leftarrow CH_4$	$[CH_4]\downarrow$	$CO_2 \rightarrow CH_4$	$[CH_4]\uparrow$

If the monitoring results indicate a decreasing trend of pollutant concentration (first line of evidence), the pattern of geochemical parameters as listed in Table 7.1 can be examined to verify the occurrence of natural attenuation (second line of evidence). On the other hand, the geochemical parameters can be used to evaluate whether the subsurface conditions are favorable for the natural attenuation of certain contaminants. Usually, oxidative conditions or aerobic conditions are favorable for the natural attenuation of petroleum products and aromatic compounds such as benzene, and reductive conditions or anaerobic conditions are desired for the natural attenuation of halogenated compounds such as PCE or TCE.

Figure 7.14 shows a theoretical parameter distribution with a contaminant groundwater plume under a natural biodegradation (AFCEE 1995). Along the plume centerline, it is presumed that the concentrations decrease with the distance. When the bacteria start working to consume organic pollutants, the inorganic elements will be consumed in a preferential sequence from nitrate to iron and manganese, and then to sulfate, and finally, when other electron acceptors are depleted, the methanogenic process starts and methane increases. If there is an inorganic distribution spatial pattern like that in Figure 7.14, it may be concluded that there is a natural attenuation, or natural biodegradation, occurring.

Monitoring natural attenuation can be conducted based on either a temporal or a spatial scenario. Namely, the parameter changes may be demonstrated temporally (increase or decrease in magnitude over time period) or spatially (increase or decrease in magnitude from location to location). The next two figures show these spatial and temporal examples, respectively.

Figure 7.15 shows an example of a spatial pattern of monitoring natural attenuation. As we can see, methane (CH_4) concentration within the plume is much higher than the numbers outside the plume. Likewise, NO_3^- and SO_4^{2-} within the plume are lower than the numbers outside the plume. By checking with Table 7.1 and Figure 7.14, we find that the aquifer has already gone through nitrate reduction and sulfate reduction, and the ORP within the plume core is low enough to support the methanogenic process, as evidenced by the increasing methane concentration. This is an indicator of a reduction situation, which is in favor of the natural attenuation of chlorinated solvent. One may notice that the oxygen number within the plume is also lower than outside the plume. This may be a situation where the aerobic bacteria work first to exhaust the oxygen, and then the anaerobic bacteria start working.

Figure 7.16 shows an example of a temporal pattern of monitoring natural attenuation, which contains monitoring data from July 2001 to January 2009. As seen (note the concentrations shown

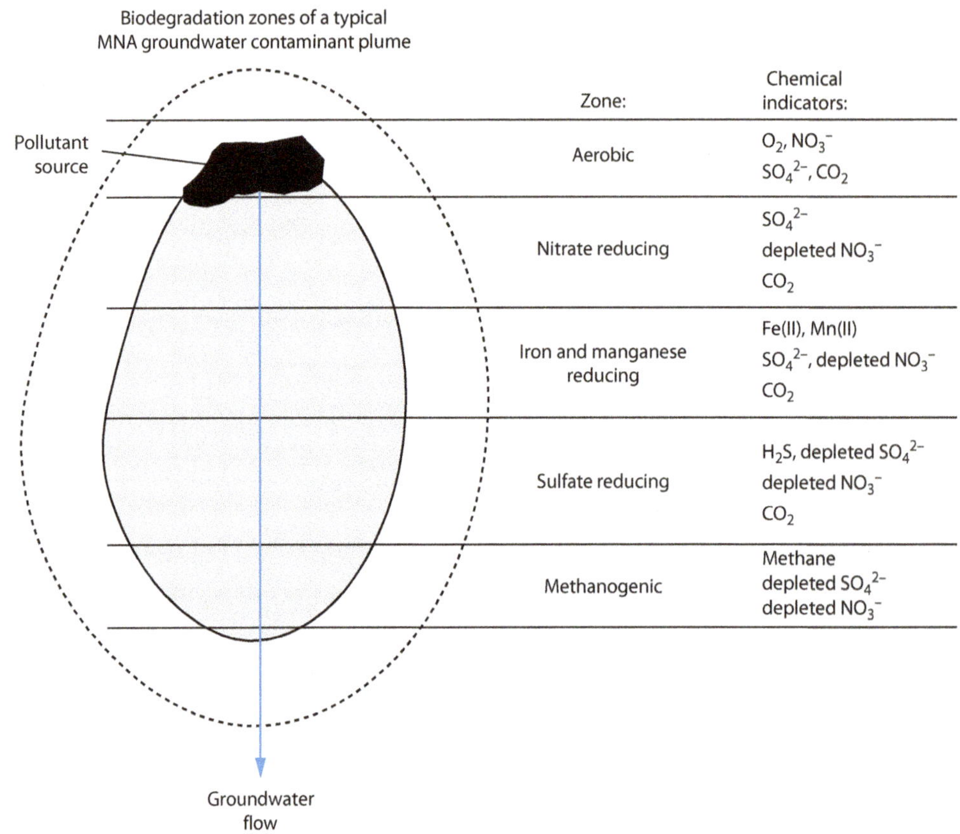

FIGURE 7.14 Natural attenuation parameter distribution within a plume.

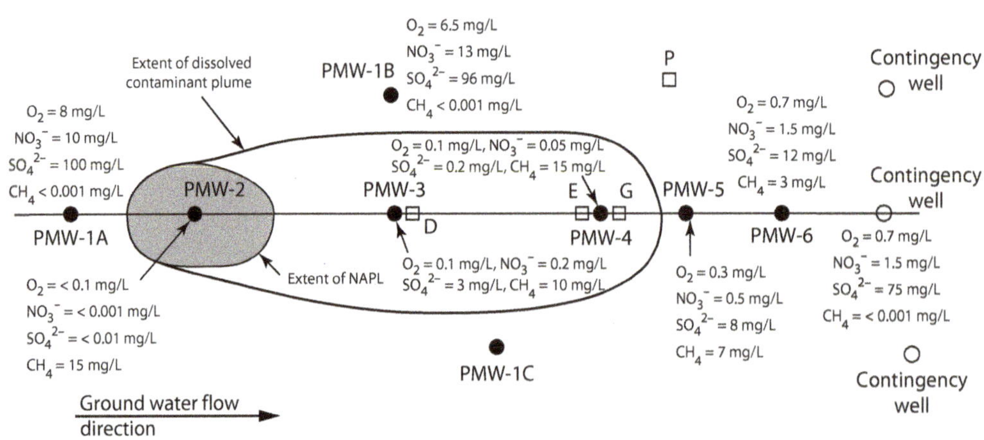

FIGURE 7.15 Natural attenuation spatial pattern.

on different scales), benzene concentration (red solid line) decreases over time, SO_4^{2-} concentration (green dashed line) shows a decreasing trend, and methane (orange solid line) increases from 2007. In addition, nitrates (gray solid line) also show an overall decreasing trend. These increasing and decreasing trends of geochemical parameters in Figure 7.16 show an oxygen-depleting scenario, where aromatic compounds such as benzene are naturally degraded, which is consistent with the

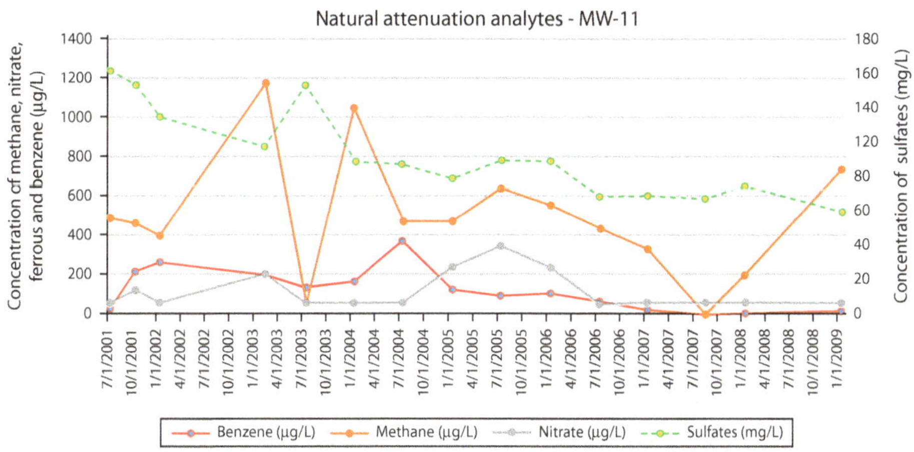

FIGURE 7.16 Natural attenuation temporal pattern.

Guidance (AFCEE 1999). Note that the trend of up and down over time is not as obvious as one might wish. However, this kind of data is often seen in real cases.

As shown, natural attenuation parameter monitoring is not always straightforward; sometimes, it is not very clearly indicative. Therefore, we should use multiple lines of evidence to make a judgment on whether natural attenuation is indeed a possibility in the subsurface environment. The multiple lines of evidence may include the trend of concentration and mass, the geochemistry parameters discussed earlier, and new technologies, such as isotope and DNA techniques. The type of contaminants, the subsurface environment, the maximum concentration and distribution, the contamination history, and the estimate of total discharge to the subsurface should also be evaluated in the framework of monitoring natural attenuation. The advantages of the natural attenuation method are that it is economical, good for polishing at the end of cleanup, and desirable for projects that would otherwise be cost prohibitive, at a polluted site that is not completely accessible, or where other remedial methods may not be feasible. The disadvantages are that it is slow, it may not work in site-specific conditions, and monitoring may become costly over time.

7.5.2 Phytoremediation

Phytoremediation is the use of natural living green plants to remove, degrade, or break down soil and/or groundwater contaminants *in situ*. This method is more dependent on species-specific characteristics, soil medium, and type of pollutants. The intake of green plants may or may not discriminate between different types of pollutants, for example, metals or organics. The advantage of phytoremediation is the low cost in most situations. The disadvantages are that this method may be very slow, the cleanup rate is dependent on the season, it is difficult to remediate the deeper zone of contamination beyond the plant root zone, and it is difficult to monitor the results of the method. In addition, the high pollutant concentrations may cause harm to the plants. The method of phytoremediation is also discussed in detail in Chapter 8.

7.6 NEW REMEDIATION TECHNOLOGIES ON THE HORIZON

Conventional remedial technology, like any other, will evolve with advancements in science and technology. Physical, chemical, and biological methods are mature and have become regular and routine in the field of site remediation. With new site assessment technologies and methods, such as three-dimensional high-resolution sampling, the subsurface formation and contaminant distribution can be understood more intelligently, which will result in better ways and methods for site remediation. Lately, green technology, sustainability, and reduction of carbon emissions have become criteria for evaluating site cleanup and remedial methods. New technologies are gradually being applied to this field, such as biological and chemical transfer methods, nanotechnology, gene analysis, and isotope analysis.

In situ injection and treatment have become common options, not only for economic reasons and for the conservation of water resources, but also for applying new technologies: for example, nanotechnology with zero-valence iron injection, and molecular diagnostic tools (MDTs) to identify specific pollutant-consuming bacteria by using gene analysis and isotope analysis. In the next several chapters, these new remedial technologies will be discussed in detail.

PRACTICE AND EXERCISES

Question 1: Site characterizations yield the following information. Groundwater is about 120 ft deep (40 m), soil contamination is most distributed about 60 ft deep in a fine-grained material layer, contaminants are primarily aromatic compounds (benzene, toluene, ethylbenzene, and xylene [BTEX]), and contamination delineation is adequate. What will the choice of soil cleanup method be?

Answer 1: The best choice is the natural attenuation method. The reasons are: (1) aromatic contaminants are naturally biodegradable (bio prone); (2) soil excavation is almost impossible or very costly to a depth of 60 ft; (3) soil vapor extraction does not work well with fine-grained material soil; (4) groundwater is relatively deep, and the chance of its being impacted may be small.

Question 2: in what circumstances is the natural attenuation method best used?

Answer 2: Natural attenuation is used most effectively when the contaminant concentrations are not very high, preferably used in the "polishing stage" of the cleanup. Another situation might be at a site having physical constraints for active cleanup methods, for example, under a building. It is also desirable where another project would be cost prohibitive, or at a polluted site that is not completely accessible, or where other remedial methods may not be feasible.

Question 3: What is the prerequisite for deciding the type of methods for site remediation?

Answer 3: A complete site assessment and characterization in terms of the type of contaminants, the extent of the contamination, the magnitude of the contamination, and the concentration distribution on-site and/or off-site. The need for an active remediation will depend on the risk and threat from the contamination to human health and the environment.

REFERENCES

Air Force Center for Environmental Excellence (AFCEE). 1995. *Technical Protocol for Implementing Intrinsic Remediation with Long-Term Monitoring for Natural Attenuation of Fuel Contamination Dissolved in Groundwater.* Technology Transfer Division, Brooks Air Force Base, San Antonio, TX.

Air Force Center for Environmental Excellence (AFCEE). 1999. *Technical Protocol for Implementing Intrinsic Remediation with Long-Term Monitoring for Natural Attenuation of Fuel Contamination Dissolved in Groundwater.* Technology Transfer Division, Brooks Air Force Base, San Antonio, TX.

U.S. Army Corps of Engineers (USACE). 1995. *Engineering Manual – Soil Vapor Extraction and Bioventing.* EM 1110-1-4001. November

U.S. Environmental Protection Agency (USEPA). 1992. *Technology Assessment of Soil Vapor Extraction and Air Sparging.* Risk Reduction Engineering Laboratory, Cincinnati, OH, EPA/600/R-92/173.

USEPA. 1995. *How to Evaluate Alternative Cleanup Technologies for Underground Storage Tank Sites: Chapter II Soil Vapor Extraction.* EPA 510-95-007.

USEPA. 1997. *Ground Water Issue – Design Guidelines for Conventional Pump-and-Treat Systems.* Office of Research and Development, and Office of Solid Waste and Emergency Response. EPA/540/S-97/504, Washington, D.C.

8 *In situ* Treatment Technologies

Xihong Zhai

CONTENTS

The term *in situ treatment* implies that the treatment process occurs primarily in place without removing large volumes of contaminated materials (e.g., extracting the contaminated groundwater or excavating the contaminated soil). *In situ* treatment either destroys the contaminants in place or concentrates the contaminants in a certain medium (gas or liquid) and subsequently collects it aboveground for treatment. Therefore, the cost and duration can be reduced substantially using *in situ* treatment when compared with the conventional treatment methods discussed in Chapter 7. The *in situ* treatment process can be chemical, physical, biological, or any combination of these three, as discussed in the following sections.

8.1 CHEMICAL TREATMENT

In situ chemical treatment uses chemical processes such as chemical oxidation or reduction to break down the organic contaminants into non-hazardous compounds such as carbon dioxide and water, or to change the oxidation-reduction state of metals into a benign state, such as from hexavalent chromium to trivalent chromium. Sections 8.1.1 and 8.1.2 discuss chemical oxidation and chemical reduction, respectively.

8.1.1 Chemical Oxidation

In situ chemical oxidation (ISCO) is the process by which an oxidant is injected into the subsurface and reacts with the contaminants, producing innocuous substances such as carbon dioxide and water. A wide variety of organic contaminants, including petroleum-related and chlorinated solvent compounds, are susceptible to ISCO. Four commonly used oxidants are hydrogen peroxide, permanganate, ozone, and persulfate. Many of these oxidants produce radicals, which are highly reactive and short-lived molecular fragments that have an unpaired electron. These radicals are especially efficient in the destruction of contaminants and are essential for successful remediation. Different oxidants may have different oxidation strength, which can be expressed as oxidation-reduction potential (ORP) in volts. Table 8.1 lists the standard ORPs of radicals and commonly used oxidants.

TABLE 8.1
Standard ORP and Relative Strength of Commonly Used Oxidants and Associated Free Radicals

Chemical Species	Standard Oxidation Potential (volts)	Relative Strength (Chlorine = 1)
Hydroxyl radical ($OH^{-}\bullet$)[a]	2.8	2
Sulfate radical ($SO_4^{-\bullet}$)	2.5	1.8
Ozone	2.1	1.5
Sodium persulfate	2	1.5
Hydrogen peroxide	1.8	1.3
Permanganate	1.7	1.2
Chlorine	1.4	1
Oxygen	1.2	0.9
Superoxide ($O^{-}\bullet$)[a]	−2.4	−1.8

Source: Data from Siegrist et al., *Guidance for In Situ Chemical Oxidation at Contaminated Sites: Technology Overview with a Focus on Permanganate Systems*, Battelle Press, Columbus, OH, 2001.

[a] These radicals can be formed when ozone and H_2O_2 decompose.

8.1.1.1 Permanganate

Permanganate has also been traditionally used as an oxidant in wastewater treatment. Its application as an *in situ* chemical oxidant for environmental remediation started in the 1990s (Siegrist et al. 2001). The oxidation power of permanganate comes from the multiple valency states of manganese. Manganese is at its most oxidized state, Mn^{+7}, in permanganate (MnO_4^-). During the ISCO reaction, manganese is reduced to lower valency states, mostly Mn^{+4} in the form of manganese dioxide (MnO_2) solids, while the contaminant(s) are oxidized. The typical half-cell reaction of permanganate is as follows:

$$MnO_4^- + 2H_2O + 3e^- \rightarrow MnO_2(s) + 4OH^- \quad (8.1)$$

Under very acidic ($pH < 3.5$) or alkaline conditions ($pH > 12$), manganese can also be reduced to the +2 or +6 valency state accordingly.

Permanganate is very effective at oxidizing organic compounds with carbon-carbon double bonds, such as aliphatic ethenes. For example, the redox reaction between permanganate and tetrachloroethylene (PCE) under normal pH conditions can be described as follows:

$$4MnO_4^- + 3C_2Cl_4 + 4H_2O \rightarrow 6CO_2 + 4MnO_2(s) + 12Cl^- + 8H^+ \quad (8.2)$$

However, for most aromatic compounds with carbon rings (e.g., benzene), the carbon-carbon double bonds are more stable and therefore, recalcitrant to permanganate oxidation. If one or more of the carbon atoms of an aromatic compound are substituted with another atom (e.g., chlorine in chlorobenzene) or organic group (e.g., CH_3 in toluene), the stability of electrons within the carbon ring is weakened, rendering it susceptible to oxidation. Permanganate is also known to have limited effectiveness on saturated aliphatic compounds (ITRC 2005a).

Two commonly available permanganate products are sodium permanganate and potassium permanganate. Sodium permanganate is soluble in water and is normally provided as a concentrated solution (30%–40%). Potassium permanganate has a relatively low solubility (4%). It is normally provided as solid particles and dissolved on site prior to implementation.

8.1.1.2 Persulfate

Persulfate is a strong oxidizing agent that is capable of breaking down many common contaminants. The standard ORP for the persulfate to sulfate reaction is relatively high (+2.0 V; refer to Table 8.1). However, unlike permanganate, the direct reaction of persulfate anion with organic compounds appears to be slow. Studies have found that free radicals can be generated by the activation or catalysis of persulfate solution; these have an even higher oxidation potential and can react with faster kinetics.

Free radicals are molecular fragments that have an unpaired electron, which are therefore highly reactive and short-lived. Reactions involving radicals can be either chain-propagating or chain-terminating. *Chain-propagating reactions* produce another radical, while *chain-terminating reactions* do not. The target compounds (e.g., organic contaminants) may or may not be involved in these reactions.

The free radicals involved in activated persulfate systems include sulfate radical (ORP of +2.5 V), hydroxyl radical (ORP of +2.8 V), and organic radicals. The persulfate activation methods, in the order of increasing strength, include metal (mostly ferrous iron), alkaline, hydrogen peroxide (dual oxidant system), and heat, as discussed in the following paragraphs.

Metal activation: Sulfate radicals may be generated by one-electron oxidation reactions, which occur more readily in the presence of metals (Liang et al. 2001; FMC 2001; Hoag et al. 2000):

$$S_2O_8^{-2}\left(\text{or } ^-O_4S - SO_4^-\right) + M^{+n} \rightarrow SO_4^{-\bullet} + SO_4^{-2} + M^{+(n+1)} \quad (8.3)$$

Ferrous iron is a commonly used metal activator for persulfate oxidation. Similarly to Fenton's reagent, acidic conditions are needed to keep iron from precipitating out of the solution. In the past several years, chelated iron has been used for persulfate activation, which maintains the ferrous ions in solution by complexation. Commonly used chelating agents include organic acids such as oxalic acid, citric acid, ethylenediamine tetra-acetic acid (EDTA), nitrilotriacetic acid (NTA), and so on (Liang et al. 2003, 2004a,b; Block et al. 2004).
Some other chain-initiating reactions involved include

$$S_2O_8^{2-} \rightarrow 2SO_4^{-\bullet} \tag{8.4}$$

$$S_2O_8^{2-} + RH \rightarrow SO_4^{-\bullet} + R^{\bullet} + HSO_4^{-} \tag{8.5}$$

Chain-propagating reactions include

$$SO_4^{-\bullet} + RH \rightarrow R^{\bullet} + HSO_4^{-} \tag{8.6}$$

$$OH^{\bullet} + RH \rightarrow R^{\bullet} + H_2O \tag{8.7}$$

$$R^{\bullet} + S_2O_8^{2-} \rightarrow SO_4^{-\bullet} + HSO_4^{-} + R \tag{8.8}$$

$$SO_4^{-\bullet} + OH^{-} \rightarrow OH^{\bullet} + SO_4^{2-} \tag{8.9}$$

Chain-terminating reactions include

$$SO_4^{-\bullet} + Fe^{2+} \rightarrow Fe^{3+} + SO_4^{2-} \tag{8.10}$$

$$OH^{\bullet} + Fe^{2+} \rightarrow Fe^{3+} + OH^{-} \tag{8.11}$$

$$R^{\bullet} + Fe^{3+} \rightarrow Fe^{2+} + R \tag{8.12}$$

$$2R^{\bullet} \rightarrow \text{Chain termination} \tag{8.13}$$

Alkaline Activation: Elevated pH (> 10.5) is also widely used to activate persulfate. An alkaline pH environment will favor the reaction shown in Equation 8.9, which increases the production of hydroxyl free radicals, as they can be generated by the reaction of OH^- and another radical. Likely because of the production of highly oxidative free radicals, alkaline activated persulfate has been shown to destroy recalcitrant saturated aliphatic compounds such as chlorinated ethanes (e.g., trichloroethane [TCA] and dichloroethane [DCA]) and chlorinated methanes (carbon tetrachloride and chloroform) (Block et al. 2004).

Alkaline activation requires high pH to be maintained, and the soil buffering capacity should generally be characterized to determine the appropriate alkaline dosing. Sodium hydroxide is a commonly used alkaline activator. In some sites where contamination is shallow and reagents can be mixed in by blending, some insoluble form of alkaline, such as quick lime (CaO), can also be used to increase soil pH. The additional benefit of using quick lime as an activator is that its hydration process generates heat and can thermally activate persulfate to a certain extent (see heat activation discussed later).

Hydrogen Peroxide Activation: When hydrogen peroxide is applied together with persulfate, the presence of hydroxyl radical generated from hydrogen peroxide enhances the chain-propagation reactions shown in Equation 8.7, and more radicals are generated with strong ORP to oxidize most organic compounds.

Heat Activation: Heat is very effective at initiating the production of sulfate radicals. The heat activation of persulfate systems requires fine tuning of the subsurface temperature. Although high temperature might thermally destroy some volatile organic chemicals (VOCs), it could also enhance the volatilization of VOCs and subsequent undesired migration prior to destruction by chemical oxidation. The optimal temperature range for persulfate activation is usually 40–60°C. Heat is normally delivered by steam injection. Some sites have applied persulfate ISCO following thermal treatment of chlorinated ethanes, using the residual heat to activate persulfate, and the thermal treatment also breaks down ethanes to some intermediate that is more susceptible to ISCO (Gu et al. 2011).

8.1.1.3 Hydrogen Peroxide

Hydrogen peroxide is commonly associated with Fenton's reagent, which is a mixture of hydrogen peroxide and ferrous iron. The generation of free radicals and the subsequent chain reaction associated with this mixture were first discovered by H. J. H, Fenton; hence the names Fenton's reaction and Fenton's reagent.

The primary free radical generated in Fenton's reaction is hydroxyl radical, a very strong oxidizing agent, as shown in Table 8.1. The subsequent chain-propagating reactions generate various other radicals, including superoxide ions ($O_2^{\bullet-}$), hydroperoxide ions (HO_2^-), and organic radicals (R•). The chain-terminating reactions convert ferrous iron to ferric iron and back again. The typical Fenton's reaction involves the following steps:
Chain-initiating reaction:

$$Fe^{2+} + H_2O_2 \rightarrow Fe^{3+} + OH^{\bullet} + OH^- \tag{8.14}$$

Chain-propagating reaction:

$$OH^{\bullet} + H_2O_2 \rightarrow HO_2^{\bullet} + H_2O \tag{8.15}$$

$$HO_2^{\bullet} \rightarrow O2^{-\bullet} + H^+ \tag{8.16}$$

$$OH^{\bullet} + RH \rightarrow R^{\bullet} + OH^- \tag{8.17}$$

$$R^{\bullet} + H_2O_2 \rightarrow ROH + OH^{\bullet} \tag{8.18}$$

Chain-terminating reactions:

$$HO_2^{\bullet} + Fe^{2+} \rightarrow O_2 + H^+ + Fe^{3+} \tag{8.19}$$

$$HO^{2\bullet} + Fe^{2+} \rightarrow HO_2^- + Fe^{3+} \tag{8.20}$$

$$Fe^{3+} + O_2^{\bullet} - \rightarrow Fe^{2+} + O_2 \tag{8.21}$$

An undesired side reaction converts two end products, ferric iron and hydroxide, into ferric hydroxide precipitate:

$$Fe^{3+} + nOH^- \rightarrow Fe(OH)_n(s) \tag{8.22}$$

This side reaction consumes iron and makes it unavailable for Fenton's reaction. Acidic conditions or a chelating agent is normally used to prevent this side reaction from occurring.

The typical Fenton's reagent consists of a low concentration of peroxide (normally around 0.03%), ferrous iron (Fe^{+2}) salt, and acidic pH (2.5–3.5). In remediation practice, a higher concentration of hydrogen peroxide (4%–20%) is normally used for cost-effectiveness, and ferrous iron is normally applied with or without a chelating agent. This is referred to as the *modified Fenton's reagent.*

A high concentration of hydrogen peroxide usually results in more free radicals and a more aggressive reaction. A vigorous redox reaction with organic compounds in the subsurface could lead to heat and gas production (ITRC 2005a), which would become a safety concern or cause undesired migration of non-aqueous phase liquid (NAPL) contaminants. Therefore, careful site characterization is often needed in the design phase to avoid excessive heat accumulation and gas generation. Another factor that needs special attention is the presence of free radical scavengers, including carbonate ion and certain metals. If they are present at high concentrations, the free radicals will be consumed by undesired reactions and exert a high oxidant demand, making hydrogen peroxide a less cost-effective option.

8.1.1.4 Ozone

Ozone is an allotrope of oxygen, but it has a higher water solubility (up to 30 mg/L) (Langlais et al. 1991) than oxygen (8 mg/L under normal conditions). The chemical reactions involved in ozone oxidation include two categories: direct oxidation and indirect oxidation, as discussed in detail in the following.

Direct Oxidation: Ozone can directly react with target compounds instead of relying on the generation of free radicals. This process has found significant use in water treatment. In a direct oxidation process, the oxygen atom is usually inserted into unsaturated carbon-carbon bonds, which results in the formation of an ozonide (Qiu et al. 1999) and oxygen:

$$O_3 + RC = CR \rightarrow RCOCR + O_2 \tag{8.23}$$

Indirect Oxidation: The indirect oxidation of ozone systems involves the generation of free radicals including hydroxyl radical ($OH^•$), which are strong oxidants that can destroy a wide variety of organic contaminants. Oxidation by hydroxyl radicals is a faster reaction than direct oxidation by the ozone itself. The oxidation products of most organic compounds are usually hydroxylated products, aldehydes, ketones, and carboxylic acids, in addition to carbon dioxide and water.

Ozone systems with neutral to basic pH ranges will favor the generation of hydroxyl radicals. Some other ways to activate ozone include ultraviolet (UV) and certain cations. In addition to the production of these radicals, a chain reaction is also initiated, causing the formation of new radicals. The hydroxyl radical chain-propagating and -terminating reactions are similar to those outlined for hydrogen peroxide and are summarized here:

Chain-initiating reactions

$$O_3 + OH^- \rightarrow O_2 + OH^• \tag{8.24}$$

$$O_3 + H_2O \rightarrow O_2 + 2OH^• \text{ (in the presence of UV light)} \tag{8.25}$$

Chain-propagating reactions

$$OH \bullet + 2H_2O \rightarrow HO_2^• + OH^- + 3H^+ \tag{8.26}$$

$$HO_2^• \rightarrow O_2^{-•} + H^+ \tag{8.27}$$

$$OH^{\bullet} + RH \rightarrow R^{\bullet} + OH^{-} \quad (8.28)$$

$$R^{\bullet} + O_3 + H_2O \rightarrow ROH + O_2 + OH^{\bullet} \quad (8.29)$$

Chain-terminating reactions

$$HO_2^{\bullet} + Fe^{2+} \rightarrow O_2 + H^{+} + Fe^{3+} \quad (8.30)$$

$$HO_2^{\bullet} + Fe^{2+} \rightarrow HO_2^{-} + Fe^{3+} \quad (8.31)$$

$$Fe^{3+} + O_2^{-\bullet} \rightarrow Fe^{2+} + O_2 \quad (8.32)$$

The typical half-life of ozone is 30 minutes at standard temperature and pressure, but it can be longer with a slow mass transfer between ozone gas and groundwater. The gaseous nature of ozone makes it suitable for vadose zone treatment, although the generation of free radicals involved in the indirect mechanism still needs to take place in water contained in soil moisture. The reaction or decomposition product of ozone is oxygen, which can be used in the subsequent aerobic biodegradation (refer to Section 8.2.1.1).

Numerous successful applications of ozonation ISCO processes have been reported using ozone injection alone as well as ozone in combination with hydrogen peroxide (Nelson and Brown 1994; Amarante 2000; Nimmer et al. 2000). Ozone–hydrogen peroxide reactions result in enhanced generation of hydroxyl radicals. This mechanism for the formation of hydroxyl radical during ozone–hydrogen peroxide treatment involves the production of hydroxyl radicals by direct hydrogen peroxide and ozone reactions and through intermediate ozone and hydrogen peroxide reactions.

Ozone and hydrogen peroxide injection is considered to be one of the most aggressive forms of ISCO technologies due to the high yields of hydroxyl radicals obtainable (Hoigne and Bader 1983; Hong et al. 1996; Kuo and Chen 1996). A variety of other oxidizing species are introduced into the subsurface, reacting with many different contaminants, and there can be significant downgradient dissolved oxygen bioremediation effects. The issues involved with the use of ozone–hydrogen peroxide to oxidize VOCs in groundwater are the same as those for the individual oxidants. One particular concern is the overdosing of hydrogen peroxide (a common practice with Fenton's reagent applications), which will serve as a scavenger, resulting in unproductive consumption of the generated hydroxyl radical.

8.1.1.5 Comparison of Commonly Used Oxidants

Among the four commonly used oxidants, permanganate and persulfate are available in solid forms (i.e., potassium permanganate and sodium persulfate). Hydrogen peroxide is available as concentrated solutions (up to 35%), and ozone is a gas. The solid forms are generally considered more stable and easier to handle.

Three of the commonly used oxidants, hydrogen peroxide/Fenton's reagent, persulfate, and ozone, involve free radicals. One of the most important characteristics of free radicals is their short life-times. It has been reported that sulfate free radicals have a half-life of about 4 seconds at an elevated temperature of about 40°C (Banerjee and Konar 1984). The half-life of hydroxyl radicals is expected to be even shorter given their faster kinetics.

All oxidants can decompose to various extents by reacting with water if they are in solution, resulting in nonproductive depletion of oxidants. The decomposition rates of permanganate and persulfate are reported to be very slow under normal conditions in the absence of species that can initiate the chain reaction of free radicals (particularly for persulfate). For persulfate, free radicals can be generated even in the absence of activators, but at a very slow rate (Kolthoff and Miller 1951; FMC 2001). Unlike the non-selective hydroxyl radicals, which react with all reduced species they encounter, sulfate radicals generated from persulfate reactions are reported to be selective and only react with certain reduced species. Therefore, these two oxidants, permanganate and persulfate, are considered long-lasting even when delivered to the subsurface.

However, if they generate short-lived free radicals, such as in ozone and hydrogen peroxide systems, the decomposition processes make these oxidants themselves relatively short-lived compared with permanganate and persulfate, and the nonproductive depletion of oxidant becomes significant. On the one hand, this exerts more oxidant dose demand and increases the treatment cost; on the other hand, the oxidants, if over-dosed, will be consumed quickly, and the subsurface redox conditions will soon be restored. In addition, the decomposition of both ozone and hydrogen peroxide can release oxygen, which can enhance the subsequent aerobic biodegradation if that is desired.

8.1.1.6 Selection of Appropriate and Effective Oxidant(s)

The selection of specific oxidant(s) needs to consider several factors, including effectiveness on contaminants of concern (COCs), soil oxidant demand (SOD), site geochemistry, site lithology, and so on, as discussed later. More discussion on oxidant selection and design consideration can be found in several published guidance documents (ITRC 2005a; NAVFAC 2015).

Effectiveness: The effectiveness of different oxidants on various COCs was discussed briefly in Sections 8.1.1.1 through 8.1.1.4 for each oxidant. Generally, ISCO is effective mostly on organic compounds, as the oxidation end products are usually carbon dioxide and water. Chlorinated ethenes with carbon-carbon double bonds are susceptible to any of the four oxidants discussed earlier. However, persulfate systems should be used with caution, as their effectiveness often depends on site-specific geochemical conditions and activation methods. Ozone and hydrogen peroxide, which both involve the generation of strong hydroxyl radicals, are considered the most aggressive oxidants. Table 8.2 summarizes the generally recognized treatment effectiveness of the four oxidants and one combination. However, site-specific treatability studies are strongly recommended.

Soil Oxidant Demand (SOD): The oxidants can react with most reduced species, including the metals and organic compounds they encounter, in addition to the target COCs. It has been estimated

TABLE 8.2
Effectiveness of Common Oxidants on Organic Contaminants

Oxidant	Amenable COCs	Reluctant COCs	Recalcitrant COCs
H_2O_2	TCA, PCE, TCE, DCE, VC, BTEX, CB, phenols, 1,4-dioxane, MTBE, TBA, explosives	DCA, CH_2Cl_2, PAHs, carbon tetrachloride, PCBs	$CHCl_3$, pesticides
Ozone	PCE, TCE, DCE, VC, BTEX, CB, phenols, MTBE, TBA, explosives	DCA, CH_2Cl_2, PAHs	TCA, carbon tetrachloride, $CHCl_3$, PCBs, pesticides
Ozone/H_2O_2	TCA, PCE, TCE, DCE, VC, BTEX, CB, phenols, 1,4-dioxane, MTBE, TBA, explosives	DCA, CH_2Cl_2, PAHs, carbon tetrachloride, PCBs	$CHCl_3$, pesticides
Permanganate	PCE, TCE, DCE, VC, BTEX, PAHs, phenols, explosives	Benzene, pesticides	TCA, carbon tetrachloride, $CHCl_3$, PCBs
Activated persulfate	PCE, TCE, DCE, VC, BTEX, CB, phenols, 1,4-dioxane, MTBE, TBA	PAHs, explosives, pesticides	PCBs

Source: ITRC, *Technical and Regulatory Guidance for In Situ Chemical Oxidation of Contaminated Soil and Groundwater*, 2nd ed. ISCO-2, Washington, DC, 2005.

that the actual amount of oxidant that is consumed by target COCs only accounts for approximately 1% of the total oxidant loading. Although some species in groundwater consume oxidant, most of the oxidant demand comes from the solid phase, or the soil; therefore, the term *soil oxidant demand* (SOD) represents the majority of the overall oxidant demand and is often used interchangeably with the term *total oxidant demand* (TOD). Metals in reduced form, such as ferrous iron, manganese (II), and chromium (III), and naturally occurring organic matter, including humic acids and fulvic acids, are all potential sources that can contribute to SOD. SOD depends on the site-specific soil and groundwater geochemical conditions and is directly related to the cost of ISCO technology. SOD can also differ for different oxidants due to the oxidants' reactivity. Therefore, SOD testing using the selected oxidant(s) is the minimum required site characterization if ISCO is the intended technology.

Site Geochemistry: In addition to its impact on SOD, soil geochemistry can impact the effectiveness of ISCO due to the potential presence of radical scavengers. The radical chain reaction can be terminated by radical scavengers, as outlined in the peroxide systems. Radical scavengers are common within soil matrices, including naturally occurring carbonates and oxidation products, mainly humic acids, and tertiary alcohols. These scavengers consume free radicals and increase the oxidant demand due to the nonselectivity of the oxidant and the free radicals. Some important soil parameters to examine include levels of bacterial biomass, total organic carbon, iron, manganese, hydrogen sulfide, and carbonates. Soils containing elevated levels of any one of these constituents should warrant some concern. It is therefore suggested that some laboratory or pilot testing be conducted to evaluate the site compatibility with ISCO from the combination of these geochemistry effects.

8.1.2 Chemical Reduction

In situ chemical reduction (ISCR) uses a reductant, or a material that can generate reductant, to detoxify the contaminants via chemical reduction in the subsurface. The specific detoxification process varies for different groups of contaminants. For organic contaminants, ISCR transforms them to potentially less toxic or nontoxic compounds. For inorganic contaminants such as metals, ISCR reduces their oxidation status and therefore immobilizes them by adsorption or precipitation, as in the case of hexavalent chromium (Cr[VI]). The commonly used ISCR reagents include zero-valent iron (ZVI), ferrous iron, sodium dithionite, calcium polysulfide, and hydrogen with a catalyst, as discussed in the following subsections.

8.1.2.1 Zero-Valent Iron

ZVI is the most commonly used ISCR reagent, effective on a wide range of organic contaminants, including halogenated ethenes and ethanes, energetics, and inorganic contaminants such as metals/metalloids (Cr[VI]), arsenic, and uranium (ITRC 2011).

Chemical reduction of organic contaminants by ZVI is reported to be via abiotic degradation pathways (i.e., pathways that do not directly involve microorganisms). For the commonly encountered chlorinated ethene compounds such as PCE and TCE, *beta elimination* is believed to be the most energetically favorable pathway. Figure 8.1 shows the different pathways of biotic and abiotic degradation for PCE and TCE. Unlike the sequential dechlorination pathway in biotic degradation (refer to Section 8.2.2.1), beta elimination literally eliminates two chlorine atoms, resulting in the formation of a triple bond across the two carbons. This unstable compound quickly breaks down to acetylene gas. The detection of acetylene compounds is generally considered as an indicator of the occurrence of beta elimination following ZVI amendment.

Although the abiotic pathway is believed to be dominant, biotic degradation after ZVI implementation has often also been observed at sites with PCE and TCE, as evidenced by the detection of biodegradation daughter products, that is, dichloroethene (DCE) isomers and vinyl chloride (VC), as illustrated in Figure 8.1.

Inorganic contaminants treatable by ZVI include metal oxyanions and divalent metals, with different mechanisms.

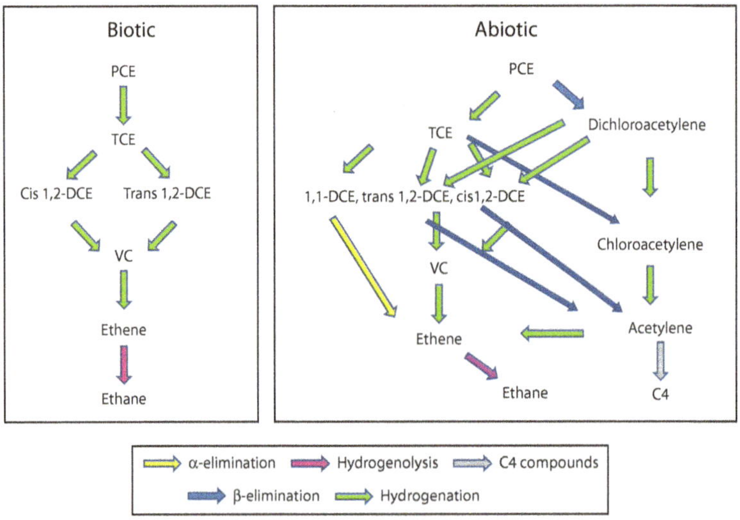

FIGURE 8.1 Typical biotic and abiotic reduction pathways of chlorinated ethene compounds. (From Department of the Navy, Moffett Field Restoration Advisory Board Progress Report – Sites 26 (EATS) and 28 (WATS), 2010.)

For metal oxyanions such as CrO_4^{2-}, Cr(VI) is reduced to Cr(III) followed by precipitation as chromium hydroxides ($Cr[OH]_3$) and oxides (Cr_2O_3) (ITRC 2011) or co-precipitation with iron phases (FeOOH or Fe_2O_3) (Yang et al. 2007; ITRC 2011).

For arsenic, the removal mechanism is mostly adsorption by ZVI corrosion products (Manning et al. 2002). The corrosion of ZVI produces various hydroxide and oxide compounds with either ferrous or ferric iron. Both arsenic (III) and arsenic (V) can form inner sphere complexes with these iron compounds and be occluded out of the water phase (Manning et al. 2002; Kanel et al. 2006).

For divalent metallic cations, the removal mechanisms include reductive precipitation, surface adsorption or complexation, or co-precipitation with iron oxyhydroxides (Cundy et al. 2008). The corrosion of the ZVI by water and dissolved oxygen also increases the pH (alkalinity) of groundwater and favors the precipitation of divalent metal cations.

ZVI is often installed as *permeable reactive barriers* (PRBs) in the subsurface. A PRB consists of a zone of reactive material, such as granular iron, intercepting the groundwater plume. As the groundwater flows through, the contaminants react with the reagent and become nontoxic or less toxic compounds. PRBs are passive systems and require little maintenance once installed. ZVI-based PRB walls have been placed using injection techniques (ITRC 2011).

While ZVI in the form of PRBs is intended for groundwater plumes, a recent evolution of ISCR has expanded the ZVI application to source zones by injecting highly reactive iron powder in a grid pattern (USEPA 2003). ZVI powder in solid form is suspended in viscous fluid for injection. The viscous fluid can be an organic material, which will also promote biological activities (ITRC 2011; ESTCP 2010; Krug et al. 2010). The injection of ZVI powder in a viscous carrier fluid targeted at low-permeability zones might require the assistance of some fracturing technology, either pneumatic or hydraulic (Forman 2010; Bures et al. 2010).

8.1.2.2 Calcium Polysulfide

Calcium polysulfide (CaS_x) is a strong reductant widely used in ISCR treatment. It is conventionally used to precipitate metals in wastewater treatment systems. It has a pH around 11. The sulfide component can precipitate with a wide variety of cations, including FeS, ZnS, PbS, CdS, and CuS. Oxidized metals such as Cr(VI) can be reduced by CaS_x to Cr(III), which then precipitates out of the water phase (Petersen and Hedquist 2006; Zawislanski et al. 2010).

Caution should be used when applying CaS_x to a Cr(VI) site if arsenic is also present abundantly in the soil because of the unique mobility of arsenic. Arsenic is stable in its most oxidized state, pentavalent arsenate (As[V]). When reduced to trivalent state arsenite (As[III]), it becomes soluble. The solubility of pentavalent arsenate also increases with increasing pH. The application of CaS_x into the subsurface creates reducing and high-pH conditions, which both make arsenic more mobile (USEPA 2011a). To conquer the arsenic mobility issue, CaS_x can be used together with an iron component, which can adsorb arsenic, as discussed in Section 8.1.2.1, in addition to reducing other contaminants.

8.1.2.3 Ferrous Iron and Sodium Dithionite

Ferrous iron is an effective reductant. However, when injected alone, ferrous iron will be oxidized to ferric iron and rapidly precipitate out of solution (USEPA 2011b). This precipitation can be prevented by acidifying the solution (USEPA 2000) as Fenton's reagent (refer to Section 8.1.1.3) or by co-injection with sodium dithionite as a buffer. Sodium dithionite is also a strong reductant conventionally used in the textile and paper industries and in wastewater treatment. An aqueous solution of sodium dithionite is acidic and decomposes to sodium thiosulfate and sodium bisulfite (Lister et al. 1959).

The combination of ferrous iron with sodium dithionite is an ISCR reagent effective at treating groundwater contaminated with oxidized metal species (Fruchter et al. 2000). Once injected, the ferrous iron readily reduces the oxidized metal, such as Cr(VI), to Cr(III). Excess ferrous iron and sulfur species from dithionite decomposition can all adsorb to the surface of solid particles, which serve as a long-term source of reducing strength to react with oxidized metal contaminants when groundwater flows through (USEPA 2011b).

8.1.2.4 Biogeochemical Reductive Dechlorination

In the past decade, a new remediation approach, *biogeochemical reductive dechlorination* (BiRD), has been developed based on field observation of sites contaminated with chlorinated solvents (primarily PCE and TCE) (Kennedy et al. 2006a,b; SERDP 2007; USEPA 2009). The BiRD approach uses the abiotic degradation pathway of organic contaminants by reductants generated from microbiological processes.

Four elements essential to the BiRD process are sulfate, iron containing minerals, organic substrate, and bacteria. Particularly, the commonly found sulfate-reducing bacteria can convert native iron minerals into iron sulfides in the presence of sulfate and organic substrate. The surface of iron sulfides can facilitate chemical reduction of many common chlorinated solvent compounds, including PCE, TCE, and DCE, similarly to ZVI (Kennedy et al. 2006a,b). Acetylene is often observed as a product of beta elimination. The most common iron sulfide mineral is pyrite. Some other minerals, including magnetite, green rust, and goethite with adsorbed iron (II), are also reported to be responsible for the BiRD process.

Some sites naturally have the four essential elements present at abundant levels to stimulate the BiRD process. For sites that do not, engineering enhancement can be performed by injection of iron and sulfate (e.g., ferrous sulfate) and/or organic substrate (Kennedy et al. 2006a,b).

8.2 BIOREMEDIATION

Bioremediation uses biological/microbiological processes to detoxify organic and inorganic contaminants in the environment by modifying environmental conditions (physical, chemical, biochemical, or microbiological) such that they are amenable to biological growth and activities. It mostly involves microorganisms. During metabolism, bacteria perform coupled redox reactions, by which the target contaminants are consumed in biological respiration processes and turned into harmless products such as carbon dioxide or ethene. A collection of guidance documents is available, from the fundamentals of *in situ* bioremediation (ISB) to its design considerations, including the U.S. Environmental Protection Agency (US EPA)'s *Introduction to In Situ Bioremediation*

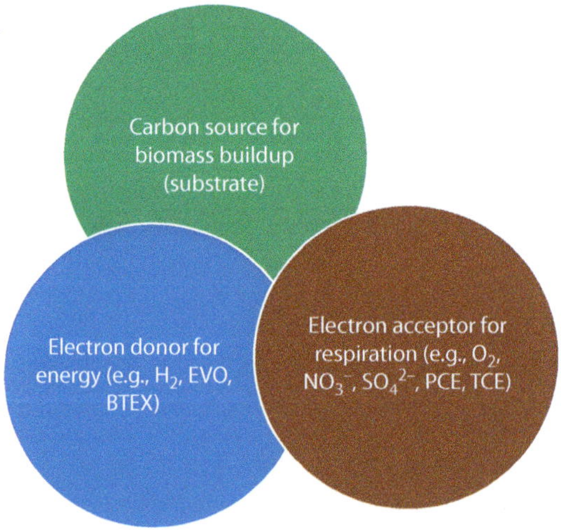

FIGURE 8.2 Three basic categories critical to microbial activities.

of Groundwater (2013), the Air Force Center for Engineering and the Environment (AFCEE)'s *Protocol for In-situ Bioremediation of Chlorinated Solvents Using Edible Oil* (2007), and the Environmental Security Technology Certification Program (ESTCP)'s *Bioaugmentation for Remediation of Chlorinated Solvents: Technology Development, Status, and Research Needs* (2005), to name a few.

Theoretically, phytoremediation, which uses plants to remediate contamination, is also a form of bioremediation. However, given its own variety and significant differences from microbiological bioremediation, phytoremediation is discussed separately in Section 8.3 as part of phytotechnology.

All microbes have basic requirements for life and growth. The basic needs of bacteria can be classified into three broad categories: a carbon source that can be used to build their biomass, an electron donor (such as hydrogen) as energy source, and a terminal electron acceptor (TEA) (e.g., oxygen) to receive the electrons the bacteria use for energy. The carbon source will also serve as the electron donor in most cases. The target contaminants can fit into any of these three categories, as shown in Figure 8.2.

The growth rates of bacteria depend on the concentration of the elements from whichever of the three categories is most limiting, which is often the substrate. Studies have shown that the specific growth rate increases directly with increasing substrate concentration, while it levels out to a constant maximum growth rate when there is sufficient substrate (Okpokwasili and Nweke 2005).

Bacteria catalyze (increase, initiate, or transform) redox reactions and use the energy released from them. As discussed in Section 8.1, a redox reaction involves the transfer of electrons from an electron donor (reductant) to an electron acceptor (oxidant). Each redox pair releases a different amount of energy, and each species of bacteria has enzymes to catalyze only certain redox pairs. Depending on the categories the target contaminants may fall into (oxidant or reductant), ISB can be classified as *oxidative bioremediation* and *reductive bioremediation*.

8.2.1 Oxidative Bioremediation

In an oxidative bioremediation process, target contaminants serve as electron donors and go through an oxidative process. The contaminants are oxidized into innocuous end products, including carbon dioxide and water. This process is applicable to a variety of organic contaminants:

- Petroleum derivatives associated with gasoline, fuel, and diesel release, such as benzene, toluene, ethyl benzene, and xylenes (BTEX)

- Gasoline additive methyl tertiary-butyl ether (MTBE)
- Solvents such as acetone
- Some simple polycyclic aromatic hydrocarbons (PAHs) such as naphthalene
- Pesticides and herbicides such as pentachlorophenol (PCP)
- Vinyl chloride (VC) and dichloroethene (DCE) isomers as daughter products from the reductive biodegradation of chlorinated solvent compounds

8.2.1.1 Respiration Processes

The microbial respiration process involved in oxidative bioremediation can be *aerobic* or *anaerobic*, depending on the TEA involved. The various TEAs that naturally exist in the subsurface are consumed in a specific preference order, according to their redox potential (and therefore energy yield), as illustrated in Figure 8.3.

Aerobic respiration uses oxygen as TEA, and anaerobic (i.e. *anoxic*) respiration uses other TEAs, such as nitrate, ferric iron, manganese, and so on. Oxygen is the most preferred TEA (therefore, aerobic respiration is dominant), and oxygen is the first TEA to be depleted. This is why in most petroleum release sites, the subsurface condition is anaerobic, due to oxygen consumption from the biodegradation of target contaminants. When oxygen is depleted, the denitrification process, which uses nitrate as TEA, kicks in. Following nitrate depletion, manganese and ferric iron become available for microbial use. Iron-reducing bacteria reduce iron (III) to iron (II), or they can use manganese (IV) and reduce it to manganese (II). Once iron and manganese have been reduced, sulfate serves as an electron acceptor and is converted by sulfur-reducing bacteria to sulfide, sulfite, or elemental sulfur.

When all external TEAs have been exhausted, bacteria can use organic molecules as both electron acceptors and donors in a metabolic pathway called *fermentation*. Fermentation generates the lowest amount of energy because of the relatively small fraction of the organic matter available as electron donor and the low redox potential of the reactions. Therefore, fermentation is of little value to bioremediation and should be the process to avoid.

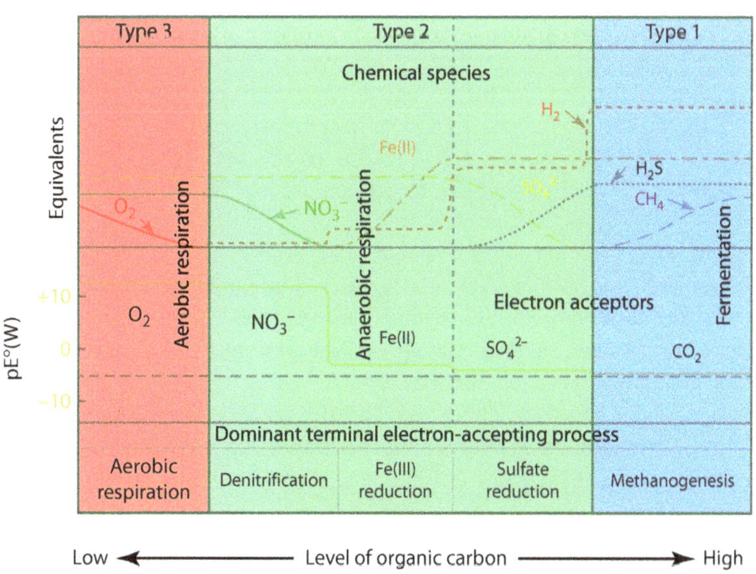

FIGURE 8.3 Dominant terminal electron accepting processes. (Modified from Bouwer, E.J., and McCarty, P.L., *Ground Water*, 22, 433–440, 1984.)

8.2.1.2 Microbial Population

The microbial population, according to the type of respiration it performs, can be generally classified as *obligate aerobic*, *facultative*, and *obligate anaerobic.*

Obligate aerobic bacteria can only thrive in aerobic conditions. These bacteria are generally considered ubiquitous in the natural environment (Bamforth and Singleton 2005). Many of the aerobic heterotrophic species can metabolize the less recalcitrant organic contaminants, including BTEX, PAHs, and pesticides, as substrates and electron donors. Some of the most common aerobic bacteria with these bioremediation capabilities, to name a few, are *Pseudomonas*, *Alcaligenes*, *Sphingomonas*, *Rhodococcus*, and *Mycobacterium* (Vidali 2001; Hendrickx et al. 2005; Bamforth and Singleton 2005).

Several microbes involved in bioremediation are *facultative*, meaning that they can perform both aerobic respiration and anaerobic respiration, but at a slower rate when oxygen is depleted. Microbes that use nitrate as an electron acceptor tend to be facultative (Firestone 1982).

Obligate anaerobic bacteria will be active only in reduced environments and will use electron acceptors with relatively low preference illustrated in Figure 8.3, such as sulfate or carbon dioxide.

Organic contaminants susceptible to oxidative bioremediation, as listed under Section 8.2.1, can be used as electron donors by either obligate aerobes, facultative bacteria, or obligate anaerobes. Supplementation of these bacteria is generally not needed, given their ubiquitous presence. However, the limiting factor in the oxidative bioremediation process is normally the electron acceptor, which often needs to be supplemented, as discussed in Section 8.2.1.3.

8.2.1.3 Electron Acceptors

Oxygen is the preferred electron acceptor in the oxidative bioremediation process due to its high energy yield from redox process. Engineering processes, such as biosparging and oxygen diffusors, have been used to directly add oxygen into the subsurface. Alternatively, oxygen-releasing compounds can be applied, which release oxygen as they dissolve or decompose in the subsurface. Common oxygen-releasing chemicals are calcium and magnesium peroxides, hydrogen peroxide, and ozone. For example, the release of oxygen from the decomposition of calcium peroxide is as shown in Equation 8.33.

$$2CaO_2 + 2H_2O \rightarrow 2Ca(OH)_2 + O_2 \tag{8.33}$$

As discussed previously in Sections 8.1.1.3 and 8.1.1.4, hydrogen peroxide and ozone are used commonly in the ISCO process. The oxygen from decomposition of residual hydrogen peroxide or ozone, while not oxidative enough to perform ISCO, may be sufficient to support aerobic biodegradation. As a result, a common practice is to use a treatment train of ISCO followed by aerobic bioremediation.

According to the TEA preference order shown in Figure 8.3, the TEAs, after oxygen depletion, include nitrate, iron (III), and sulfate sequentially. These inorganic species are often naturally present in aquifers but with limited quantities available for the anoxic biodegradation of contaminants. Oxidative bioremediation often involves the application of commercial products containing these TEAs to drive the anaerobic oxidation process.

Nitrate is highly soluble in water, and it provides the second highest free energy for microbial action. However, nitrate is generally not recommended as a supplement for anoxic biodegradation because of regulatory standards (its maximum contaminant level [MCL] is 10 mg/L, above which it will have adverse effects on humans and animals).

Iron(III) generally presents in the subsurface in solid form. The reduction product of iron (III) is iron(II), which is much more soluble in water. Iron(II) concentration in groundwater is often used as one of the indicators when evaluating the subsurface redox condition. Iron(III) is not a preferred TEA supplement due to the challenge associated with delivering a solid-form reagent and the low water quality threshold of the reaction by-product iron(II), with a secondary MCL of 0.3 mg/L.

Sulfate salts are generally soluble in water, will not sorb to a solid phase, and are generally unreactive. The end product of sulfate reduction is sulfide, which can precipitate with iron(II) and be effectively immobilized. A sulfate salt such as gypsum is a commercially available ISB product. However, care should be taken when applying it in acidic environments, as sulfide can react with hydrogen ion and produce hydrogen sulfide gas, which is toxic to the bacterial population.

The selection of a TEA to promote oxidative bioremediation needs to consider the contaminant and the favorable ORP state of the existing microbial population. For example, if a site with BTEX contamination is depleted of oxygen and nitrate, sulfate-reducing bacteria may have thrived in the sulfate-reducing subsurface conditions. Then, a sulfate-containing product should be considered to enhance the sulfate reduction process.

8.2.2 Reductive Bioremediation

In the reductive bioremediation process, the target contaminants are reduced; for example, PCE is reduced to ethene. As shown in Figure 8.4, the contaminants play the role of TEAs, and the naturally occurring TEAs, including oxygen, nitrate, and sulfate, become competing species. Electron donors/organic substrates are normally supplemented in the reductive bioremediation process.

8.2.2.1 Applicable Contaminants

The reductive bioremediation process is applicable to contaminants in their oxidative states, as listed in several groups in this subsection:

Chlorinated solvent compounds, including PCE, TCE, TCA, and so on. These compounds go through a sequential dechlorination process by losing one chlorine atom in each step, often referred to as *reductive dechlorination*, as illustrated in Figure 8.5.

Take the fully chlorinated ethene, PCE, as an example. It becomes TCE, DCE isomers, VC, and finally ethene sequentially. The partially chlorinated compounds derived from this process, the *daughter products*, are generally also considered toxic, some even with more stringent regulatory standards (e.g., VC). However, the reductive dechlorination of PCE and TCE often stops or becomes slow after the formation of DCE isomers, which is termed *stall*. It is speculated that several factors may contribute to the stalling of the reductive dechlorination process.

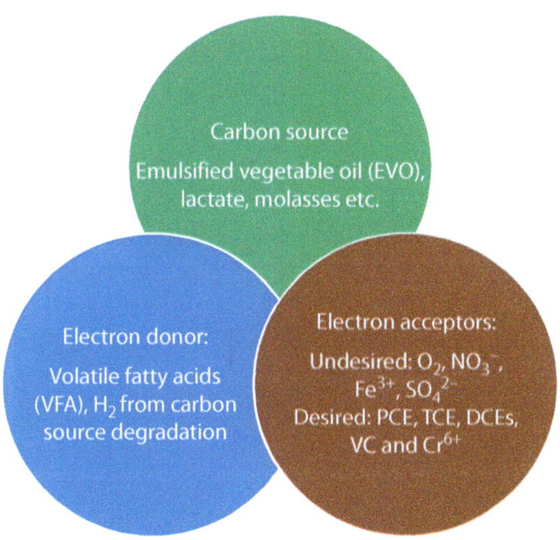

FIGURE 8.4 The elements in the three categories of reductive bioremediation process.

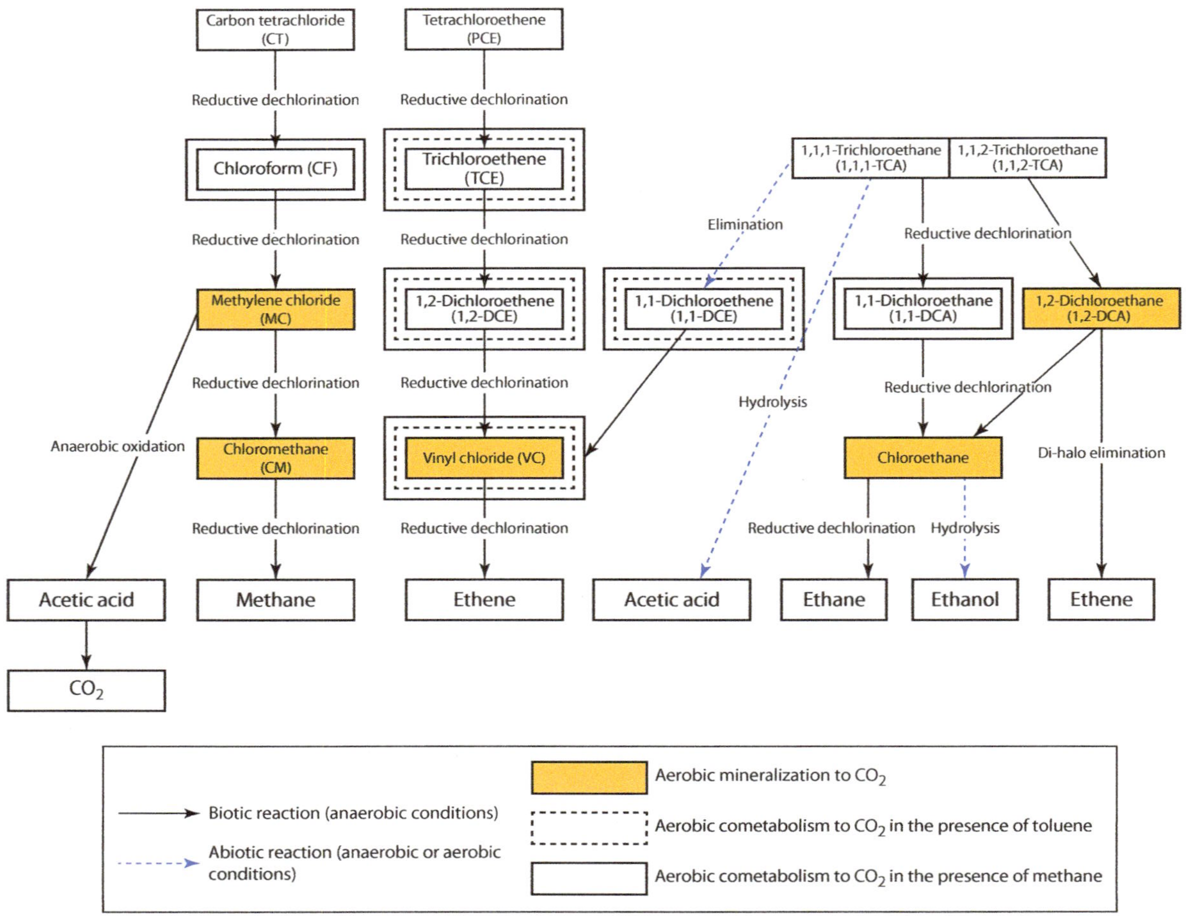

FIGURE 8.5 Degradation pathway of chlorinated compounds. (Modified from Agency for Toxic Substances and Disease Registry (ATSDR), Properties and Degradation Pathways of Common Organic Compounds in Groundwater, by Lawrence, S. J., Atlanta, Georgia, 2007.)

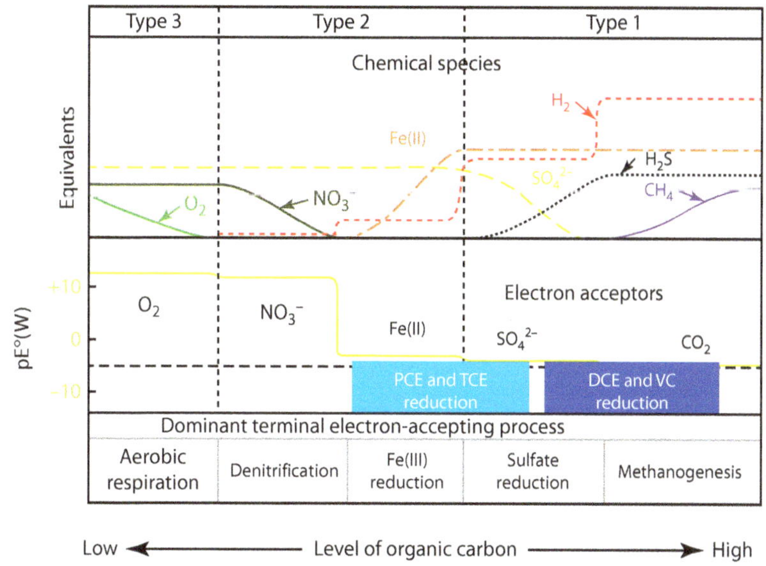

FIGURE 8.6 Reductive dechlorination process of chlorinated ethenes. (Modified from Bouwer, E.J., and McCarty, P.L., *Ground Water*, 22, 433–440, 1984.)

- Further dechlorination of DCE and VC requires more reductive conditions beyond the commonly observed sulfate reducing environment, as shown in Figure 8.6.
- The energy yield from the dechlorination of parent products (i.e., PCE and TCE) is often greater than that from daughter products; therefore, the former becomes a preferred process (AFCEE 2004).
- The bacterial species known to achieve complete dechlorination of chlorinated ethenes, *Dehalococcoides mccartyi* (DHC), is not present ubiquitously (Maymo-Gatell et al. 1997; Ernst 2009; Löffler et al. 2013). *Bioaugmentation*, that is, supplementation with this bacterium, is often needed to sustain complete dechlorination.
- Other geochemical conditions can inhibit the microbial growth activity. Most bacteria thrive in a circum-neutral pH (between 6 and 8) environment. Acidic conditions are harmful to the growth of DHC, particularly when a high level of sulfide is generated from sulfate reduction activities. The hydrogen sulfide from the combination of sulfide and hydrogen ion in a low-pH environment is toxic to bacteria.

Perchlorate can be treated using anaerobic reductive bioremediation. Perchlorate is reduced sequentially to chlorate, chlorite, and finally, chloride and oxygen. Slightly reducing conditions, such as nitrate reducing, are sufficient for perchlorate reduction to occur at a significant rate (ITRC 2005b).

Demolition Explosive (RDX) Anaerobic biodegradation of nitroaromatic compounds such as 2,4,6-trinitrotoluene (TNT) by amendment with a carbon source has been shown to be an effective remediation strategy (Barnes et al. 2001; Downey et al. 2005, 2011). The nitro substituents of the nitroaromatic compounds are highly electronegative, which makes them available as electron acceptors. The reduction of nitro groups can occur via biotic and abiotic reduction pathways (Rieger and Knackmuss 1995). The nitro substituents in TNT are reduced to amino substituents, forming compounds such as 2-amino (A)-4,6-dinitrotoluene (DNT) and 4-A-2,6-DNT. Consequently, the complete anaerobic reduction of TNT into 2,4,6-triaminotoluene (TAT) requires reducing conditions less than −0.2 V. TAT as a product of complete reduction of TNT is irreversibly bound to the soil matrix through several possible mechanisms, including adsorption, polymerization, and humification with naturally presented humic acid in soil. ISB of TNT has been successfully applied at the West Virginia Ordinance Work site (Downey et al. 2005, 2011).

Some pesticides and herbicides are susceptible to anaerobic reductive bioremediation. Organochlorine pesticides, such as toxaphene and dieldrin, which adsorb strongly to soils but do not readily decompose in the natural environment, were treated effectively with the application of solid organic carbon and ZVI to groundwater in a field study (Seech et al. 2008).

Semi-volatile organic compounds (SVOCs) from wood treating wastes include creosote and PCP, which are typically treated using bioremediation in the soil matrix, in sediments, or in mulch. In groundwater, typical treatment includes anaerobic reductive dechlorination followed by anaerobic oxidation. However, there are examples of full-scale *in situ* implementation that report success using a carbon source supplemented with oxygen (Fields and Wickramanayake 2010). Since PAHs are the main components in creosote, bioremediation technologies for PAHs can also be considered for creosote contamination (Zhang 2010), although PAHs above three rings are recalcitrant to bioremediation under any scenario because of their typically low solubility.

Metals, unlike organic compounds, cannot be destroyed through remediation technologies. Rather, remediation processes alter the physical or chemical characteristics of metals and therefore, their mobility and toxicity. The reductive bioremediation of chromium may be the most common application of bioremediation to metals; in this process, microbes reduce soluble Cr(VI) to insoluble Cr(III) and precipitate it out of solution. Selenium is another metal suitable for reductive bioremediation, with the same mechanism as chromium reduction, whereby the highly soluble oxidized form of selenium is biologically reduced to insoluble elemental selenium (CLU-IN 2008).

Another mechanism of metal immobilization is by microbial alteration of the redox state of naturally occurring metals, including iron and manganese, which bind the target contaminants, that is, heavy metals, so that they adsorb to soil (Lovley and Coates 1997).

An indirect metal remediation mechanism involves the use of anaerobic reductive bioremediation to promote bacterial sulfate reduction, which produces sulfide. Sulfide can consequently precipitate with various metals to form insoluble metal sulfides, as demonstrated at the Stoller Chemical Site in Jericho, South Carolina (CLU-IN 2006). This mechanism is used in the BiRD process, as discussed in Section 8.1.2.4.

8.2.2.2 Microbial Population

Unlike those involved in oxidative bioremediation, bacterial species used in reductive bioremediation can be contaminant specific, and their presence is often not ubiquitous. Therefore, *bioaugmentation*, amendment of the subsurface by specific microorganism populations (i.e., strains of bacteria) for the target contaminants in addition to organic substrates, is widely used in reductive bioremediation.

The most frequently supplemented bacterium is probably DHC, the only bacterium known to completely dechlorinate PCE and TCE (Maymo-Gatell et al. 1997). DHC is found absent at many sites, or at very low densities if present. The application of anaerobes, such as DHC, requires careful planning to ensure an oxygen-free subsurface environment and no exposure of the bacterial consortium to ambient air. The subsurface geochemical conditions need to be evaluated for any potential harm to the bacteria (refer to Section 8.2.2.1) and the need for pre-conditioning (i.e., adjustment of pH, dissolved oxygen [DO], etc.). The ESTCP guidance (ESTCP 2005) gives an in-depth discussion about the planning of DHC supplementation.

Dehalobacter spp. are capable of converting 1,1,1-TCA to chloroethane. 1,1,1-TCA is known to adversely impact the ability of DHC to dechlorinate TCE and daughter products (Duhamel et al. 2002). This inhibitory effect can be relieved by conversion of 1,1,1-TCA to 1,1-DCA (Chan et al. 2011). Therefore, when chlorinated ethene contaminants are comingled with chlorinated ethanes, it is recommended to inoculate mixed consortia of DHC and *Dehalobacter* spp.

Carbon tetrachloride can be reduced to methane by either *Pseudomonas stutzeri* KC, *Methanosarcina barkeri*, *Desulfobacterium autotrophicum*, *Moorella thermoacetica*, or *Methanobacterium thermoautotrophicum*.

Perchlorate-reducing microorganisms are generally ubiquitous in the environment, but bioaugmentation is sometimes needed to reach the population densities required for treatment (ITRC 2005b). Perchlorate reducers include the species *Dechloromonas aromatic* (Salinero et al. 2009), *Moorella perchloratireducens* (Balk et al. 2008), and *Sporomusa* sp. (Balk et al. 2010). These microbes produce an enzyme that can lower the activation energy for perchlorate reduction.

8.2.2.3 Carbon Source/Electron Donors

In the anaerobic reductive biodegradation process, the carbon source is also the electron donor in most cases. The materials commonly used as electron donors can be either *quick-release substrates* or *slow-release substrates*, depending on the rate at which they release bioavailable ingredients into the aqueous phase. In some cases, *hydrogen and propane gases* are used as electron donors, but to a much lower extent.

Quick-Release Substrates: Quick-release substrates normally consist of relatively simple molecules, including lactate compounds, organic acids, methanol, ethanol, molasses, and high fructose corn syrup (AFCEE 2004). These compounds are readily soluble in water and can move with groundwater flow, which allows wide distribution into the subsurface. Quick-release substrates generally are consumed quickly by the microbes, and therefore, they require more frequent replenishment. Another potential disadvantage of quick-release substrates is that the redox condition might become an overly reducing environment (more than sulfate reducing), which leads to the unproductive consumption of substrates by methanogens (Smatlak 1996).

Slow-Release Substrates Slow-release substrates include some oil products or solid organic matters such as mulch, compost, and chitin. Either these products are insoluble or they have low solubility and high viscosity, which make them generally immobile, and they sorb to the aquifer matrix. They are slowly broken down to soluble products such as fatty acids, which are then available for bacteria involved in biodegradation processes.

The low-solubility and high-viscosity nature of slow-release substrates makes them challenging to deliver into the subsurface. Emulsion technology was used to make vegetable oil mixable with water for injection, termed *emulsified vegetable oil* (EVO). The emulsion can be quickly broken once injected into the subsurface and forms small oil droplets adhering to solid particles. These oil droplets are slowly fermented into bioavailable fatty acids over time. The relatively slow consumption rates of EVO substrates result in increased longevity and less frequent need for replenishment. Guidance documents are currently available for the design and addition of specific substrate types, such as edible oil (AFCEE 2007).

Solid substrates, such as mulch, compost, and chitin, can last even longer (approximately 5–10 years) (AFCEE 2004) than EVO. These substrates are delivered by trenching, excavation, or surface application, targeting shallow contamination.

Hydrogen and Propane Gas: The microbial breakdown of fatty acids releases hydrogen, which is then used in the reduction of contaminants. The implementation of hydrogen gas is therefore the most direct approach for stimulating anaerobic reductive bioremediation (AFCEE 2004). Some hydrocarbon gases, such as propane, can also be used. These gaseous substrates can be delivered by biosparging or permeable membrane diffusers, just like oxygen for aerobic bioremediation. Special care must be taken over health and safety during the handling of these materials, given their combustible nature.

The selection of an appropriate substrate for a specific site depends on various factors, such as contaminant distribution and concentration levels, soil retention capacity for substrate, site hydrogeology, and the project timeline. In some cases, a mixture of quick-release and slow-release substrate is used, so that anaerobic conditions can be established quickly by the quick-release substrate and the slow-release substrate can sustain the reductive bioremediation over an extended period. Substrate dosing needs to take into account all available electron acceptors, including oxygen, nitrate, sulfate, and so on. Multiple tools are available to estimate the substrate dosing based on site geochemical conditions (ESTCP 2005).

8.2.3 Natural Attenuation

Natural attenuation refers to a set of naturally occurring processes, including biodegradation, dispersion, dilution, sorption, volatilization, and radioactive decay, as well as chemical or biological stabilization, transformation, or destruction of contaminants, which can reduce contaminant concentrations. These processes typically occur at all sites but to varying degrees of effectiveness, depending on the types and concentrations of contaminants present and the physical, chemical, and biological characteristics of the soil and groundwater.

Natural attenuation can be used to reduce contamination to cleanup goals at sites where contaminant concentrations are relatively low (i.e., close to the cleanup goal), which means that more aggressive treatment is not cost effective, and there are no immediate threats to human health and the environment. For most sites under natural attenuation, destruction, particularly biological destruction, is the only process that destroys contaminants and is therefore the preferred process. Several lines of evidence need to be collected to verify the occurrence of natural attenuation. For example, US EPA (1998) specified three lines of evidences to support natural attenuation of chlorinated ethenes in groundwater:

- Concentration trends of contaminants, such as decrease of PCE and TCE, and increase of daughter products, including DCE isomers and VC
- Decrease of contaminant mass across the groundwater plume and hydrogeologic and geochemical data that indirectly indicate natural attenuation (US EPA 1999)
- Microbiological laboratory data such as a microcosm study, which proves the occurrence of biological destruction of contaminants

Natural attenuation can be used as the polishing step after the more active treatment methods, such as ISB or ISCO, discussed previously. The monitoring requirement is similar to that for these active treatment technologies.

8.3 PHYTOTECHNOLOGIES

The term *phytotechnologies* refers to a set of technologies that use plants to remediate or contain contaminants in soil, groundwater, surface water, or sediments. These technologies have become attractive in comparison to conventional cleanup technologies due to the relatively low capital costs and the inherently aesthetic nature of planted sites. The following sections focus on the mechanism and application forms involved in phytotechnologies. More in-depth information can be found in ITRC's document *Phytotechnology Technical and Regulatory Guidance and Decision Trees* (2009).

8.3.1 Phytotechnology Mechanisms

The basic mechanisms involved in phytotechnology processes include sequestration, degradation, volatilization, and hydraulic containment, as discussed in more detail later.

Sequestration can occur either in the root zone (rhizosphere) or in the transpiration stream (stems and leaves, etc.). Some literature (e.g., ITRC 2009) defines *phytosequestration* as the fixation of contaminants at the rhizosphere and *phytoextraction* as the uptake of contaminants by plant stems and leaves. While harvesting can be applied to remove contaminants collected in the stems and leaves, it is not applicable to those sequestered in the rhizosphere.

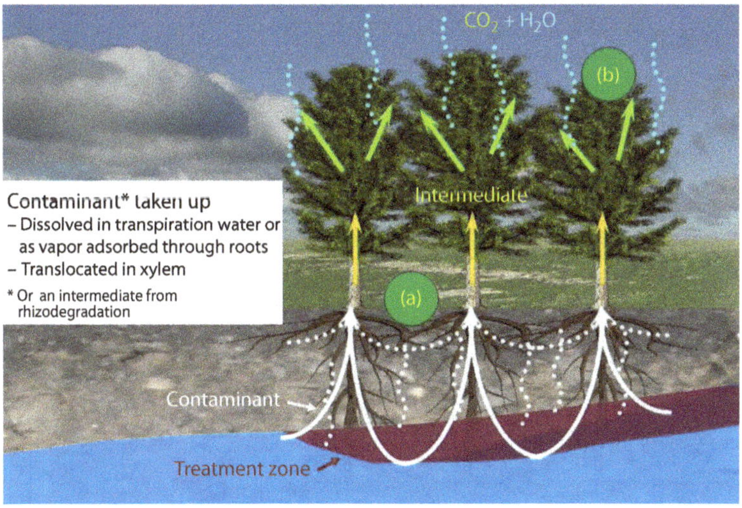

FIGURE 8.7 Phytodegradation mechanisms. (a) Plant enzymatic activity, (b) photosynthetic. (Source: ITRC 2009, *Phytotechnology Technical and Regulatory Guidance and Decision Trees.*)

Degradation can also occur in either the rhizosphere (rhizodegradation), or plant stems and leaves after extraction (phytodegradation). *Rhizodegradation* is the enhanced breakdown of a contaminant by increasing bioactivity using the plant rhizosphere environment to stimulate microbial populations. Plant roots can exude specific phytochemicals, including sugars, alcohols, carbohydrates, and so on, that can be used as carbon sources by bacteria and therefore enhance the bioactivity in the rhizosphere. This enhanced bioactivity represents the primary means through which organic contaminants can be remediated.

Phytodegradation uses the biological processes occurring within the plant itself following the phytoextraction into the transpiration stream. It involves various internal enzymatic reactions and metabolic processes. Figure 8.7 depicts these mechanisms. In addition, the redox cycle involved in photosynthesis produces strong oxidants and reductants in the plant system (from +1.1 to −1.3 V), which offer additional contaminant breakdown potential. These oxidants and reductants are normally stronger than those commonly available in biodegradation.

Volatilization of contaminants from the plant can occur either from the leaf stomata or from plant stems. The tendency of a chemical to volatilize depends on its Henry's constant and vapor pressure. While some contaminants might be not readily volatile, its breakdown product(s) derived from rhizodegradation or phytodegradation might be (Chappell 1998). Once volatilized into the atmosphere, many chemicals, while recalcitrant in the subsurface environment, can react rapidly with hydroxyl radicals, an oxidant formed during the photochemical cycle in air.

Hydraulic containment can be achieved by *phytohydraulics*, that is, the ability of plants to capture and evaporate water off the plant and take up and transpire water through the plant. Plants can significantly affect local hydrology. The vertical migration of water involved in evapotranspiration of deep-rooted species such as prairie plants and trees, through the root system, can form a cone of depression of groundwater elevation, which in turn, can contain or control the horizontal migration of groundwater (USEPA 2000).

8.3.2 Phytotechnology Applications

The term *phytotechnology* is sometimes misperceived as phytoremediation. The difference between these two can be viewed from the regulatory perspective: cleanup goals may be remediation, containment, or both. Phytotechnologies include containment strategies in addition to (phyto)remediation

TABLE 8.3
Summary of Phytotechnology Applications, Potential Mechanisms, and Applicable Contaminants

Application		Mechanisms	Applicable Contaminants
Phytostabilization	Phytostabilization covers and tree hydraulic barriers	Phytohydraulics Phytosequestration Phytoextraction	BTEX, PAHs, PCBs, PCE, TCE, DCE, VC, TPH, metals, acetone, MIBK (waste oil), refinery leachate, salinity, sodium (sodic soils), metals tailings, acid tar
Phytoremediation	Groundcover	Rhizodegradation Phytoextraction Phytodegradation Phytovolatilization	Organic compounds, TPH, TCE, PCE, PAHs, PCBs, BTEX, MTBE, energetics, cyanide, metals, radionuclides
	Tree stand	Phytohydraulics Rhizodegradation Phytoextraction Phytodegradation Phytovolatilization	BTEX, MTBE, naphthalene, toluene, EDC, TDA, TPH, TCE and PCE, metals (Cu, Cd, Pb), salinity, pH
Constructed wetlands		Phytohydraulics Rhizodegradation Phytoextraction Phytodegradation Phytovolatilization	As, Cr, F, N, P, Se, cyanide, phenols, TPH, BTEX, chlorinated hydrocarbons, municipal waste water, alkalinity, gray water, crude oil

Source: Modified from ITRC, *Phytotechnology Technical and Regulatory Guidance and Decision Trees, Revised.* PHYTO-3, Washington, DC, 2009.

strategies; that is, phytoremediation is a subset of phytotechnologies. The two broad categories of phytotechnologies are phytoremediation and phytostabilization. Phytostabilization is generally considered to comprise the "nonremediation" aspects of phytotechnologies. Constructed wetlands use many of the mechanisms of phytoremediation, and wetland technology itself is mature enough to have its own discipline. Table 8.3 illustrates the various forms of phytotechnology applications, the mechanisms involved, and their applicable contaminants.

8.3.2.1 Phytoremediation

Phytoremediation uses plants to remove, transfer, stabilize, and destroy contaminants in soil and groundwater. The mechanisms include rhizodegradation, phytodegradation, and/or phytovolatilization to reduce contaminant concentrations at the site. It can also involve phytoextraction if harvesting and contaminant removal are planned. Depending on the target depth, phytoremediation can be applied as groundcover for surface soil or tree stands for deep soil and/or groundwater.

Phytoremediation groundcovers have been widely applied to soils impacted with recalcitrant compounds such as PAHs, PCBs, and other persistent organic pollutants that are typically immobile, insoluble, and not very bioavailable. Phytoremediation groundcovers have been used to extract certain inorganic contaminants such as metals, salts, and radionuclides.

Phytoremediation tree stands, with the capability to reach a certain depth into the subsurface, have been widely applied to soil or groundwater impacted with soluble contaminants such as petroleum products (BTEX, MTBE, aliphatics, gasoline range organics [GRO], diesel range organics (DRO), and total petroleum hydrocarbon (TPH)) and chlorinated hydrocarbons (PCE, TCE, DCE, VC, etc.).

8.3.2.2 Phytostabilization

The main mechanism in phytostabilization is phytohydraulics, while phytosequestration and phytoextraction can also play a role in contaminated media.

Similarly to phytoremediation, phytostabilization can also be applied to different depths, that is, ground surface and subsurface. *Phytostabilization groundcover* targets surface soil or sediment. It uses vegetation with fibrous root systems to control bulk soil migration (erosion control) and/or prevent contaminant migration through phytosequestration if the soil or sediment is impacted. *Tree hydraulic barriers* target the subsurface and/or groundwater (plume control). The tree species are generally deep rooted and high transpiring, such that they actively tap into the groundwater to create the barrier. A relatively large number of trees (and the associated area) are generally required to extract sufficient volume of groundwater to achieve containment.

8.3.2.3 Constructed Wetlands

Constructed wetlands (also referred to as *constructed treatment wetlands*) are manmade wetlands to remove various types of contaminants in water. They possess a rich microbial community in the soil or sediment to effect the biochemical transformation of contaminants. They are biologically productive and most importantly, self-sustaining. The mechanisms involved in constructed wetlands include phytosequestration, rhizodegradation, phytoextraction, phytodegradation, and phytovolatilization. The hydrology of the constructed wetlands is affected by the vegetation with the same mechanisms driving phytohydraulics.

Besides surface water treatment, constructed wetlands can be used to treat saturated sediments and even shallow groundwater *in situ*. Riparian buffer strips are a type of constructed wetland that protects adjacent water resources from non–point source pollution. The *Technical and Regulatory Guidance for Constructed Treatment Wetlands* (ITRC 2003) provides more detail on this technology.

8.3.3 Advantages and Limitations

Phytotechnologies embrace a broad set of technologies that can be applied to a wide range of environmental conditions. In comparison with other *in situ* technologies, they are among the few that can be applied to both organic and inorganic contaminants. The applicable media include soil/sediment, surface water, and groundwater. In some cases, they can be applied to various combinations of contaminant types and impacted media simultaneously. Most importantly, they are a green and sustainable technology.

The major limitation of phytotechnologies is the depth they can reach. In general, 70%–80% of the root structure is within the top 1–2 ft of soil with exploratory roots sent deeper. For wetland species, typical depths are less than 1 ft due to oxygen limitations. For trees, typical depths are 10–15 ft, but they often require special culturing practices. The maximum practical depth of phytoremediation is down to 25 ft, although the influence of phytotechnology at 40 ft below ground surface has been measured. A general rule of thumb, however, is that trees will not access deeper than 5 ft into the saturated zone (ITRC 2009).

8.4 PHYSICAL TREATMENT

In situ physical treatment technologies include some conventional methods, including air sparging (AS), soil vapor extraction (SVE), multi-phase extraction (MPE), and some "innovative" methods such as thermal treatment and soil flushing.

AS uses a series of injection wells in which air is injected throughout a contaminated aquifer. Injected air travels through the saturated zone and removes contaminants via volatilization. An SVE system is typically installed in the vadose zone in conjunction with AS to capture the contaminated vapor. AS also aerates the aquifer, which promotes the aerobic biodegradation of contaminants.

AS is similar to ozone sparging, since both processes deliver gas streams to the subsurface, but ozone sparging provides the oxidation capacity to destroy contaminants *in situ*, whereas AS typically requires above-ground treatment of the volatilized contaminants.

MPE involves simultaneous extraction of soil vapor, groundwater, and NAPL if present to remediate multiple contaminated media.

The more conventional treatment methods, AS/SVE and MPE, are discussed in Chapter 7. The following sections focus on thermal treatment and soil flushing.

8.4.1 Thermal Remediation

In situ thermal remediation (ISTR) uses heat to enhance the volatilization of organic contaminants for subsequent collection via SVE. ISTR is considered the most aggressive and effective of all *in situ* treatment methods and a significant expenditure. Most *in situ* thermal technologies were originally developed and applied in the petroleum industry to enhance oil recovery. ISTR is more efficient at treating the source zone (or hot spots) within the saturated or unsaturated zones than at addressing larger, less contaminated areas of soil or groundwater plumes.

A detailed discussion of ISTR can be found in several publications, including US EPA's *Engineering Paper: In Situ Thermal Treatment Technologies: Lessons Learned* (US EPA 2014), and U.S. Army Corps of Engineers (USACE)'s *Design: In Situ Thermal Remediation, Engineer Manual* (USACE 2014), among others. The following sections discuss the principles governing the removal efficiency in ISTR and the three commonly used heat delivery methods: thermal conductive heating (TCH), electrical resistance heating (ERH), and steam enhanced extraction (SEE).

8.4.1.1 Fundamental Principles

Regardless of the different heat delivery methods, temperature increase in the subsurface will have similar effects on the physicochemical properties, such as vapor pressure, viscosity, and aqueous solubility, of organic contaminants. Changes in physicochemical properties as temperatures increase are illustrated in Figure 8.8.

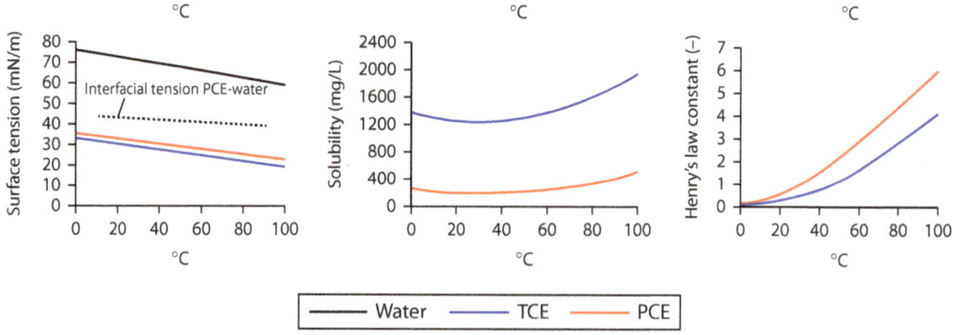

FIGURE 8.8 Change of physicochemical properties of common chlorinated solvents with increasing temperature. (From Heron, G., R.S. Baker, J.M. Bierschenk, and J.C. LaChance, Heat it All the Way—Mechanisms and Results Achieved Using In-Situ Thermal Remediation. *Proceedings of the Fifth International Conference on Remediation of Chlorinated and Recalcitrant Compounds*, Monterey, CA, 2006.)

TABLE 8.4
Characteristics of the Three Commonly Used Heat Delivery Methods

	Heat Delivery Mechanism	Upper Bound Temperature	Fluid Delivered	Applicable Contaminants	Applicable Geology
Thermal conductive heating	Thermal conduction	750–800°C	None	VOCs, SVOCs, tar, creosote	All except for high-groundwater flow zones
Electrical resistance heating	Ohmic resistance	100°C	Water, saline	VOCs, partial treatment of SVOCs	Not limited by permeability, particularly effective at clay and silt zones, requires water
Steam enhanced extraction	Steam injection	170°C	Steam	VOCs, partial treatment of SVOCs	Permeable zones

Source: Data from Hero, *In Situ Thermal Remediation*, 2017, and USACE *Design: In Situ Thermal Remediation*, Engineer Manual, May, EM 200-1-21, 2014.

These changes of physicochemical properties impact the transport behavior of organic contaminants in all phases, including NAPL, solid, water, and vapor.

Although the specific contribution of each process in ISTR treatment varies based on different compounds and different site conditions, the most important process in ISTR is probably volatilization. The volatilization of an organic compound from the liquid or solid phase into the vapor phase is controlled by its vapor pressure, that is, the pressure exerted by the vapor phase in equilibrium with its liquid or solid phase. Enhanced dissolution as a result of increased aqueous solubility has also been reported to play a significant role in ISTR (USACE 2014; Smith 2000).

For the NAPL phase, dissolution into water and evaporation into vapor are both enhanced by higher aqueous solubility and higher vapor pressure at elevated temperature. For contaminants in the water phase, vaporization is expedited with higher Henry's Law constants. For contaminants sorbed on the solid phase, desorption into the water and vapor phases are both enhanced, as the sorption coefficients of soil-water and soil-gas decrease with increasing temperature.

For less volatile contaminants such as oils, tars, creosote, and SVOCs, viscosity reduction and subsequent mobilization may also be an important mechanism. MPE is coupled with ISTR to remove the mobilized fluids. Although the heating can be done using any of the three major ISTR methods, complete removal of these less volatile contaminants requires TCH to achieve temperatures above the boiling point of water (refer to Table 8.4).

In addition to accelerating the rate and degree of contaminant mass recovery from the target treatment area, thermal treatment can also catalyze *in situ* destruction, such as hydrolysis (decomposition in the presence of water), pyrolysis (decomposition in the absence of oxygen), and oxidation. Up to a certain temperature range before sterilization occurs, bioremediation can also be enhanced by higher microorganism activity and increased bioavailability of contaminants (resulting from enhanced dissolution and desorption).

8.4.1.2 Common Heat Delivery Methods

The most widely used heat delivery methods include thermal conductive heating (TCH), electrical resistance heating (ERH), and steam enhanced extraction (SEE). Table 8.4 presents the characteristics of these three methods, and Figure 8.9 illustrates the schematic of the system associated with each method.

Thermal Conduction Heating (TCH) (also referred to as *electrical conductive heating* or *in situ thermal desorption*) supplies heat to the soil through steel wells or with a blanket that covers the

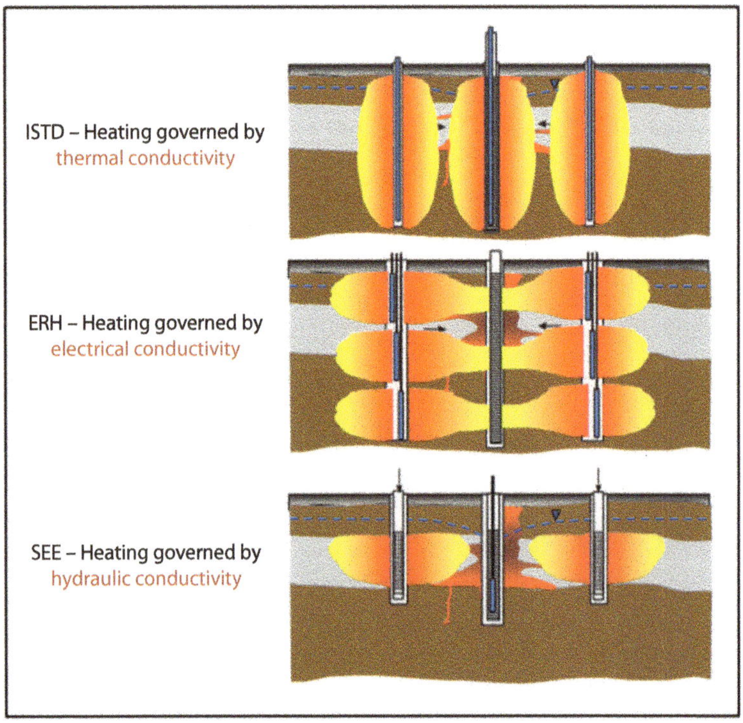

FIGURE 8.9 Schematics of the three commonly used ISTR methods. (From Heron, G., In Situ Thermal Remediation, 2017.)

ground surface. It uses an extraction and treatment system that collects the contaminants and vapors out of the ground for treatment. The soil is heated by simple thermal conduction via heated metal pipes in the ground. Typical TCH systems use circuits of electrically powered heaters. Alternatively, gas may also be used to heat the boreholes. As soil is heated, contaminants are vaporized or destroyed by several mechanisms, including evaporation, oxidation, and pyrolysis.

TCH can achieve a wide range of temperatures up to 800°C, giving it the capacity to treat less volatile contaminants such as SVOCs, oil and tar, and so on. It can be applied at low (<100°C), moderate (about 100°C), and high (>100°C) temperature levels depending on the target contaminant(s). Although it is applicable to both the vadose zone and the saturated zone, removal of groundwater (by pumping or vaporization) is required if a temperature higher than the boiling point of water is desired. It is therefore not very effective for zones with high groundwater flow.

Electrical Resistance Heating (ERH) heats the subsurface by passing electricity through it. It uses arrays of electrodes installed around a central neutral electrode to create a concentrated flow of current toward the central point. Electrical current is passed through the thin layer of moisture that coats the soil particles. Resistance to current flow in the soil generates heat. Due to its reliance on moisture to conduct the electrical current, the maximum temperature achievable by ERH corresponds to the boiling point of water at the local pressure conditions (generally from approximately 100–120°C). Volatized contaminants are collected by vacuum extraction wells and delivered to the surface for recovery or treatment.

Because ERH delivers heat by electrical resistance, it is applicable both above and below the water table. However, when moisture content is low, vadose zone ERH may require an electrode wetting system, with water or a salt solution supplied to each electrode. Unlike other *in situ* treatment technologies that deliver a fluid, ERH is not limited by permeability. It is especially advantageous and well suited for sites with low permeability, such as clays and silts. However, sites with very fast groundwater flow rates (e.g., greater than 1 ft/day) may be problematic, as the water acts as a heat sink.

Steam Enhanced Extraction (SEE) uses steam injection to heat the subsurface. A typical SEE system consists of a steam generator, a thermal well field, an MPE, and an effluent treatment system. Steam is normally generated on-site through the combustion of fuel oil or gas (e.g., propane) in a boiler system. SEE is governed by steam propagation and heat transfer to the surrounding soil formation. It relies on wells to inject steam and extract heated fluids. Subsurface heating occurs initially through the transfer of heat carried in the steam to the subsurface during condensation near the injection points. Then, the condensation front radiates out and eventually reaches extraction wells with continued steam injection. Thermal conduction and convection may also help distribute the heat into the less permeable zones.

The injected steam enhances contaminant removal via the following mechanisms: (1) increased pressure gradient for recovery of fluids, (2) reduced viscosity, (3) increased vapor pressure for accelerated vaporization, and (4) enhanced dissolution, destruction, and removal with the extracted water.

SEE is most effective at sites where the hydraulic conductivities of the bulk formation are sufficient to allow steam to be efficiently injected. It can be used in higher–groundwater flow zones where ERH or TCH is not as effective. SEE has been used in combination with other ISTR methods, with steam applied to permeable portions and other technologies, such as TCH or ERH, applied to less permeable zones.

Other less commonly used ISTR technologies are hot water injection, radio-frequency heating, microwave heating, and vitrification (USACE 2014). A new technology has used smoldering, that is, controlled combustion in the contaminated zone with very high concentrations of hydrocarbons as energy source (e.g., creosote and coal tars), which has been studied at pilot scale (USEPA 2014).

8.4.1.3 Advantages and Limitations

In most subsurface materials subject to *in situ treatment*, the variation of thermal conductivity is within an order of magnitude. In comparison to the variation of hydraulic conductivities over several orders of magnitude, the relatively narrow range of thermal conductivities allows a more uniform heat distribution and subsequent treatment within the contaminated zone than other *in situ* methods that rely on delivery of reagents. This advantage is particularly dominant for TCH and to a lesser extent, for ERH and SEE. The multiple mechanisms involved in ISTR also lead to a more complete removal of contaminant in a shorter period than is possible with other techniques.

ISTR apparently cannot be used to remove inorganic contaminants except for volatile metals such as mercury. Organic contaminants with very low volatility, including pesticides, some PAHs, dioxins, and PCBs, are not susceptible to ISTR. Other site conditions that may exclude ISTR include high groundwater flow, the presence of ordnance or explosive substances, or critical subsurface facilities or utilities. And what is more, the cost of ISTR can sometimes become a limiting factor, depending on the magnitude of contamination and site-specific situations.

8.4.2 Soil Flushing

In situ soil flushing is virtually a modified pump-and-treat technology with the addition of a carrier to increase the recovery of water-immiscible solvents, precipitated metals, contaminants that have diffused into small pore spaces, or those that adhere strongly to soils. *In situ* soil flushing differs from the pump-and-treat technology in that the former can be applied to vadose zone soil, whereas the latter is mostly for groundwater in the saturated zone. In contrast to soil washing, which involves excavation of the contaminated soil and above-ground treatment in a soil washer, soil flushing involves an injection/recirculation process in place. The soil flushing carriers include surfactant and cosolvent, as discussed in the following subsections.

8.4.2.1 Surfactant Flushing

Surfactants are surface active agents that can alter the physical-chemical properties at the solution interface. Surfactant molecules consist of a *hydrophilic* (water-loving) head and a *hydrophobic* (water-hating) tail, as illustrated in Figure 8.10 for a typical surfactant. The hydrophilic head group often includes anions or cations such as sodium, chloride, or bromide as a counter-charge balancing ion. The hydrophobic portion of a surfactant molecule is typically a long hydrocarbon chain with strong affinity for NAPL.

This amphiphilic nature leads to the accumulation of surfactant monomers at NAPL-water interfaces with the hydrophobic tail residing in the NAPL and the hydrophilic head facing toward the aqueous phase (Figure 8.11). The molecular weight of surfactants generally used in the environmental remediation area ranges from 200 to 2000 g/mol (AATDF 1997). Depending on the nature of various head groups, surfactants are typically classified as cationic, anionic, nonionic, or zwitterionic (having both cationic and anionic groups).

The amphiphilic nature is a characteristic that surfactants share in common with cosolvents. Surfactants have a unique feature, that is, the formation of micelles, that distinguishes them from cosolvents. A surfactant molecule that exists as a single unit is called a *surfactant monomer*. With increasing surfactant concentrations, the monomer concentration increases up to a threshold value, above which micelles form. The threshold value is referred to as the *critical micelle concentration* (CMC). When the surfactant concentration is above the CMC, the number of monomers remains constant, and the excess surfactant molecules aggregate to form micelles (Figure 8.11). Micelles in aqueous solutions will have their hydrophobic tail pointing toward the interior of the micelle and the hydrophilic head oriented toward the aqueous solution. CMC is an important property

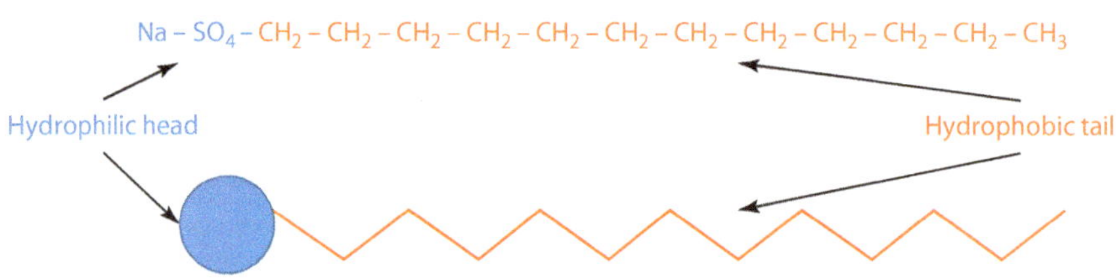

FIGURE 8.10 Structure of a typical surfactant, sodium dodecyl sulfate.

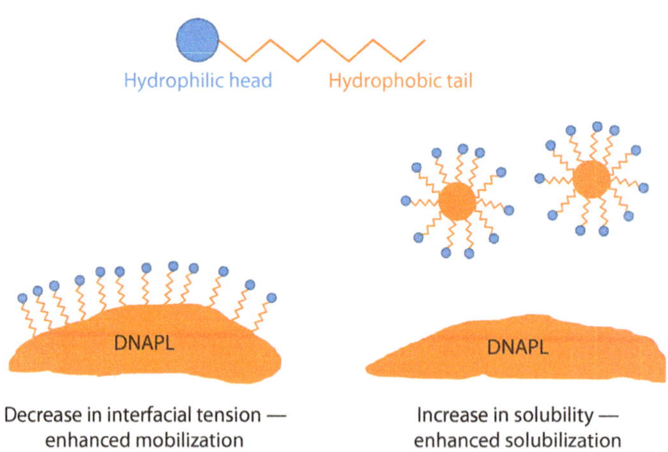

FIGURE 8.11 Mobilization versus solubilization of DNAPL by surfactant. (Modified from Lee, L. S., X. Zhai, and J. Lee, INDOT Guidance Document for In-Situ Soil Flushing. Publication FHWA/IN/JTRP-2006/28-2. Joint Transportation Research Program, Indiana Department of Transportation and Purdue University, West Lafayette, Indiana, 2007.)

of surfactants, and the CMC values of commonly used surfactants are available in the literature (Lee et al. 2007).

Hydrophile-lipophile balance (HLB) is another characteristic feature of surfactants. HLB is an indication of the relative strength of the hydrophilic and hydrophobic portions of the surfactants. A high HLB value indicates higher water solubility and lower affinity for the NAPL (Sabatini et al. 1996), and vice versa. When choosing a surfactant for soil flushing, one of the factors to consider is that the HLB of the surfactant should be as close as possible to that of the contaminants (Rosen 1989).

The hydrophobic nature inside the micelle makes it an amenable place for NAPL to reside (Figure 8.11). This NAPL removal mechanism is defined as *solubilization*, as opposed to *mobilization*. Mobilization is mainly caused by decreased interfacial tension (IFT) between the NAPL and the surfactant. For example, NAPL can be trapped in soil pores due to capillary forces, and this is typically referred to as *residual NAPL*. These capillary forces are proportional to the IFT at the NAPL-water interface (West and Harwell 1992). During surfactant flushing, surfactants accumulate at the NAPL-water interface, and the IFT is reduced between the two phases because of the amphiphilic nature of surfactant (refer to total trapping number, discussed by Pennell et al. 1996 and Lee et al. 2007). When buoyancy and viscous forces overcome capillary forces, the NAPL migrates in the direction of the net force, and the phenomenon is termed *mobilization*. However, the mobilization of DNAPL into deeper zones is apparently not desired, as it causes the contamination to spread, making it even harder to clean up.

8.4.2.2 Cosolvent Flushing

Cosolvents are organic compounds with a hydrophobic part, usually hydrocarbon chains, and a hydrophilic functional group, such as hydroxyl, carboxylic, and aldehyde groups. The amphiphilic and hydrophobic groups, respectively, enable cosolvents to be miscible in both the water and the NAPL phase. Alcohols are most commonly used as *in situ* flushing cosolvents.

Cosolvents increase contaminant solubility in water by making the aqueous phase less polar. The cosolvent-enhanced solubility is exponentially correlated to the volume fraction of the cosolvent in the cosolvent-water mixture by the cosolvency power (Yalkowsky and Roseman 1981):

$$S_m = S_W \cdot 10^{\sigma f} \tag{8.34}$$

where:

- S_m is the solubility in cosolvent-water mixture (mass or moles/L)
- S_w is the solubility in water (mass or mol/L)
- σ is known as the cosolvency power (dimensionless)
- f is the cosolvent volume fraction

Cosolvency power is characteristic of the specific cosolvent selected. To achieve the cosolvency effect, the volume fraction of cosolvent should generally be greater than 10% (Schwarzenbach et al. 2002; Yalkowsky and Roseman 1981).

Cosolvents enhance the removal of NAPLs by two processes: solubilization and mobilization. The mechanisms responsible for mobilizing NAPL contaminants include

- creation of a single phase condition
- decrease in the water-NAPL interfacial tension
- swelling of the NAPL by solubilization of the cosolvents within this phase

On the other hand, the addition of cosolvent can also lower the IFT to an extent that causes NAPL mobilization. For sites with DNAPL, downward movement of DNAPL to previously clean regions would worsen the environmental risk. Caution must be practiced to optimize solubilization while minimizing the risk of DNAPL mobilization. There should be a maximum level of cosolvent that can be used at a given site where DNAPL solubility is sufficiently increased without enough decrease in IFT to induce mobilization. Such a system can be designed on the basis of careful characterization of the system by considering various factors, as discussed in detail in the *Technology Practices Manual for Surfactants and Cosolvents* (AATDF 1997).

8.4.2.3 Advantages and Limitations

Soil flushing with surfactant or cosolvent solution can significantly improve the contaminant mass recovery efficiency compared with conventional pump-and-treat. However, soil flushing technologies have not been widely used in the past two decades except for limited research and field-scale demonstration projects, due to several drawbacks (NRC 2012).

The first limiting factor of soil flushing is the mobilization potential when IFT is lowered, especially in the case of DNAPLs. Second, soil flushing is a mass recovery technology, and there is still the need for subsequent extraction and treatment of the effluent waste stream (contaminant and surfactant or cosolvent). Additionally, the selection of a surfactant formulation that is both safe and effective requires laboratory and/or pilot-scale treatability tests. The up-front costs for active ingredients and effluent treatment systems associated with soil flushing can be substantial.

In addition to these limitations for both soil flushing carriers, cosolvent has its own extra challenges. First, the composition of cosolvent solution with more than half (70%–90%) of alcohol results in a significant density difference relative to water. The subsequent density override effect needs to be conquered by gradient injection (gradually increasing the concentration of cosolvent delivered over time) (Rao et al. 1997) and strategic placement of injection wells to promote upward migration of the miscible effluent in unconfined systems (Jawitz et al. 2000). Finally, cosolvents are highly flammable and require special safety procedures during handling.

PRACTICE AND EXERCISES:

Question 1: The total mass of COCs at a contaminated site was estimated to be 10 kg of PCE, 12 kg of TCE, 15 kg of DCE isomers, and 2 kg of VC.

a. Calculate the stoichiometric demand of sodium permanganate for ISCO treatment.
b. A bench-scale test was conducted to estimate the total oxidant demand (TOD) using site-specific soil and groundwater samples. Based on the actual measured TOD, the calculated sodium permanganate dosing should be 2100 kg. Is this higher or lower than the calculated stoichiometric demand in a? Explain the factors that caused this difference.
c. If sodium permanganate is supplied as 40% (by mass) solution, with a density of 1.36 g/mL, estimate the volume of sodium permanganate solution needed based on the demand of 2100 kg.
d. This site originated from a release of PCE at a former dry cleaner's facility. Explain why other compounds such as TCE, DCE isomers, and VC are detected.
e. What other *in situ* technology will you consider? And what site information do you need to know to evaluate the feasibility of these technologies?

Question 2: The following table presents the groundwater monitoring data, including COC concentrations and geochemical parameters, observed in a monitoring well near the center of a plume. A water injection test was conducted at the site, and an injection rate of 30 L/minute was achieved, indicating that the site soil is permeable. Based on the given information, which *in situ* technology will you choose for this site? State your rationale.

COC Concentrations (μg/L)		Geochemical Parameters			
TPH (g)	1100	DO (mg/L)	0.19	Total organic carbon (mg/L)	76
Benzene	15	ORP (mV)	38	DHC bacteria (cells/mL)	6.00×10^{-1}
Toluene	ND	pH (S.U.)	6.67	Methane (μg/L)	24.1
Ethyl benzene	38	Temperature (°C)	22.9	Ethane (μg/L)	ND
Xylenes	16	Specific conductance (μS/cm)	4924	Ethene (μg/L)	ND
TCE	840	Turbidity (NTU)	5	Iron, ferrous (mg/L)	ND
PCE	ND	Total dissolved solids (mg/L)	4200	Arsenic (mg/L)	ND
trans-1,2-DCE	ND	Nitrate (as N) (mg/L)	1.3	Boron (mg/L)	0.963
cis-1,2-DCE	800	Nitrite (as N) (mg/L)	ND	Iron (mg/L)	ND
1,2-DCA	26	Sulfate (mg/L)	1900	Manganese (mg/L)	1.61
Vinyl chloride	2.7	Chloride (mg/L)	340		

Question 3: A bench-scale treatability study was conducted for Cr(VI) contamination at a site. The following data were obtained:

Treatment	Treatment Time (Days)	pH (S.U.)	ORP (mV)	Cr(VI) (mg/L)	Total Cr (mg/L)	Arsenic (mg/L)	Manganese (mg/L)	Nitrate (mg/L)	Nitrite (mg/L)	Sulfate (mg/L)
Cleanup goal		–	–	–	0.1	0.01	–	–	–	–
Control	1	7.2	207.9	1.02	1.15	0.04	<0.016	13	< 3.8	2030
	7	7.32	−32.3	1.38	1.13	0.04	<0.016	11.8	< 3.8	2090
	21	7.26	288.1	1.43	1.06	0.04	<0.016	15.2	< 3.8	2010
	84	6.7	233.7	1.04	1.05	0.03	<0.016	16.7	< 3.8	1900
EVO	7	6.9	512.1	0.80	0.77	0.18	0.24	< 2.8	< 3.8	2220
	21	6.46	248	0.75	0.46	0.23	0.57	< 2.8	< 3.8	1940
	84	6.62	154.6	< 0.0125	0.23	0.24	0.47	< 2.8	< 3.8	1950
ZVI and organic carbon	7	6.35	−192	0.02	0.072	0.30	4.23	< 2.8	< 3.8	2220
	21	6.13	−194.3	0.01	< 0.026	0.13	4.37	< 2.8	< 3.8	1850
	84	6.41	−196.3	< 0.0125	< 0.026	0.01	2.14	< 2.8	< 3.8	1770
CaS_5	1	9.24	−464.8	NA	< 0.026	0.03	<0.016	12.4	< 3.8	1970
	7	9.26	−465.8	0.06	< 0.026	0.13	0.03	11.4	< 3.8	2090

a. Based on the given data, which treatment(s) is/are shown to be effective? What percentage of Cr(VI) reduction was achieved at the end of the effective treatment(s)?
b. Which factors led to the ineffectiveness of the other treatment(s)?
c. If you are to design this treatability study, which species listed in the table do you need to consider to calculate the dosing rate of treatment reagents?

Question 4: An ISB treatability study was conducted for a site with chlorinated solvent compounds. The treatment results are presented in the following two figures.

a. Which two substrates were tested in this treatability study? What are the major differences between these two substrates?
b. What competing electron acceptors were present in the system? How did they respond differently to the treatment with the two substrates?
c. Do you observe a stall? Based on what?

d. What was the purpose of adding ferrous chloride at the later stage of treatment? Was it effective?
e. What would you recommend for field implementation?

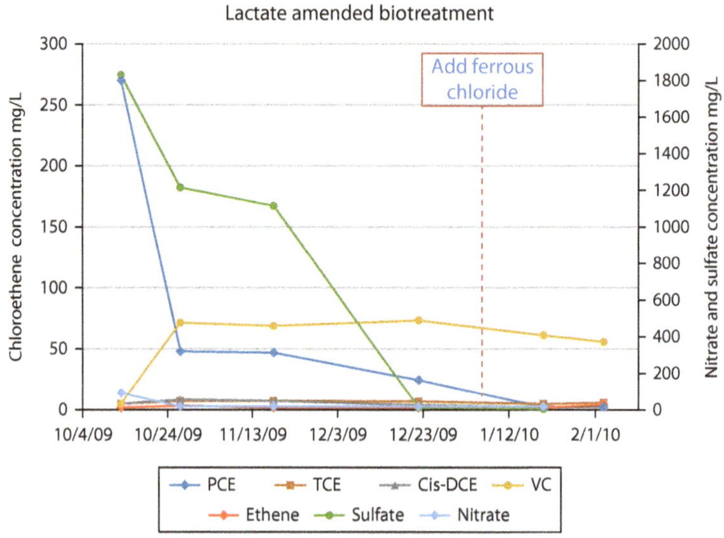

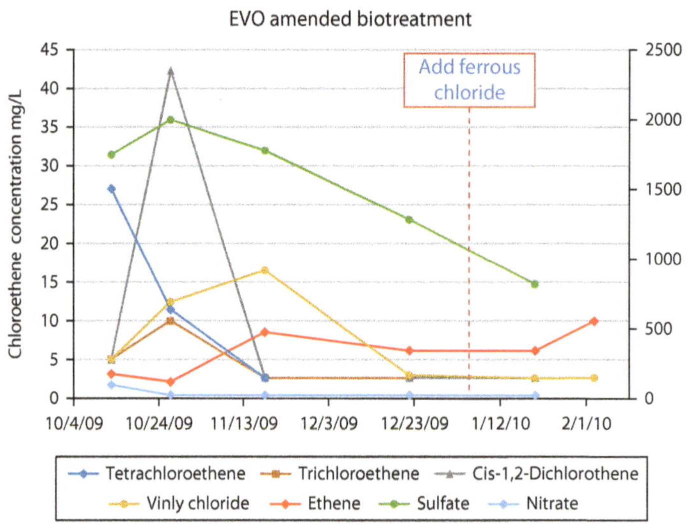

Question 5: A site is contaminated with heavy metal (primarily lead) and TPH in both soil and groundwater. The groundwater plume extends to the seepage face to a creek, where contamination was delineated at the depth interval of 10–15 ft bgs. Will you consider phytotechnology to protect the creek? Which one and why?

Question 6: For a pump-and-treat application at a site with TCE contamination, it was estimated that a total of 5,000,000 L of groundwater needed to be pumped to reduce the TCE mass in the source zone. Assume TCE solubility is approximately 1100 mg/L.

a. Estimate the total mass of TCE that can be removed by 5,000,000 L of water, assuming the extractant has an average TCE concentration of 1% of its solubility.
b. If ethanol is to be used as a cosolvent to enhance the groundwater pumping, calculate the volume of pumping required to achieve the same mass reduction as in Problem 6a. Assume that ethanol is to be applied at 90% by volume. The cosolvency power of ethanol is 3.62 (unitless).

c. What factors do you need to consider if cosolvent flushing is to be implemented?

Question 7: The historical COC concentration changes are shown in the following table for a site with gasoline spill.

		Concentrations (μg/L)							
Sampling Date	**Depth to Groundwater (feet)**	**TPH as Gasoline**	**MTBE**	**Tert-Butyl Alcohol**	**Benzene**	**Toluene**	**Ethylbenzene**	**Total Xylenes**
May 5, 2000	13.63	120,000	<500	NA	13,000	18,000	4,100	16,000
January 5, 2001	17.81	78,000	<500	NA	8,900	10,000	2,900	10,000
May 24, 2001	14.57	94,000	1,100	NA	9,400	11,000	3,300	13,000
November 17, 2001	18.22	70,000	<100	NA	7,200	7,600	2,800	10,000
March 9, 2002	17.48	73,000	<50	<2,500	7,700	8,900	3,200	11,900
June 15, 2002	18.6	54,000	<100	<1,000	7,200	7,200	3,100	10,700
August 26, 2002	19.09	59,000	<100	<1,000	7,900	5,400	3,300	10,400
December 21, 2002	16.85	72,000	<100	<1,000	7,900	7,700	3,400	12,200
March 2, 2003	15.18	52,000	<100	<1,000	8,900	10,000	3,100	13,700
May 23, 2003	15.3	73,000	<50	<500	7,500	9,900	3,900	17,000
October 4, 2003	16.83	53,000	<10	<100	6,300	7,300	3,000	12,000
December 13, 2003	15.88	64,000	<100	<1,000	6,600	9,100	3,800	15,800
February 28, 2004	13.23	54,000	<100	<1,000	5,800	7,900	3,100	12,600
June 20, 2004	14.13	54,000	<100	<1,000	6,100	7,800	2,900	13,100
September 19, 2004	15.45	70,000	<50	<500	6,600	8,000	4,000	14,900
December 20, 2004	12.73	96,000	<50	<500	6,600	9,400	3,500	15,800
March 12, 2005	7.62	61,000	<100	<1,000	6,200	9,300	3,300	15,500
November 20, 2005	10.66	140,000	<25	<250	4,600	4,200	2,800	11,300
December 16, 2006	10.29	54,000	<25	<250	4,000	2,500	2,700	9,600

The current groundwater geochemical parameters are

- Nitrate < 0.1 mg/L
- Sulfate < 1 mg/L
- Iron 2.47 mg/L
- Manganese 1.25 mg/L

Will you consider natural attenuation as an effective method without any additional active treatment? Why?

Question 8: If oxidative bioremediation is to be implemented for the site described in Problem 7, answer the following questions:

a. What needs to be supplemented to the subsurface? Electron donor, electron acceptor, or organic substrate?
b. Calculate the dosing rate of calcium peroxide to treat the contaminants in groundwater.
c. Calculate the dosing rate of calcium peroxide to satisfy the demand from naturally presented species given in Problem 7.
d. Is there any other potential demand for calcium peroxide that needs to be considered to decide the dosing?

Question 9: An accidental release of TCE at a metal processing facility resulted in DNAPL on top of a clay lens at a depth of 50 ft bgs. Thermal treatment is considered for expedited cleanup. Which heat delivery method will you use and why?

ANSWERS:

Answer 1: Oxidant dose calculation for ISCO treatment of chlorinated compounds.

a. The calculation needs to consider the molecular weight of each contaminant and its reaction equation with permanganate. The total stoichiometric demand is estimated to be approximately 111 kg.

COC	Mass of COC (kg)	Molecular weight (g/mole)	Moles	Reaction ratio (moles of MnO_4: moles of COC)	Moles of $NaMnO_4$ required	Mass of $NaMnO_4$ required (kg)
PCE	10	166	60.24	1.33	80.32	11.41
TCE	12	131.5	91.25	2.00	182.51	25.92
DCE	15	97	154.64	2.67	412.37	58.56
VC	2	62.5	32.00	3.33	106.67	15.15
Total						111.03

b. The actually measured TOD is generally much higher than the stoichiometric demand from contaminants due to the demand from naturally presented organic matter and reduced species such as ferrous iron and manganese.

c. $2100\text{ kg} / 40\% / 1.36\text{ kg} / \text{L} = 3860\text{ L}$.

d. The reductive dechlorination of PCE does not require highly reducing conditions, and it can occur naturally. However, complete dechlorination to ethene requires strong reducing conditions and an abundance of DHC bacteria.

e. Given the observed biodegradation daughter products, ISB can also be used to treat the COCs. As an ISB alternative, other sources of electron acceptors, including nitrate, sulfate, ferric iron, and so on, need to be considered. DHC may need to be bioaugmented to provide a large enough microbial population.

ISCR can also treat the COCs observed at the site. Based on background iron and sulfate, the BiRD process can be used. Thermal treatment can also be effective if expedited cleanup is desired.

Site lithology, primarily soil permeability, needs to be investigated for ease of delivery of *in situ* treatment reagent.

Answer 2: Given the relatively permeable soil lithology, *in situ* technologies that involve reagent delivery are feasible. ISCO is effective on both the chlorinated compounds and BTEX compounds detected at the site.

Answer 3: Bench-scale treatability study for Cr(VI) contamination at a site.

a. The treatment using a mixture of ZVI and organic carbon appears to be effective, as the other two both caused increasing concentration of arsenic. Percentage reduction is calculated using the Cr(VI) concentration for treatment and control after 84 days of treatment:

b. $(1.04 - 0.0125) / 1.04 \times 100\% = 98.8\%$

c. The reduced form of arsenic, As(III), is more mobile than its oxidized form, As(V). The reducing conditions resulting from EVO or CaS_5 treatment converted As(V) to As(III). In the treatment with a mixture of ZVI and organic carbon, iron from ZVI is able to precipitate arsenic even in its reduced form.

d. The reducible species listed in the table include Cr(VI), arsenic, nitrate, nitrite, and sulfate, which should all be considered when calculating the treatment dose.

Answer 4: ISB treatability study for a site with chlorinated solvent compounds.

a. Lactate and EVO are used as two treatment reagents. Lactate is a fast-releasing substrate and will establish reducing conditions quickly. EVO is a slow-releasing substrate; it will last longer and establish reducing conditions gradually.

b. Nitrate and sulfate are competing electron acceptors in the system. Nitrate was reduced to a low level quickly by both treatments. The sulfate reduction rate was faster in lactate treatment than in EVO treatment.
c. Yes, in the system with lactate treatment, cis-DCE and VC are both detected throughout the course of treatment.
d. Ferrous sulfate was added likely to precipitate the sulfide generated from sulfate reduction. VC still persisted after ferrous sulfate addition to lactate treatment, indicating that it is not effective. For EVO treatment, it is hard to say based on the given information, as the situation was not as bad (cis-DCE and VC were already reduced) before its addition.
e. EVO treatment appears to perform better than lactate treatment and is therefore recommended for field implementation.

Answer 5: Yes, phytotechnology is a feasible option given that the impacted depth of groundwater is within its working range. Constructed wetland and a phytoremediation tree barrier both have the capacity to protect the creek. However, a phytoremediation tree barrier might have a better capacity to reach the depth of groundwater contamination.

Answer 6: Cosolvent flushing of TCE contamination.

a. $1{,}100\ \text{mg/L} \times 1\% \times 5{,}000{,}000\ \text{L} \times 0.001\ \text{g/mg} = 55{,}000\ \text{g}$ of TCE removed.
b. Solubility of TCE in 90% by volume ethanol solution:

$$1{,}100\ \text{mg/L} \times 10^{(0.9 \times 3.62)} = 28{,}597\ \text{mg/L}.$$

It is likely that the cosolvency flushing will also achieve 1% of the solubility after cosolvency enhancement. To extract 55,000 g of TCE, the volume required is

$$55{,}000\ \text{g}/(28{,}597\ \text{mg/L} \times 1\%) \times 1{,}000\ \text{mg/g} = 192{,}326\ \text{L}.$$

c. The density of 90% ethanol solution is significantly greater than that of natural groundwater. The density override needs to be overcome by some strategy such as gradient flushing. DNAPL is likely present at the site. So, the design of ethanol flushing should take caution to avoid DNAPL mobilization.

Answer 7: The data indicated that the benzene concentration was reduced approximately threefold in six years; the other contaminants showed even lower reduction. The last known concentrations of BTEX compounds are still much higher (about 1000 times) than their MCLs. BTEX biodegradation takes the oxidative pathway, and electron acceptors are needed. Naturally occurring electron acceptors, such as nitrate and sulfate, are not detected, and the reduction products ferrous iron and manganese are at significant levels. The groundwater geochemical condition is not favorable for oxidative degradation of BTEX. So, natural attenuation is not likely to clean up the contamination in a reasonable time frame.

Answer 8: Oxidative bioremediation for BTEX contamination:

a. Electron acceptors need to be added to the subsurface to induce oxidative degradation of BTEX contamination.
b. Dose calculation can take the reaction of contaminants with oxygen (O_2) and then calculate the mass of calcium peroxide needed to release the needed O_2. The last measured concentrations in December 2006 are used in the following calculation. The molecular weights of BTEX and their oxidation reactions are used to calculate the stoichiometric demand (mass of O_2 per unit mass of contaminant).

Contaminant	Concentration of Contaminants (mg/L)	Stoichiometric Demand (Mass of O_2/Mass of Contaminant)	Oxygen Demand (mg/L O_2)
Benzene	4	3.08	12.31
Toluene	2.5	3.12	7.79
Ethylbenzene	2.7	3.16	8.54
Xylenes	9.6	2.99	28.74
	Total		57.38

Based on Equation 8.33 in Section 8.2.1.3, 2 moles (144 g) of CaO_2 can release 1 mole (32 g) of O_2. To get 57.38 mg/L of O_2, the mass of CaO_2 required is $57.38/32 \times 144 = 258.22$ mg/L.

c. Naturally presented species that compete for electron acceptors include ferrous iron and manganese.

Species	Concentration (mg/L)	Stoichiometric Demand (Mass of O_2/Mass of Species)	Oxygen Demand (mg/L O_2)
Ferrous iron	2.47	0.14	0.35
Manganese	1.25	0.29	0.36
	Total		0.72

Similarly to step (b), to get 0.72 mg/L of O_2, the mass of CaO_2 required is $0.72/32 \times 144 = 3.23$ mg/L.

d. These calculations do not take account of natural organic matter in soil that can be oxidized. Also, the contaminants do not just exist in groundwater. A significant portion is sorbed to solid phase. Soil/water partitioning effects need to be considered to estimate the oxygen demand from sorbed contaminants.

Answer 9: TCE has a fairly low boiling point (87.2°C) and high vapor pressure and is susceptible to thermal treatment. A temperature of 100°C or lower is sufficient to vaporize TCE. Steam injection and ERH can both achieve the desired temperature range. ERH is better suited, given the low permeability associated with a clay lens.

REFERENCES

Advanced Applied Technology Demonstration Facility for Environmental Technology Program (AATDF). 1997. *Technology Practices Manual for Surfactants and Cosolvents*, https://clu-in.org/PRODUCTS/AATDF/toc.html.

Air Force Center for Energy and the Environment (AFCEE). 2004. Principles and practices of enhanced anaerobic bioremediation of chlorinated solvents. August. Available at: www.costperformance.org/remediation/pdf/principles_and_practices_bioremediation.pdf

AFCEE. 2007. Final technical protocol for in situ bioremediation of chlorinated solvents using edible oil. *Technical Directorate. Environmental Science Division*. October. Available at: www.clu-in.org/download/remed/Final-Edible-Oil-Protocol-October-2007.pdf

Agency for Toxic Substances and Disease Registry (ATSDR). 2007. Properties and degradation pathways of common organic compounds in groundwater, Atlanta, Georgia.

Amarante, D. 2000. Applying in situ chemical oxidation. *Pollution Engineering*, 32(2): 40–42.

Balk, M., F. Mehboob, A. H. van Gelder, W. I. Riipstra, J. S. Damste, and A. J. M. Stams. 2010. (Per)chlorate reduction by an acetogenic bacterium, Sporomusa sp., isolated from an underground gas storage. *Applied Microbiology and Biotechnology*. 88: 595–603.

Balk, M., T. van Gelder, S. A. Weelink, and A. J. M. Stams. 2008. (Per)chlorate reduction by the thermophilic bacterium Moorella perchloratireducens sp. nov., isolated from underground gas storage. *Applied and Environmental Microbiology*. 74: 403–409.

Bamforth, S. M. and I. Singleton. 2005. Bioremediation of polycyclic aromatic hydrocarbons: Current knowledge and future directions. *Journal of Chemical Technology & Biotechnology*. 80: 723–736.

Banerjee, M. and R. S. Konar. 1984. Comment on the paper "polymerization of acrylonitrile initiated by K2S2O8-Fe(II) redox system". *Journal of Polymer Science: Polymer Chemistry*. 22: 1193–1195.

Barnes, P., M. Heaston, and J. Compton. 2001. Treatment of explosives-contaminated groundwater by in situ cometabolic reduction. In V. Magar, G. Johnson, S. K. Ong, and A. Leeson (Eds.), *Bioremediation of Energetics, Phenolics, and Polynuclear Aromatic Hydrocarbons*. San Diego, CA: Battelle.

Block, P. A., R. A. Brown, and D. Robinson. 2004. Novel activation technologies for sodium persulfate in situ chemical oxidation, *Proceedings of the Fourth International Conference on the Remediation of Chlorinated and Recalcitrant Compounds*, Monterey, CA.

Bouwer, E. J. and P. L. McCarty. 1984. Modeling of trace organics biotransformation in the subsurface. *Ground Water.* 22(4): 433–440.

Bures, G., J. A. Skog, D. Swift, J. Rothermel, R. Starr, and J. Moreno. 2010. Fracture-emplacement and 3-D mapping of a microiron/carbon amendment in TCE-impacted sedimentary bedrock (abstract), *Seventh International Conference on Remediation of Chlorinated and Recalcitrant Compounds*, May, Monterey, CA.

Chan, W. W., A. Grostern, F. E. Löffler, and E. A. Edwards. 2011. Quantifying the effects of 1,1,1-trichloroethane and 1,1-dichloroethane on chlorinated ethene reductive dehalogenases. *Environmental Science & Technology.* 45(22): 9693–9702.

Chappell, J. 1998. Phytoremediation of TCE in groundwater using Populus. Status report prepared for the U.S. Environmental Protection Agency Technology Innovation Office. Available at: http://clu-in.org/products/phytotce.htm

CLU-IN. 2006. *Technology news and trends.* February. Available at: www.clu-in.org/products/newsltrs/tnandt/view.cfm?issue=0206.cfm

CLU-IN. 2008. *Technology news and trends.* January. Available at: www.clu-in.org/products/newsltrs/tnandt/view.cfm?issue=0108.cfm#4

Cundy, A., L. Hopkinson, and R. Whitby. 2008. Use of iron-based technologies in contaminated land and groundwater remediation: A review. *Science of the Total Environment.* 400: 42–51.

Department of the Navy. 2010. Moffett Field Restoration Advisory Board Progress Report- Sites 26 (EATS) and 28 (WATS). January. Available at: www.slideshare.net/

Downey, S., T. Ladaa, and D. Mullendore. 2005. In situ enhanced bioremediation to treat nitroaromatics in groundwater. *Federal Facilities Environmental Journal*, 16(1). www.interscience.wiley.com, DOI:10.1002/ffej.20043.

Downey, S., X. Zhai, and R. Meadows. 2011. Field-scale treatability study for enhanced in situ bioremediation of explosives in groundwater: BioBarrier installation and hot spot treatment using DPT injection, *Environment, Energy Security & Sustainability (E2S2) Symposium & Exhibition*, May, New Orleans, LA.

Duhamel, M., S. D. Wehr, L. Yu, H. Rizvi, D. Seepersad, S. Dworatzek, E. E. Cox, et al. 2002. Comparison of anaerobic dechlorinating enrichment cultures maintained on tetrachloroethene, trichloroethene, cis-dichloroethene and vinyl chloride. *Water Research.* 36: 4193–4202.

Environmental Security Technology Certification Program (ESTCP). 2005. Bioaugmentation for remediation of chlorinated solvents: Technology development, status, and research needs. www.serdp-estcp.org/Tools-and-Training.

ESTCP. 2010. Emulsified zero-valent nano-scale iron treatment of chlorinated solvent DNAPL source areas, ER-200431, 61 p.

Ernst, T. 2009. Use of "Dehalococcoides" to bioremediate groundwater contaminated with chlorinated solvents. *MMG 445 Basic Biotechnology.* 5(1): 72–77.

Fields, K. A. and G. B. Wickramanayake (Chairs). 2010. Remediation of chlorinated and recalcitrant compounds—2010, *Seventh International Conference on Remediation of Chlorinated and Recalcitrant Compounds.* Battelle Memorial Institute, May, Columbus, Ohio.

Firestone, M. K. 1982. Biological denitrification. In F. J. Stevenson (Ed.), *Nitrogen in Agricultural Soils.* Madison, WI: American Society of Agronomy. pp. 289–326.

FMC. 2001. Persulfate technical information. Available at: www.peroxychem.com/media/90826/aod_brochure_persulfate.pdf

Forman, K. 2010. Chloroform and TCE reduction using pneumatically injected Microscale Zero-Valent Iron (ZVI), *Seventh International Conference on Remediation of Chlorinated and Recalcitrant Compounds*, May, Monterey, CA.

Fruchter, J. S., C. R. Cole, M. D. Williams, V. R. Vermeul, J. E. Amonette, J. E. Szecsody, J. D. Istok, et al. 2000. Creation of a subsurface permeable treatment zone for aqueous chromate contamination using in situ redox manipulation. *Ground Water Monitoring and Remediation.* 20(2): 66–77.

Gu, X., S. Lu, L. Li, Qiu , Q. Sui, K Lin, and Q. Luo. 2011. Oxidation of 1,1,1-trichloroethane stimulated by thermally activated persulfate. *Industrial & Engineering Chemistry Research.* 50(19): 11029–11036.

Hendrickx, B., W. Dejonghe, W. Boënne, M. Brennerova, M. Cernik, T. Lederer, M. Bucheli-Witschel, et al. 2005. Dynamics of an oligotrophic bacterial aquifer community during contact with a groundwater plume contaminated with benzene, toluene, ethylbenzene, and xylenes: An in situ mesocosm study. *Applied and Environmental Microbiology.* 71(7): 3815–3825.

Heron, G, R. S. Baker, J. M. Bierschenk, and J. C. LaChance. 2006. Heat it all the way—Mechanisms and results achieved using in-situ thermal remediation. *Proceedings of the Fifth International Conference on Remediation of Chlorinated and Recalcitrant Compounds*, May, Monterey, CA.

Heron, G. 2017. In situ thermal remediation. Available at: www.environmentalrestoration.wiki

Hoag, G. E., P. V. Chheda, B. A. Woody, and G. M. Dobbs. 2000. *Chemical Oxidation of Volatile Organic Compounds*, Patent No. 6,019,548, February 1, 2000.

Hoigne, J. and H. Bader. 1983. Rate constants of reactions of ozone with organic and inorganic compounds in water—II, dissociating organic compounds. *Water Research*. 17: 185–194.

Hong, A., M. Zappi, C. Kuo, and D. Hill. 1996. Modeling the kinetics of illuminated and dark advanced oxidation processes. *Environmental Engineering*. 122(1): 58–62.

Interstate Technology & Regulatory Council (ITRC). 2003. *Technical and Regulatory Guidance Document for Constructed Treatment Wetlands*. Washington, DC: Interstate Technology & Regulatory Council, Phytotechnologies Team, Tech Reg Update. Available at: www.itrcweb.org

ITRC. 2005a. *Technical and Regulatory Guidance for In Situ Chemical Oxidation of Contaminated Soil and Groundwater*, 2nd ed. ISCO-2. Washington, DC: Interstate Technology & Regulatory Council, In Situ Chemical Oxidation Team. Available at: www.itrcweb.org

ITRC. 2005b. *Perchlorate: Overview of Issues, Status, and Remedial Options*. Washington, DC: Interstate Technology & Regulatory Council, Perchlorate Team. Available at: www.itrcweb.org

ITRC. 2009. *Phytotechnology Technical and Regulatory Guidance and Decision Trees, Revised*. PHYTO-3. Washington, DC: Interstate Technology & Regulatory Council, Phytotechnologies Team, Tech Reg Update. Available at: www.itrcweb.org

ITRC. 2011. *Permeable Reactive Barrier: Technology Update. PRB-5*. Washington, DC: Interstate Technology & Regulatory Council, PRB: Technology Update Team. Available at: www.itrcweb.org

Jawitz, J. W., R. K. Sillan, M. D. Annable, P. S. C. Rao, and K. Warner. 2000. In-situ alcohol flushing of a DNAPL source zone at a dry cleaner site. *Environmental Science & Technology*. 34(17): 3722–3729.

Kanel, S. R., J. M. Greneche, and H. Choi. 2006. Arsenic(V) removal from groundwater using nano-scale zero-valent iron as a colloidal reactive barrier material. *Environmental Science & Technology*. 40: 2045–2050.

Kennedy, L. G., J. W. Everett, E. Becvar, and D. DeFeo. 2006a. Field-scale demonstration of induced biogeochemical reductive dechlorination at Dover Air Force Base, Dover, Delaware. *Journal of Contaminant Hydrology*. 88(1-2): 119–136.

Kennedy, L. G., J. W. Everett, and J. Gonzales. 2006b. Assessment of biogeochemical natural attenuation and treatment of chlorinated solvents, Altus Air Force Base, Altus, Oklahoma. *Journal of Contaminant Hydrology*. 83: 221–236.

Kolthoff, I. M. and I. K. Miller. 1951. The chemistry of persulfate. I. The kinetics and mechanism of the decomposition of the persulfate ion in aqueous medium. *Journal of the American Chemical Society*. 73(7): 3055–3059.

Krug, T., S. O'Hara, M. Watling, and J. Quinn. 2010. *Final Report: Emulsified Zero-Valent Nano-Scale Iron Treatment of Chlorinated Solvent DNAPL Source Areas*, ESTCP Project CU-0431 763 p.

Kuo, C. and S. Chen. 1996. Ozonation and peroxone oxidation of toluene in aqueous solutions. *Industrial and Engineering Chemistry Research*. 35(11): 1206–1213.

Langlais, B., D. Reckhow, and D. Brink. 1991. *Ozone in Water Treatment*. Chelsea, MI: Lewis.

Lee, L. S., X. Zhai, and J. Lee. 2007. *INDOT Guidance Document for In-Situ Soil Flushing*. Publication FHWA/IN/JTRP-2006/28-2. Joint Transportation Research Program, Indiana Department of Transportation and Purdue University, West Lafayette, Indiana.

Liang, C., C. J. Bruell, M. C. Marley, and K. Sperry. 2001. Kinetics of thermally activated persulfate oxidation of trichloroethylene (TCE) and 1,1,1-trichloroethane (TCA). *The First International Conference on Oxidation and Reduction Technologies for In-Situ Treatment of Soil and Groundwater*, June 25–29, Niagara Falls, Ontario, Canada.

Liang, C., C. J. Bruell, M. C. Marley, and K. L. Sperry. 2003. Thermally activated persulfate oxidation of TCE and 1,1,1-TCA in aqueous systems and soil slurries. *Soil and Sediment Contamination*. 12(2): 207–228.

Liang, C., C. J. Bruell, M. C. Marley, and K. L. Sperry. 2004a. Persulfate oxidation for in situ remediation of TCE: I. Activated by ferrous ion with and without a persulfate-thiosulfate redox couple. *Chemosphere*. 55(9): 1225–1233.

Liang, C., C. J. Bruell, M. C. Marley, and K. L. Sperry. 2004b. Persulfate oxidation for in situ remediation of TCE: II. Activated by chelated ferrous ion. *Chemosphere*. 55(9): 1225–1233.

Lister, M. W. and R. C. Garvie. 1959. Sodium dithionite, decomposition in aqueous solution and in the solid state. *Canadian Journal of Chemistry*. 37: 1567–1574.

Löffler, F., K. M. Ritalahti, and S. H. Zinder. 2013. Dehalococcoides and reductive dechlorination of chlorinated solvents. In H. Stroo et al. (Eds.), *Bioaugmentation for Groundwater Remediation*, SERDP/ESTCP Remediation Technology Monograph Series. New York, NY: Springer Science & Business Media. pp. 39–88. 361 p.

Lovley, D. R. and J. D. Coates. 1997. *Bioremediation of metal contamination. Biotechnology.* 8: 285–289.

Manning, B. A., M. Hunt, C. Amrhein, and J. A. Yarmoff. 2002. Arsenic(III) and arsenic(V) reactions with zerovalent iron corrosion product. *Environmental Science & Technology.* 36: 5455–5461.

Maymo-Gatell, X., Y. Chien, J. M. Gossett, and S. H. Zinder. 1997. Isolation of a bacterium that reductively dechlorinates tetrachloroethene to ethene. *Science.* 276(5318): 1568–1571.

National Research Council (NRC). 2012. *Alternatives for Managing the Nation's Complex Contaminated Groundwater Sites.* The National Academies Press, ISBN 978-0-309-27874-4 I DOI: 10.17226/14668.

Naval Facilities Engineering Command (NAVFAC). 2015. *Design Considerations for In Situ Chemical Oxidation*, Engineering and Expeditionary Warfare Center, TM-NAVFAC-EXWC-EV-1502.

Nelson, C. and R. Brown. 1994. Adapting ozonation for soil and groundwater cleanup. *Chemical Engineering* (November): EE18–EE22.

Nimmer, M., B. Wayner, and A. Morr. 2000. In situ ozonation of contaminated groundwater. *Environmental Progress.* 19(3): 183–196.

Okpokwasili, G. C and C. O. Nweke. 2005. Microbial growth and substrate utilization kinetics. *African Journal of Biotechnology.* 5(4): 305–317.

Pennell, K. D., G. A. Pope, and L. M. Abriola. 1996. Influence of viscous and buoyancy forces on the mobilization of residual tetrachloroethylene during surfactant flushing. *Environmental Science & Technology.* 30(4): 1328–1335.

Petersen, S. and K. Hedquist. 2006. *Treatability Test Report for Calcium Polysulfide in the 100-K Area*, DOE, DOE/RL-2006-17, 138 p.

Qiu, Y., M. Zappi, C. Kuo, and E. Fleming. 1999. A kinetic and mechanistic study of the ozonation of dichlorophenols in aqueous solutions. *Journal of Environmental Engineering.* 125(5): 441–450.

Rao, P. S. C., M. D. Annable, R. K. Sillan, D. Dai, K. Hatfield, W. D. Graham, A. L. Wood, and C. G. Enfield. 1997. Field-scale evaluation of in situ cosolvent flushing for enhanced aquifer remediation. *Water Resources Research.* 33: 2673–2686.

Rieger, P.-G. and H. J. Knackmuss. 1995. Basic knowledge and perspectives on biodegradation of 2,4,6-trinitrotoluene and related nitroaromatic compounds in contaminated soil. In J. C. Spain (Ed.), *Biodegradation of Nitroaromatic Compounds.* New York: Plenum Press. pp. 1–18.

Rosen, M. J. 1989. *Surfactants and Interfacial Phenomena.* New York: John Wiley & Sons. Roundtable, F. R. T. "Abstracts of Remediation Case Studies."

Sabatini, D. A., R. C. Knox, and J. H. Harwell. 1996. Surfactant-enhanced DNAPL remediation: Surfactant selection, hydraulic efficiency, and economic factors. Prepared for U.S. Environmental Protection Agency.

Salinero, K. K., K. Keller, W. S. Feil, H. Feil, S. Trong, G. Di Bartolo, and A. Lapidus. 2009. Metabolic analysis of the soil microbe Dechloromonas aromatic str. RCB: indications of a surprisingly complex lifestyle and cryptic anaerobic pathways for aromatic degradation. *BMC Genomics.* 10: 351.

Schwarzenbach, R. P., P. M. Gschwend, and D. M. Imboden. 2002. *Environmental Organic Chemistry.* John Wiley & Sons, Hoboken, NJ.

Seech, A., K. Bolanos-Shaw, D. Hill, and J. Molin. 2008. In situ bioremediation of pesticides in soil and groundwater. *Remediation.* Winter Edition 19(1): 87–98.

Siegrist, R. L., M. A. Urynowicz, O. R. West, M. L. Crimi, and K. S. Lowe (eds.). 2001. *Guidance for In Situ Chemical Oxidation at Contaminated Sites: Technology Overview with a Focus on Permanganate Systems.* Columbus, OH: Battelle Press.

Smatlak, C. R., J. M. Gossett, and S. H. Zinder. 1996. Comparative kinetics of hydrogen utilization for reductive dechlorination of tetrachloroethene and methanogenesis in an anaerobic enrichment culture. *Environmental Science & Technology.* 30(9): 2850–2858.

Smith, G., D. Fleming, V. Jurka, and T. Adams. 2000. Closure of a trichloroethene and 1,1,1- trichloroethane DNAPL remediation using thermal technologies. *Proceedings of the Second International Conference on Remediation of Chlorinated and Recalcitrant Compounds*, Monterey, CA.

Strategic Environmental Research and Development Program (SERDP). 2007. Sustainability of long-term abiotic attenuation of chlorinated ethenes. SERDP Project ER-1369. September.

U.S. Army Corps of Engineers (USACE). 2014. *Design: In Situ Thermal Remediation*, Engineer Manual, May, EM 200-1-21. Available at: www.publications.usace.army.mil

U.S. Environmental Protection Agency (USEPA). 1998. *Technical Protocol for Evaluating Natural Attenuation of Chlorinated Solvents in Ground Water*, Office of Research and Development, EPA/600/R-98/128.

USEPA. 1999. *Use of Monitored Natural Attenuation at Superfund, RCRA Corrective Action, and Underground Storage Tank Sites*, Office of Solid Waste and Emergency Response, Directive 9200.4-17P.

USEPA. 2000. *In Situ Treatment of Soil and Groundwater Contaminated with Chromium*, EPA 625/R-00/005, 98 p.

USEPA. 2003. *Cost and Performance Summary Report in Situ Chemical Reduction at the Marshall Space Flight Center, Source Area 2*. Huntsville, Alabama: Office of Solid Waste and Emergency Response Technology Innovation Office. Available at: https://frtr.gov/costperformance/

USEPA. 2009. *Identification and Characterization Methods for Reactive Minerals Responsible for Natural Attenuation of Chlorinated Organic Compounds in Ground Water*. Ada, Oklahoma Office of Research and Development, National Risk Management Research Laboratory, EPA 600/R-09/115, December.

USEPA. 2011a. *Five Year Review: Coast Wood Preserving*. Ukiah, CA.

USEPA. 2011b. *Research Snapshot: Reducing Subsurface Hexavalent Chromium to Harmless Trivalent Chromium through Injection of Ferrous Iron*.

USEPA. 2013. *Introduction to In Situ Bioremediation of Groundwater*, EPA-542-R-13-018, Office of Solid Waste and Emergency Response.

USEPA. 2014. *Engineering Paper: In Situ Thermal Treatment Technologies: Lessons Learned*, May, EPA 542-R-14-012. Available at: www.epa.gov/superfund

Vidali, M. 2001. Bioremediation. An overview. *Pure and Applied Chemistry*. 73(7): 1163–1172.

West, C. C. and J. H. Harwell. 1992. Surfactants and subsurface remediation. *Environmental Science & Technology*. 26(12): 2324–2330.

Yalkowsky, S. H. and T. J. Roseman. 1981. Solubilization of drugs by cosolvents. In S. H. Yalkowsky (Ed.), *Techniques of Solubilization of Drugs*. New York, NY: Dekker. pp. 91–134.

Yang, J. E., J. S. Kim, Y. S. Ok, and K. R. Yoo. 2007. Mechanistic evidence and efficiency of the Cr(VI) reduction in water by different sources of zero-valent irons. *Water Science and Technology*. 55(1–2): 197–202.

Zawislanski, P., J. Horst, J. Gillow, and D. Liles. 2010. Challenges associated with arsenic remediation in media affected by multiple metals, Seventh International Conference on Remediation of Chlorinated and Recalcitrant Compounds, May, Monterey, CA.

Zhang, B. 2010. *Bioremediation of Creosote-Treated Wood Waste*. Master Thesis, Dalhousie University, Halifax, Nova Scotia. October.

9 Nanotechnology in Site Remediation

Zhong Xiong and Jun Lu

CONTENTS

9.1 INTRODUCTION

Nanotechnology is defined by the National Nanotechnology Initiative as understanding and controlling matter at dimensions between approximately 1 and 100 nm, where unique phenomena enable novel applications. A nanometer is one billionth of a meter (10^{-9} m), approximately one hundred thousand times smaller than the diameter of a human hair, a thousand times smaller than a red blood cell, or about half the size of the diameter of DNA. Nanoscale materials that have macroscale counterparts frequently display different or enhanced properties (such as electrical, catalytic, magnetic, mechanical, thermal, or imaging features) compared with the macroscale form.

Nanoscale materials are being used in a variety of applications within the scientific, environmental, industrial, and medical arenas. And over the past two decades, an increasing variety of nanoscale materials with environmental applications have been developed. Nanotechnology is of interest for environmental applications, such as contaminated site remediation, because of the large specific surface area of nanoscale materials, as a result of which their reactivity in chemical or biological surface-mediated reactions can be greatly enhanced in comparison with the same material at larger sizes. Early treatment remedies for groundwater were primarily pump-and-treat operations. Due to the lengthy operating periods for these remedies and the associated high costs, the use of *in situ* treatment technologies is increasing. One example of *in situ* groundwater treatment technologies is the installation of a trench filled with macroscale zero-valent iron, that is, ZVI or Fe(0) or Fe^0, which has demonstrated effectiveness for treating chlorinated solvents and heavy metals. Recent research indicates that nanoscale zero-valent iron (nZVI) may prove more effective and less costly than macroscale ZVI under similar environmental conditions. For example, nZVI has a much larger specific surface area compared with macroscale ZVI, and laboratory tests have shown that nZVI can degrade trichloroethene (TCE), a common groundwater contaminant at Superfund sites, more rapidly and completely than larger ZVI particles. Also, nZVI can be injected directly into the subsurface where contaminants reside, therefore eliminating the need to dig a trench and related costs (USEPA, 2008).

Most of the bench-scale research and field applications of nanotechnology for contaminated site remediation have focused on nZVI and related products. The reminder of this chapter focuses on the nZVI technology.

9.2 nZVI FOR SITE REMEDIATION

nZVI is a relatively new groundwater remediation material that has received significant attention over the past decade in research and field applications (Wang and Zhang, 1997; Gavaskar et al., 2005; He and Zhao, 2005; Liu et al., 2005; Saleh et al., 2005; Xiong et al., 2007), largely because of its potential for rapid destruction of recalcitrant compounds, such as TCE and tetrachloroethene (PCE). Laboratory studies have shown that nZVI particles typically have surface areas up to 30 times greater than larger-sized granular iron particles and are up to 1000 times more reactive for degrading chlorinated solvents (Zhang, 2003). Coating nZVI with a small quantity of palladium as a catalyst to form Fe-Pd bimetallic nanoparticles enhances the rates of nZVI reaction and accelerates the dechlorination process, and thereby limits the formation of cis-1,2-dichloroethene (cDCE) and vinyl chloride (VC) from TCE and PCE reduction (Wang and Zhang, 1997; He and Zhao, 2005).

9.2.1 Development History of nZVI

The study of ZVI for environmental applications started in the 1990s, when the effectiveness of ZVI for the degradation of chlorinated ethane, ethene, and methane was discovered. In addition, ZVI was found to be effective for the remediation of a variety of priority pollutants, including metals, polychlorinated biphenyls (PCBs), chlorinated pesticides, nitro aromatic compounds, and nitrates. This research was followed by the application of ZVI for groundwater remediation, primarily in the form of a permeable reactive barrier (PRB), in which a trench is installed in an alignment perpendicular to the groundwater flow direction, and the trench is filled with ZVI as a reactive medium. When the contaminated groundwater flows through the PRB, ZVI reacts with the contaminants, and the groundwater is treated. PRB is an effective remediation technology, and hundreds of PRBs have been installed. However, PRB is a passive remedy that is not designed to target contaminant sources and has limited application due to the achievable maximum installation depth.

A breakthrough was reported by Wang and Zhang (1997), who developed a method to prepare nanoscale ZVI. nZVI provides several advantages, as detailed in the following sections, and immediately attracted much attention from academia and professionals in the site remediation field. Follow-up research has demonstrated that nZVI exhibits much higher specific surface area and greater reaction rate than larger-size ZVI. A secondary metal, such as palladium or nickel, can be added to the nZVI as a catalyst to form bimetallic nanoparticles, which have been shown to be more effective for contaminant degradation.

9.2.2 Preparation Methods

As with many other nanotechnologies, nZVI may be produced via a "top down" approach, such as the milling or grinding of macroscale material, or, most commonly, via a "bottom up" approach, such as a wet chemistry synthesis method that creates nZVI materials from component atoms or molecules (EPA, 2008).

With the "top down" approach, nZVI can be prepared by grinding large iron or steel particles to smaller-size particles using a mill or a grinder. One reported method involves adding iron or steel beads that have a diameter of 100–250 μm to a device called an *attritor.* Such a device is a stirred-media mill in which the grinding media (iron or steel beads) are moved by a series of rods. Shear fields and impacts are generated by the motion of the rotating rods, which lead to more efficient power transmission, finer grinding, and a shorter grinding time. The ZVI particles produced using this method have a diameter of less than 100 nm.

With the "bottom up" approach, nZVI can be synthesized by a number of methods, including the gas phase reduction method, the liquid phase reduction method, the sonochemical method, and the electrochemical method. Among these, gas phase reduction and liquid phase reduction are the most common methods for synthesizing nZVI for remediation purposes. Reactive nanoscale iron particles (RNIP), commercially available from Toda Kogyo Corp., are produced by reducing goethite and hematite particles at high temperatures with hydrogen gas. The reduction of ferrous iron by sodium borohydride in a liquid solution is the most common way of synthesizing nZVI particles.

9.2.3 Structure of nZVI

nZVI particles typically exhibit a typical core-shell structure, as illustrated in Figure 9.1. The core is composed primarily of ZVI. As the ZVI reacts with water or dissolved oxygen in the aqueous phase, a layer of iron (hydr)oxide forms outside the ZVI core. The composition of the actual iron oxide layer is complicated due to the lack of long-range order and its amorphous structures. The exact composition of the iron oxide shell may also depend on the preparation method and environmental conditions. However, it has been reported that the iron oxide shell consists of maghemite (γ-Fe_2O_3), partially oxidized magnetite (Fe_3O_4), wustite (FeO), or a combination of these compounds. The thin and distorted oxide layer allows electron transfer from the ZVI core through three different mechanisms: (1) directly through defects such as pits and pinholes on the shell; (2) indirectly via the oxide conduction band, impurity, or localized band; (3) from sorbed or structural Fe^{2+}, thus sustaining the capacity of the particles for the reduction of contaminants (O'Carroll et al., 2013). The outer (hydr) oxide layer may also act as an efficient adsorbent for various contaminants, such as heavy metals.

nZVI particles are mostly spherical in shape. Figure 9.2 shows the transmission electron microscopy (TEM) images of nZVI synthesized using the sodium borohydride method in the presence of a stabilizer, sodium carboxymethyl cellulose (CMC). The stabilized nZVI particles appear as discrete nanoparticles and remain dispersed in water solution. Analysis of the TEM images similar to the one shown on Figure 9.2 indicated that the stabilized nZVI particles had a particle size of around 15 nm, and the surface area of the nZVI particles was approximately 55 m^2/g (Xiong et al., 2007).

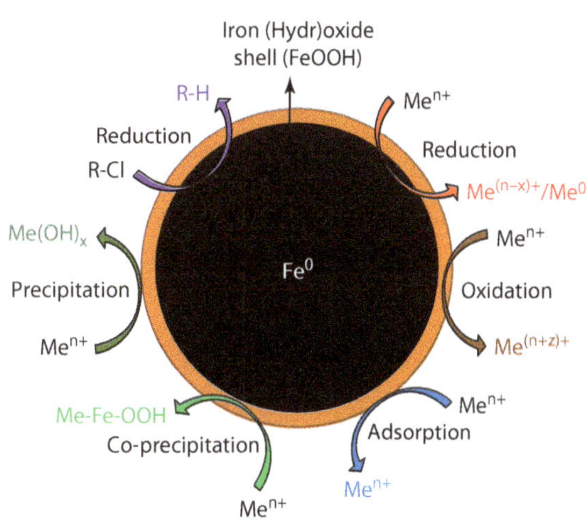

FIGURE 9.1 Core-shell structure of nZVI depicting various mechanisms for the removal of chlorinated compounds and metals. (From O'Carroll, D., et al., *Advances in Water Resources,* 51, 104–122, 2013.)

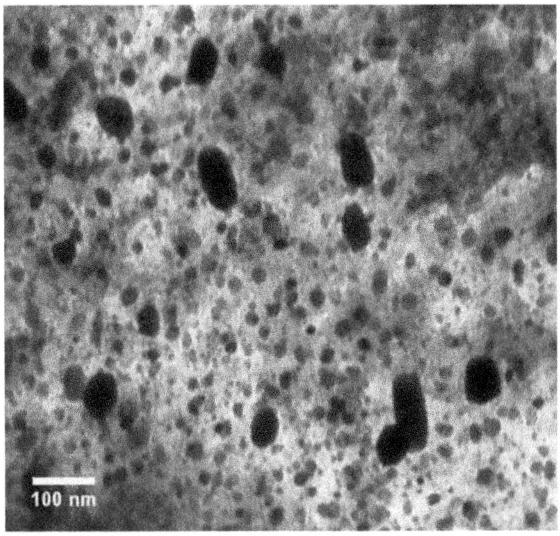

FIGURE 9.2 TEM image of nZVI particles prepared in the presence of a stabilizer. (From Xiong, Z., et al., *Water Research*, 41, 3497–3505, 2007.)

9.2.4 Advantages of nZVI

The advantages of nZVI compared to larger size of ZVI are threefold: (1) larger specific surface area or surface area to mass ratio; (2) faster reaction rate; and (3) mobility in the subsurface.

The specific surface area is defined as the surface area per unit of mass. It has a unit of square meters per gram (m^2/g). Assuming the ZVI particles are in a spherical shape, the surface area and the mass of a ZVI particle can be described with the following equations:

$$\text{Surface area} = \pi \cdot d^2 \tag{9.1}$$

$$\text{Mass} = \rho \cdot \frac{\pi \cdot d^3}{6} \tag{9.2}$$

then

$$\text{Specific surface area} = \left(\pi \cdot d^2\right) \bigg/ \left(\rho \cdot \frac{\pi \cdot d^3}{6}\right) = \frac{6}{\rho \cdot d} \tag{9.3}$$

where:
- d is the diameter of ZVI particles (m)
- ρ is the density of ZVI (g/m^3)

As shown in Equation 9.3, the specific surface area of ZVI particles is inversely proportional to the particle diameter. The smaller the particle, the higher is the specific surface area. It is reported that the specific surface area of nZVI is up to 40 m^2/g (He and Zhao, 2005; Sakulchaicharoen et al., 2010), while micron-scale ZVI has a specific surface area on the order of less than 1 m^2/g (Wang and Zhang, 1997; Matheson and Tratnyek, 1994).

ZVI-related reactions occur in the surface of ZVI, and small particle size results in a larger number of reactive sites on ZVI particles, thus increasing the removal rate of nZVI. For example, the kinetics of halogenated organic compound degradation by iron metal can be described with the following equation (Johnson et al., 1996):

$$\frac{dC}{dt} = -k_{SA} \cdot a_s \cdot C_{ZVI} \cdot C \tag{9.4}$$

where:

C is the concentration of contaminant (g/L)
k_{SA} is the rate constant (m/h)
a_s is the specific surface area (m^2/g)
C_{ZVI} is the concentration of zero-valent iron (g/m^3)
t is the reaction time (h)

As shown in Equation 9.4, the degradation rate of a halogenated organic compound (e.g., TCE) is proportional to the specific surface area. High contaminant degradation rates can be achieved by increasing the specific surface area. Various laboratory studies have proved that nZVI exhibits higher contaminant degradation rates for various contaminants compared with micron-scale ZVI.

Another expected advantage of nZVI is that the particles are so small that they are likely to be mobile in the aquifer system and can be easily injected into the subsurface (Wang and Zhang, 1997). Therefore, groundwater could be treated *in situ* by direct injection, converting chlorinated ethenes to environmentally benign reaction products, such as ethane, ethane, and chloride (Lien and Zhang, 2001; Liu et al., 2005). In the absence of an effective stabilizer, nZVI often aggregates rapidly (within a few minutes) in water due to Van der Waal forces and magnetic attraction, resulting in micron- or even millimeter-scale aggregates. The aggregated particles are not transportable through the aquifer, because they are too large, and thus, they are difficult to deliver to the targeted source zones. Saleh et al. (2008) reported that non-stabilized nZVI is essentially immobile in porous media. This process has been mitigated by coating nanoparticles with a negatively charged polymer with large molecular weight to stabilize the nanoparticles through electrostatic and electrosteric repulsion (Saleh et al., 2008; He et al., 2007).

9.3 REACTION MECHANISMS WITH nZVI

Laboratory studies have established that nZVI is effective in treating a variety of common environmental contaminants, including chlorinated organic compounds and heavy metal ions (Lien, 2000; Lien and Zhang, 1999, 2001; Zhang et al., 1998). Examples of contaminants that can be treated by nZVI are given in Table 9.1 (Zhang, 2003). nZVI can react with halogenated hydrocarbon as a reductant, while it functions as a reductant and sorbent for heavy metals. Laboratory evidence also indicates that iron-based materials have been successful in the transformation of other contaminants, such as anions (e.g., NO_3^-, $Cr_2O_7^{2-}$) and radionuclides (e.g., UO_2^{2+}).

9.3.1 Reaction of Chlorinated Solvents with nZVI

Iron is a strong reductant with a reduction potential of −0.44 V and can reduce chlorinated compounds by reductive dehalogenation while it is oxidized:

$$Fe^0 \rightarrow Fe^{2+} + 2e^- \tag{9.5}$$

There are two main pathways involved in reductive dehalogenation: hydrogenolysis and β-elimination (Arnold and Roberts, 2000). Hydrogenolysis of chlorinated compounds, such as trichloroethene, involves the replacement of a chlorine atom by a hydrogen atom and requires an electron donor as well as a proton donor (hydrogen). The general equation for reduction of chlorinated compounds by hydrogenolysis is (O'Carroll et al., 2013)

$$ClHC = CCl_2 + 2e^- + H^+ \rightarrow ClHC = CHCl + Cl^- \tag{9.6}$$

TABLE 9.1
Common Environmental Contaminants that can be Transformed by nZVI

Chlorinated methanes
- Carbon tetrachloride (CCl_4)
- Chloroform ($CHCl_3$)
- Dichloromethane (CH_2Cl_2)
- Chloromethane (CH_3Cl)

Chlorinated benzenes
- Hexachlorobenzene (C_6Cl_6)
- Pentachlorobenzene (C_6HCl_5)
- Tetrachlorobenzenes ($C_6H_2Cl_4$)
- Trichlorobenzenes ($C_6H_3Cl_3$)
- Dichlorobenzenes ($C_6H_4Cl_2$)
- Chlorobenzene (C_6H_5Cl)

Pesticides
- DDT ($C_{14}H_9Cl_5$)
- Lindane ($C_6H_6Cl_6$)

Organic dyes
- Orange II ($C_{16}H_{11}N_2NaO_4S$)
- Chrysoidine ($C_{12}H_{13}ClN_4$)
- Tropaeolin O ($C_{12}H_9N_2NaO_5S$)
- Acid Orange
- Acid Red

Heavy metal ions
- Mercury (Hg^{2+})
- Nickel (Ni^{2+})
- Silver (Ag^+)
- Cadmium (Cd^{2+})

Trihalomethanes
- Bromoform ($CHBr_3$)
- Dibromochloromethane ($CHBr_2Cl$)
- Dichlorobromomethane ($CHBrCl_2$)

Chlorinated ethenes
- Tetrachloroethene (C_2Cl_4)
- Trichloroethene (C_2HCl_3)
- *cis*-Dichloroethene ($C_2H_2Cl_2$)
- *trans*-Dichloroethene ($C_2H_2Cl_2$)
- 1,1-Dichloroethene ($C_2H_2Cl_2$)
- Vinyl chloride (C_2H_3Cl)

Other polychlorinated hydrocarbons
- PCBs
- Dioxins
- Pentachlorophenol (C_6HCl_5O)

Other organic contaminants
- N-nitrosodimethylamine (NDMA) ($C_4H_{10}N_2O$)
- TNT ($C_7H_5N_3O_6$)

Inorganic anions
- Dichromate ($Cr_2O_7^{2-}$)
- Arsenic (AsO_4^{3-})
- Perchlorate (ClO_4^-)
- Nitrate (NO_3^-)

Source: Zhang, W., *Journal of Nanoparticle Research*, 5, 323–332, 2003.

In β-elimination, chlorine atoms are released by the chlorinated compound without addition of hydrogen, resulting in a decrease in the degree of saturation of the carbon-carbon bond. An example of β-elimination of TCE to chloroacetylene is shown as follows:

$$ClHC = CCl_2 + 2e^- \rightarrow HC \equiv CHCl + 2Cl^- \quad (9.7)$$

One of the significant concerns of any remediation technology is the formation of daughter products, which can be more toxic than the parent compounds. For example, TCE can be reduced through hydrogenolysis to cis-1,2-dichloroethene (cDCE) and vinyl chloride (VC), which are more toxic than TCE itself. Although hydrogenolysis can be an important dehalogenation pathway for many chlorinated solvents, the formation of VC following the hydrogenolysis of PCE, TCE, and cDCE is undesirable. In contrast, VC is not formed in the β-elimination of TCE.

9.3.2 Reaction of Heavy Metals with nZVI

The environmental characteristics (e.g., mobility, toxicity, and solubility) of heavy metals are governed by redox reactions, precipitation/dissolution reactions, and adsorption/desorption phenomena. Remediation strategies for heavy metal contaminants typically involve manipulating these mechanisms to control the availability and toxicity of metal contaminants to environmental receptors. The reaction mechanisms involved in heavy metals and ZVI depend on the standard redox potential (E^0) of the heavy metal contaminant (O'Carroll, 2013). Metals that have an E^0 more negative than, or similar to, that of Fe^0 (e.g., Cd and Zn) are removed purely by adsorption to the iron (hydr)oxide that is formed outside the ZVI particles. Metals with E^0 much more positive than Fe^0 (e.g., Cr, As, Cu, U, and Se) are preferentially removed by reduction and precipitation. Metals with slightly more positive E^0 than Fe^0 (e.g., Pb and Ni) can be removed by both reduction and adsorption. Oxidation and co-precipitation by iron oxides are the other possible reaction mechanisms, depending on the prevailing geochemical conditions.

9.4 APPLICATION METHODS

nZVI is typically applied in the field as an *in situ* technology. The method of application is usually site specific and is dependent on the type of geology and hydrogeology in the treatment zone and the form in which the nZVI is injected. The most direct route of injection uses existing monitoring wells, piezometers, or injection wells. Recirculation is a technique that involves injecting remediation reagents such as nZVI in hydraulically upgradient wells while extracting groundwater from downgradient wells. The extracted groundwater is mixed with additional remediation reagents and reinjected into the injection wells. A hydraulic gradient is established between the injection well and the extraction well, which enhances the delivery and transport of nZVI. The wells also keep the water in the aquifer in contact with nZVI and prevent the larger iron particles from settling out, allowing continuous contact with the contaminant.

Other application methods and processes include direct push, pressure-pulse technology, liquid atomization injection, pneumatic fracturing, and hydraulic fracturing. The direct-push method involves driving direct push rods, similar to small drilling augers, and injection tools progressively deeper into the ground. This method allows materials to be injected without having to install permanent injection wells. Pressure-pulse technology uses large-amplitude pulses of pressure to insert the nZVI mixture into porous media at the water table; the pressure then excites the media and increases fluid level and flow. Liquid atomization injection involves injecting an nZVI-fluid mixture into the subsurface using a carrier gas. The nZVI liquid-gas combination aerosolizes, allowing more effective distribution; this method is reported to be used in lower-permeability geologic formations. Fracturing injection (pneumatic or hydraulic) is a high-pressure injection technique using compressed air (pneumatic) or a water-based, highly viscous slurry containing sand (hydraulic) that fractures rock and allows liquids and vapors to be transported quickly through the channels created.

9.5 FIELD APPLICATIONS OF nZVI FOR SITE REMEDIATION

The U. S. Environmental Protection Agency (USEPA) compiled a list of contaminated sites using or testing nanoparticles for remediation at pilot or full scale. The list is periodically updated, and details of the sites are available at https://clu-in.org/products/nanozvi/. As of November 2011, there were about 70 sites that were using or testing nanoparticles. The majority of these sites used nZVI, while others used bimetallic nanoparticles, emulsified zero-valent iron (EZVI), and nanoscale calcium ions with a noble metal catalyst. The most frequently treated contaminants of concern were chlorinated solvents, such as TCE, PCE, TCA, and VC. Details of three example projects are provided in the following list.

1. *Use of nanoscale bimetallic particles at a former manufacturing site in Trenton, New Jersey*: This was one of the earliest field applications of nanoparticles for site remediation. A "proof of concept" field demonstration was performed, in which nanoscale bimetallic (iron-palladium; Fe-Pd) particles were delivered to the subsurface through gravity feed to treat groundwater contaminated with chlorinated compounds. The contaminated site

selected for the pilot-scale test is an active manufacturing facility, and the test area is located within a relatively well-characterized plume downgradient of a former aboveground trichloroethene (TCE) storage tank. The test area includes one injection well and three nested pairs of piezometers at 1.5-m intervals downgradient of the injection well along the principal direction of groundwater flow. The injection system included two 625-L polyethylene storage tanks, one of which was used to feed the nanoparticle suspension while the other was on standby or was used to store recirculation volume. The feed tank was fitted with a mechanical agitator and a nitrogen gas feed system to maintain low dissolved oxygen conditions in the tank. Recirculation was accomplished by pumping groundwater from two piezometers into the storage tank or the feed tank. The bimetallic nanoparticles were prepared using the sodium borohydride reduction method, and the palladium to iron ratio was controlled at less than 1:300 (w/w). Approximately 1.7 kg of the nanoparticles was fed into the test area over a 2-day period, resulting in minimal clogging of the injection well. Despite the low nanoparticle dosage, TCE reduction efficiencies of up to 96% were observed over a 4-week monitoring period, with the highest reduction values observed at the injection well and adjacent piezometers (Elliott and Zhang, 2001).

2. *Use of nZVI at a former manufacturing facility in Alabama*: A pilot test was conducted at a former manufacturing facility in Alabama in which carboxymethyl cellulose (CMC)–stabilized ZVI nanoparticles (with a trace amount of Pd catalyst) were used for *in situ* destruction of chlorinated ethenes, such as PCE and TCE, and polychlorinated biphenyls (PCBs), which had been in the groundwater for decades. The test area was located in a well-characterized secondary source zone of PCBs and chlorinated ethenes. Four test wells were installed along the groundwater flow direction (spaced 5 ft apart), including one injection well, one upgradient monitoring well, and two downgradient monitoring wells. A stabilized nanoparticle suspension was prepared on-site and injected into the 50-ft deep, unconfined aquifer in two phases. During the first phase, approximately 150 gallons of 0.2 g/L Fe-Pd nanoparticle suspension was gravity-fed through the injection well over a 4-h period. One month later, during the second phase, another 150 gallons of 1.0 g/L Fe-Pd nanoparticle suspension was injected into the same injection well at a well head injection pressure <5 psi. Post-injection groundwater monitoring results showed that rapid degradation of PCE and TCE was observed in downgradient monitoring wells following each injection, with the maximum degradation being observed during the first week of the injections. The chlorinated ethene concentrations gradually returned to their pre-injection levels after 2 weeks, indicating exhaustion of the ZVI's reducing power. However, the injection of CMC-stabilized nanoparticles and the abiotic reductive dechlorination process appeared to have boosted a long-term *in situ* biological dechlorination thereafter, which was evidenced by the fact that PCE and TCE concentrations showed further reduction after 2 weeks. After 596 days from the first injection, the total chlorinated ethene concentration decreased by up to 61% in the downgradient monitoring wells, and a PCB 1242 reduction of up to 87% was observed in a monitoring well. During the 596 days of testing, the total concentrations of cis-DCE (dichloroethylene) and VC (vinyl chloride) decreased by 20% and 38%, respectively (He et al., 2010).
3. *Application of nZVI at Operable Unit 2 in Hill Air Force Base, Utah*: A field demonstration of stabilized Fe-Pd nanoparticles for *in situ* destruction of chlorinated solvents in soil and groundwater was performed in Operable Unit 2 in Hill Air Force Base, Utah. The demonstration site is a known hot spot where elevated concentrations of solvents (mainly trichloroethylene [TCE] at concentrations up to 47,400 μg/L) have been measured in monitoring wells. The low-permeability aquifer contains mainly clay and silt with thin sand seams. Five hundred and fifty-five gallons of CMC-stabilized Fe-Pd nanoparticle suspension was prepared on-site using readily available chemicals and equipment to

ensure maximum reactivity. The nanoparticles, with an average hydrodynamic particle size of 178 nm, were demonstrated to be stable; no agglomerates or precipitates were observed. The nanoparticle suspension was successfully injected into the formation using low pressures (6–9 psi) under a flow rate ranging from 0.25 to 0.67 gallons per minute (gpm). The nanoparticles were mobile in the subsurface, traveling a distance at least 5 ft from the injection well. The nanoparticle injection created a highly reducing environment with an oxidation reduction potential (ORP) as low as −606 mV and dissolved oxygen as low as 0.02 mg/L. In addition, groundwater pH temporarily increased by approximately two standard units (s.u.) in the injection well and in a 5-ft downgradient monitoring well. The stabilizer, CMC, increased the total organic carbon (TOC) concentration in groundwater, which stimulated reductive biological dechlorination processes over an area larger than that to which the nanoparticles traveled. VOC degradation was observed in a well 15 ft downgradient and in a side-gradient well, demonstrating a much greater radius of influence than that created by the abiotic chemical reactions produced by the ZVI. A key finding of the demonstration was that both abiotic chemical degradation and reductive biological dechlorination processes were shown to be underway at the demonstration site following nanoparticle injection. The highly reducing environment produced by the ZVI and the increase of TOC concentrations produced by the injection of biodegradable CMC stimulated the biological degradation processes. The nanoparticles completely degraded the chlorinated solvents to non-toxic products, such as ethene and ethane, at the demonstration site. High concentrations (up to 1100 μg/L) of ethane and ethene, generated during the demonstration, support an estimate that 28% of the total volatile organic compound (VOC) mass present in the baseline groundwater and soil was degraded to non-toxic products, that is, ethane and ethene. However, groundwater concentrations of TCE in demonstration wells were reduced by up to 92% 10 months after the last nanoparticle injection (Xiong et al, 2013).

9.6 LIMITATIONS OF nZVI TECHNOLOGY

It has been more than two decades since the first article regarding nZVI as a site remediation technology was published. Much research on nZVI has been published in the literature, and a number of contaminated sites have used nZVI as the selected remedy. However, the nZVI technology has not yet achieved widespread commercial application for site remediation. Among practitioners and environmental regulators, a prevailing view on the nZVI-based remedy is that nZVI is less competitive in terms of its cost-effectiveness compared with other alternative remedies or products. This perception is largely based on field observations that nZVI has not shown superior treatment effectiveness over other *in situ* amendments that would justify its significantly higher cost. In many cases, other alternatives have shown similar, if not better, treatment performance than nZVI. For example, micron ZVI (mZVI)–based amendments are currently the most widely used *in situ* chemical reduction (ISCR) reagents, which are much cheaper than nZVI and have shown more effective treatment performance when coupled with enhanced biological reductive dechlorination (Fan et al., 2016).

The low cost-effectiveness of nZVI observed in the field appears to contradict the conventional wisdom (obtained from numerous bench studies) that nZVI has much higher reactivity with contaminants than ZVI with larger particle size (Zhang and Elliott, 2006; Tratnyek et al., 2010). This discrepancy is largely due to the effects of different time scales reflected at the bench and field scale. Short-term effects of nZVI are predominant in most bench studies, because the experimental conditions are controlled and optimized to achieve this purpose, whereas field studies reflect the long-term effects more, as contaminant destruction in the subsurface is limited by other rate-limiting processes, such as desorption from soil grains, dissolution of dense non-aqueous phase liquid (DNAPL), or diffusion (Fan et al., 2016).

The main limitations facing the nZVI technology can be summarized as follows:

- *Low selectivity*: nZVI is a strong reductant and can react with both contaminants and natural reductant demand (NRD). NRD quantifies the consumption of the reducing capacity due to reactions with naturally occurring oxidants as opposed to contaminants. In the case of nZVI, the two major components that contribute to NRD are dissolved oxygen and water. Oxidation of Fe(0) by water actually represents the major fraction of NRD, because Fe(0) corrodes in water and water is unlimited. As a result of high NRD, nZVI has shown low selectivity toward reaction with contaminants. For example, Fan et al. (2016) examined the selectivity of nZVI for abiotic dechlorination based on bench-scale laboratory results and found that less than 2% of the reducing capacity is used for abiotic dechlorination. This is expected, because water is much more abundant than contaminants and therefore reacts faster. In addition, the large surface area of nZVI further favors fast reaction with water, leading to less reducing capacity for abiotic contaminant reduction.
- *Limited longevity*: As a result of the small particle size of nZVI, each nZVI particle has a limited amount of Fe(0) and related reducing capacity, which can be consumed by contaminants and NRD in a short period of time. Laboratory and field evidence has indicated that the nZVI can last for a few days to several weeks once it is applied to the subsurface. This is a short period of time, considering it typically takes months to years for the contaminants to desorb from the soil surface or dissolve from a pure phase source such as DNAPL.
- *Limited mobility in the subsurface*: The early development of nZVI suffered a significant challenge, that is, agglomeration of nZVI particles. The nZVI particles tend to stick together to form large agglomerates and then lose the benefits of small particles, such as being mobile in the subsurface. Several field applications of nZVI for site remediation have suffered the challenge of nZVI agglomerates clogging the injection wells. However, with recent developments of nZVI, this challenge has been significantly mitigated by using a stabilizer. Various stabilization reagents, such as carboxymethyl cellulose (CMC), have been found to successfully keep nZVI particles separated from each other and retaining their small size and mobility in the subsurface.
- Difficulties in obtaining representative data for the properties of media that affect the behavior of nanoparticles. The properties include soil matrix, ionic strength of the groundwater, hydraulic properties of the media, presence of organic matter, and other geochemical properties, such as pH, dissolved oxygen, oxidation reduction potential (ORP), and concentrations of competing oxidants. For example, chlorinated solvents at most sites are in fine-grained materials with low permeability, but the site data are primarily collected from permeable zones (aquifers).
- *Concern over nanotoxicity*: There is a growing concern in the regulatory community over the potential toxicity of manufactured nanomaterials. The concern is that the increased surface area and larger number of reactive sites of nanomaterials may equate to greater biological activity per unit mass than micro- or macroscale particles of the same composition. Substances considered non-toxic at macroscale may have negative impacts on human health when nanoscale particles are inhaled, absorbed through skin, or ingested. Because of the minute size of nanomaterials, the particles have the potential to migrate to or accumulate in places that larger particles cannot, such as the alveoli in the lungs, thereby potentially increasing toxicity (USEPA, 2008). The current challenge is that there are no reliable methods developed to monitor the nanoparticles so that the concerns can be addressed with confidence.

9.7 RECENT DEVELOPMENTS IN nZVI RESEARCH AND APPLICATION

To overcome some of the limitations listed in the previous section, different strategies have been explored, and a few of them have become commercially available. The main strategies to improve the selectivity and longevity of nZVI include ways to modify nZVI by various additives or supports. Three promising approaches, that is, emulsified zero-valent iron (EZVI), activated carbon (AC)-supported nZVI, and sulfidation, are described by Fan et al. (2016) and summarized as follows:

- EZVI combines nZVI (or mZVI) with vegetable oil and small amounts of other additives to create oil emulsions that encapsulate ZVI particles (Quinn et al., 2005; Hara et al., 2006). The encapsulation limits ZVI contact with water, which slows down the corrosion reaction of Fe(0). Inhibited corrosion is evidenced by moderately negative ORP values measured in the field, as mentioned earlier. Another advantage proposed for EZVI is that EZVI is more effective in treating the source area of DNAPL because EZVI and DNAPL have similar hydrophobic properties. Compared with other types of nZVI that are primarily applied at the pilot scale, full-scale application of EZVI has been carried out at a number of sites since 2005, because of the favorable treatment performance observed and relatively low costs. However, distinguishing between abiotic and biotic degradation for EZVI is still difficult, but the general consensus is that nZVI mainly serves to maintain the reducing conditions to favor biodegradation. Therefore, the reducing capacity of nZVI may still not be efficiently used for abiotic degradation, although better electron efficiency for TCE might be expected than with unmodified nZVI due to limited corrosion. EZVI represents an example that has already been successfully commercialized and applied in the field.
- The AC-nZVI composite has several variations depending on the particle size of AC. A commercial product called BOS 100™, developed by Remediation Product Inc. (RPI) in 2004, uses granular activated carbon (GAC), while more recently a similar composite was developed using colloidal AC as the support (Bleyl et al., 2012). The former has already been widely applied at contaminated sites, mostly in the form of a permeable reactive barrier, while the latter has shown satisfactory mobility in packed columns and is currently tested in the field within the scope of NanoRem (Busch et al., 2014, 2015). For both materials, Fe(III) oxide was first seeded inside the internal pores of activated carbon and then reduced to Fe(0) by exposing the composite to a reducing atmosphere at high temperature. Although limited information is available regarding the longevity of nZVI for these two specific products, data obtained for similar materials suggested that AC plays favorable roles in inhibiting nZVI reactions with water.
- Compared with the previous two examples, sulfidation is less well known to practitioners. To date, sulfidation has mainly been studied in the laboratory. Sulfidation involves treating nZVI with reducing sulfur compounds, primarily sodium sulfide and sodium dithionite (Xie and Cwiertny, 2010; Fan et al., 2013; Rajajayavel and Ghoshal, 2015; Nunez et al., 2016). These treatments result in the formation of iron sulfide (FeS) on the surface of nZVI due to the high affinity between iron and sulfide. An earlier study showed that doping nZVI with a small amount of aqueous sulfide (S/Fe=0.1) significantly inhibited the corrosion of Fe(0) (Fan et al., 2013). Recent research showed that sulfidation can limit the loss of nZVI reducing capacity due to reduction of water. Because FeS is also known for its ability for abiotic dechlorination, sulfidation does not passivate the reactivity of Fe(0) as much as iron (III) oxides. Therefore, sulfidation provides a promising solution that may selectively favor abiotic contaminant reduction.

9.8 SUMMARY

nZVI technology is an innovative technology for contaminated site remediation that has been demonstrated to be effective for a wide range of contaminants. The unique characteristics of nZVI, such as small particle size and large specific surface area, make it effective for fast reaction with contaminants, as shown in controlled laboratory experiments. Some pilot-scale and full-scale field applications of nZVI for soil and groundwater remediation have also been reported with mixed results. But widespread commercial application has not been achieved for a number of reasons. nZVI has low selectivity toward reactions with contaminants, and as a result, much of the reducing capacity of nZVI is consumed by reaction with water and dissolved oxygen. The reactive lifetime of nZVI is limited due to the small amount of ZVI mass in each nZVI particle. nZVI particles have limited mobility once they are injected into the subsurface, as nZVI particles tend to agglomerate to form big particles, although this problem is addressed to some extent by using a stabilizer. In addition, there is a growing concern among the regulatory community over the potential toxicity of nanomaterials, which also affects the widespread application of nZVI in site remediation.

Additional research and development are needed to address the limitations of nZVI mentioned earlier. In the meantime, the technology is evolving, and positive developments have been achieved in recent years. EZVI and AC-ZVI are good examples of recent developments. Both technologies are related to nZVI and have attracted much attention from practitioners, with growing commercial applications for contaminated site remediation.

PRACTICE AND EXERCISES

Question 1: What are the main advantages of nZVI?

Answer 1: The main advantages of nZVI include (1) larger specific surface area or surface area to mass ratio; (2) faster reaction rate; and (3) mobility in the subsurface.

Question 2: What are the application methods for using nZVI for site remediation?

Answer 2: The application methods include (1) injection using existing monitoring wells, piezometers, or injection wells, (2) recirculation, (3) direct push, (4) pressure-pulse technology, (5) liquid atomization injection, (6) pneumatic fracturing, and (7) hydraulic fracturing.

Question 3: What is the specific surface area (m^2/g) for ZVI particles with a diameter of 50 nm and 50 μm? Note that the density of ZVI is 7.8 g/L.

Answer 3: Use Equation 9.3,

$$\text{Specific surface area} = \left(\pi \cdot d^2\right) \Big/ \left(\rho \cdot \frac{\pi \cdot d^3}{6}\right) = \frac{6}{\rho \cdot d}$$

where:

d is the diameter of ZVI particles (m)

ρ is the density of ZVI (g/m^3) or 7800 g/m^3

If $d = 50$ nm or 5×10^{-8} m, specific surface area $= 6/(7800\ g/m^3 \times 5 \times 10^{-8}\ m) = 15{,}384\ m^2/g$

If $d = 50$ μm or 5×10^{-5} m, specific surface area $= 6/(7800\ g/m^3 \times 5 \times 10^{-5}\ m) = 15.4\ m^2/g$.

REFERENCES

Arnold, W.A. and Roberts, A.L. 2000. Pathways and kinetics of chlorinated ethylene and chlorinated acetylene reaction with Fe(0) particles. *Environmental Science and Technology,* 34: 1794–1805.

Bleyl, S., Kopinke, F.-D., and Mackenzie, K. 2012. Carbo-iron®—synthesis and stabilization of Fe (0)-doped colloidal activated carbon for in situ groundwater treatment. *Chemical Engineering Journal*, 191: 588–595.

Busch, J., Meissner, T., Potthoff, A., and Oswald, S. 2014. Investigations on mobility of carbon colloid supported nanoscale zero-valent iron (nZVI) in a column experiment and a laboratory 2D-aquifer test system. *Environmental Science and Pollution Research*, 21(18): 10908–10916.

Busch, J., Meissner, T., Potthoff, A., Bleyl, S., Georgi, A., Mackenzie, K., Trabitzsch, R., Werban, U., and Oswald, S. 2015. A field investigation on transport of carbon-supported nanoscale zero-valent iron (nZVI) in groundwater. *Journal of Contaminant Hydrology*, 181: 59–68.

Elliott, D.W. and Zhang, W.-X. 2001. Field assessment of nanoscale bimetallic particles for groundwater treatment. *Environmental Science and Technology*, 35: 4922–4926.

Environmental Protection Agency (EPA), 2008. Nanotechnology for Site Remediation Fact Sheet. doi.org/10.1002/rem.20194.

Fan, D., R.P. Anitori, B.M. Tebo, P.G. Tratnyek, J.S. Lezama Pacheco, R.K. Kukkadapu, M.H. Engelhard, M.E. Bowden, L. Kovarik, and B.W. Arey. 2013. Reductive sequestration of pertechnetate (99TcO4–) by nano zero-valent iron (nZVI) transformed by abiotic sulfide. *Environmental Science and Technology*, 47(10): 5302–5310.

Fan, D., O'Carroll, D.M., Elliott, D.W., Xiong, Z., Tratnyek, P.G., Johnson, R.L., and Garcia, A.N. 2016. Selectivity of nano zerovalent iron in in situ chemical reduction: Challenges and improvements. *Remediation Journal*, 26(4): 27–40.

Gavaskar, A., Tatar, L., and Condit, W., 2005. Cost and performance report: Nanoscale zerovalent iron technologies for source remediation; Naval Facilities Engineering Service Center: Port Hueneme, California.

Hara, S.O., Krug, T., Quinn, J., Clausen, C., and Geiger, C. 2006. Field and laboratory evaluation of the treatment of DNAPL source zones using emulsified zero-valent iron. *Remediation Journal*, 16(2): 35–56.

He, F. and Zhao, D. 2005. Preparation and characterization of a new class of starch-stabilized bimetallic nanoparticles for degradation of chlorinated hydrocarbons in water. *Environmental Science and Technology*, 39: 3314–3320.

He, F., Zhao, D., Liu, J., and Roberts, C. B., 2007. Stabilization of Fe-Pd nanoparticles with sodium carboxymethyl cellulose for enhanced transport and dechlorination of trichloroethylene in soil and groundwater. *Industrial and Engineering Chemistry Research*, 46: 29–34.

He, F., Zhao, D., and Paul, C. 2010. Field assessment of carboxymethyl cellulose stabilized iron nanoparticles for in situ destruction of chlorinated solvents in source zones. *Water Research*, 44: 2360–2370.

Johnson, T.L., Scherer, M.M., and Tratnyek, P.G. 1996. Kinetics of halogenated organic compound degradation by iron metal. *Environmental Science and Technology*, 30(8): 2634–2640.

Lien, H. 2000. Nanoscale Bimetallic Particles for Dehalogenation of Halogenated Aliphatic Compounds. Dissertation, Lehigh University, Bethlehem, PA.

Lien, H. and Zhang, W. 1999. Reactions of chlorinated methanes with nanoscale metal particles in aqueous solutions. *Journal of Environmental Engineering*, 125(11): 1042–1047.

Lien, H. and Zhang, W. 2001. Complete dechlorination of chlorinated ethenes with nanoparticles. *Colloids Surfaces A*, 191: 97–105.

Liu, Y., Majetich, S.A., Tilton, R.D., Sholl, D.S., and Lowry, G.V. 2005. TCE dechlorination rates, pathways, and efficiency of nanoscale iron particles with different properties. *Environmental Science and Technology*, 39: 1338–1345.

Matheson, L.J. and Tratnyek, P.G. 1994. Reductive dehalogenation of chlorinated methanes by iron metal. *Environmental Science and Technology*, 28: 2045–2053.

Nunez, A., Boparai, H.K., and O'Carroll, D.M. 2016. Enhanced dechlorination of 1, 2-dichloroethane by coupled nano iron-dithionite treatment. *Environmental Science and Technology*, 50(10): 5243–5251.

O'Carroll, D., Sleep, B., Krol, M., Boparai, H., and Kocur, C. 2013. Nanoscale zero valent iron and bimetallic particles for contaminated site remediation. *Advances in Water Resources*, 51: 104–122.

Quinn, J., Geiger, C., Clausen, C., Brooks, K., Coon, C., O'Hara, S., Krug, K., Major, D., Yoon, W.-S., and Gavaskar, A. 2005. Field demonstration of DNAPL dehalogenation using emulsified zero-valent iron. *Environmental Science and Technology*, 39(5): 1309–1318.

Rajajayavel, S.R.C. and Ghoshal, S. 2015. Enhanced reductive dechlorination of trichloroethylene by sulfidated nanoscale zerovalent iron. *Water Research*, 78: 144–153.

Saleh, N., Phenrat, T., Sirk, K., Dufour, B., Ok, J., Sarbu, T., Matyiaszewski, K., Tilton, R.D., and Lowry, G.V. 2005. Adsorbed triblock copolymers deliver reactive iron nanoparticles to the oil/water interface. *Nano Letter*, 5: 2489–2494.

Saleh, N., Kim, H.-J., Phenrat, T., Matyjaszewski, K., Tilton, R.D., and Lowry, G.V., 2008. Ionic strength and composition affect the mobility of surface-modified Fe0 nanoparticles in water-saturated sand columns. *Environmental Science and Technology*, 42: 3349–3355.

Sakulchaicharoen, N., O'Carroll, D.M., and Herrera, J.E. 2010. Enhanced stability and dechlorination activity of pre-synthesis stabilized nanoscale FePd particles. *Journal of Contaminant Hydrology*, 118: 117–127.

Tratnyek, P.G., Salter, A.J., Nurmi, J.T., and Sarathy, V. 2010. Environmental applications of zerovalent metals: Iron vs. zinc. *Nanoscale Materials in Chemistry: Environmental Applications*. L.E. Erickson, R.T. Koodali and R.M. Richards, American Chemical Society, Washington, DC. 1045: 165–178.

United States Environmental Protection Agency (USEPA). 2008. Nanotechnology for Site Remediation Fact Sheet. October. EPA 542-F-08-009.

Wang, C.B. and Zhang, W.X. 1997. Synthesizing nanoscale iron particles for rapid and complete dechlorination of TCE and PCBs. *Environmental Science and Technology*, 31: 2154–2156.

Xie, Y. and Cwiertny, D.M. 2010. Use of dithionite to extend the reactive lifetime of nanoscale zero-valent iron treatment systems. *Environmental Science and Technology*, 44 (22): 8649–8655.

Xiong, Z., Zhao, D., and Pan, G. 2007. Complete degradation of perchlorate by stabilized zero valent iron nanoparticles. *Water Research*, 41: 3497–3505.

Xiong, Z., Kaback, D., and Zhao, D. 2013. A field-observed synergistic effect between abiotic and biological degradation processes induced by stabilized nZVI injection, *Presentation at the Second International Symposium on Bioremediation and Sustainable Environmental Technologies*. Jacksonville, FL. June 10–13.

Zhang, W. 2003. Nanoscale iron particles for environmental remediation: An overview. *Journal of Nanoparticle Research*, 5: 323–332.

Zhang, W. and Elliott, D.W. 2006. Applications of iron nanoparticles for groundwater remediation. *Remediation Journal*, 16(2): 7–21.

Zhang, W., Wang, C., and Lien, H. 1998. Catalytic reduction of chlorinated hydrocarbons by bimetallic particles. *Catalyst Today*, 40(4): 387–395.

10 Diagnostic Tools and Processes Employed in Bioremediation

Stephen Koenigsberg and Aaron Peacock

CONTENTS

10.1 PREAMBLE: INDUSTRY OVERVIEW AND THE ROLE OF ADVANCED DIAGNOSTIC TECHNOLOGIES IN BIOREMEDIATION

To begin this overview, we will examine the genesis of the environmental remediation industry as a basis for understanding how and why we have harnessed various diagnostic technologies and processes to facilitate our work as environmental physicians. This analogy is especially meaningful in light of the fact that we are actually using certain products of medical science to aid in diagnosing bioremediation needs and managing intervention.

In a broader context, environmental remediation is fraught with complications and uncertainties that accompany the fact that we are addressing multi-variable circumstances, often on a fairly large scale. When an impacted environment needs to be addressed and repaired, these complex factors may have weighty operational costs associated with the remedial processes. One of the important features of environmental diagnostics is that they are intended to help form more accurate conceptual site models and also guide the designs and actions that follow. This can save time and money in achieving the goal of site resolution. The word *resolution* has special meaning, as opposed to clean-up, because not all sites can actually be readily restored to regulatory standards.

When the requisite science and engineering skill sets came together to form the initial structures for responding to contaminated sites, it was an imperfect beginning, and while progress has been made, there are still significant challenges in doing this work today. Again, in a proper context, the diagnostics can make their mark. The good news is that our understanding of problems, and the strategies and methods used to intervene to fix them, has evolved. The latter part of this chapter is a reflection of some of the most sophisticated and useful techniques and processes available in the realm of environmental diagnostics that now play an increasingly important role in site mitigation. That word *mitigation* is carefully chosen, because it is defined as the lessening of the force or intensity of something unpleasant. Nowhere in that statement is there an implication as to the degree to which that the unpleasantness is resolved.

When there came a time to confront environmental insults, in many cases when negative consequences to human health manifested, mitigation started out with what was at hand, and that was largely in the domain of the civil and mechanical engineers. To be more specific, depending on the site, the first obvious solution was to dig up the problem and haul it away. Clearly, this is one of the more capital- and labor-intensive remedial actions and served as an impetus to begin to do things "better, faster and cheaper." The problem with the "dig and haul" solution is that, in the words of a founding thinker in remediation, Dr. Wesley (Wes) Eckenfelder, "we can't just keep digging up the stuff in one place and move it to another."

In essence, that did not seem like much of a best practice for resolving problems with contaminated media, and that set the stage for bioremediation and the diagnostics that came along in support. We can look to the mid-1970s as a time when the field of bioremediation started. In the simplest terms, the original industrial targets were petroleum hydrocarbons, and the problem was linked to the basic production, product formation, and distribution sequence from the oil fields themselves to the refineries and ultimately to the retail gas stations. Soil contamination was typically managed in "biopiles," whereby soil was excavated and, rather than being hauled away to landfills, was treated with nutrients and oxygen that supported native microbes, which performed the remediation. When clean to regulatory standards, the soil could then be returned to the excavation from which it came. There are variations on this theme that will be mentioned, but first, so as not to get too far ahead, we must stop and consider the scientific underpinnings of what is involved here.

Beyond this Preamble, the following section addresses the fine details of the biology and chemistry behind how certain microorganisms carry out the work of bioremediation. In the area of petroleum hydrocarbons to begin with, we are focused on aerobes that use oxygen and are supported by certain nutrients to become the agents of hydrocarbon bioremediation. Briefly, for this overview, hydrocarbons are food for these aerobes, but they need oxygen to serve as an electron acceptor, which is an endpoint in the cellular energetic process. Also, the hydrocarbons provide basic chemical building blocks to generate molecular building blocks for their own growth and reproduction. The reader may already see the parallel with metabolism in our own selves, for which we also need food and oxygen, but the essential difference here is that we do not have the enzymes and metabolic ability to use hydrocarbons in the way microorganisms do. For us, it is the baked potato or loaf of bread that constitutes our carbon and energy source, and further, we can also reference edible fats and oils, other than petroleum oils, and their derivatives.

At this point, the reader should anticipate that diagnostics that identify key microbes and their enzymes have an important role to play in bioremediation. The microbes, in most cases, were already adapted to the polluting hydrocarbons, at which point they were able to metabolize them. Diagnostic methods are reasonably effective in sorting out the organisms that are doing the heavy lifting, and in some cases, microbial cultures of these organisms are added to the contaminated media in a process we now call *bioaugmentation*.

Current convention defines the first operations involving the addition of nutrients and oxygen as *biostimulation*, in contrast to bioaugmentation, which defines the use of the organisms themselves. In addition, if there is space available and the inclination to do so, an alternative to forming and managing a biopile is to *landfarm* the contaminated soil. As the name implies, this involves spreading the soil out on a surface and mechanically disking in the nutrients and providing oxygen. Then, there was also the notion that one could actually perform these tasks within the ground (*in situ*) by injecting the supportive reagents and air/oxygen.

Back to Dr. Eckenfelder. He was on the faculty of Vanderbilt University and a luminary in the field of industrial wastewater management, and in that arena, microbes are harnessed as degraders of waste products. This component of sanitation engineering, a subset of civil engineering, helped establish the concept that microbes could play an important role in environmental clean-up. Since organic industrial wastes are, well, organic wastes—in the same vein as what comes from municipal sewage and food processing—a basis to move laterally into industrial remediation realms developed. Consequently, practical applications, some of which preceded Dr. Eckenfelder's academic framing, had surfaced.

One widely recognized contributor from industry, Richard (Dick) Raymond Sr. of Sun Oil, went after fugitive petroleum hydrocarbons in the subsurface that now had the status of being pollutants. He did this with injections of oxygen and nutrients to support microbial metabolism. In addition, in parallel with Wes, other academics, such as Dr. Perry McCarty at Stanford University, also set forth key scientific and engineering principles for other classes of contaminants, such as chlorinated hydrocarbons. These agents were used ubiquitously as industrial solvents for grease and oil, and in their misuse, environmental impacts showed up wherever they were used for cleaning machinery or washing circuit boards and even fabrics, with emphasis on use in the dry-cleaning industry.

The twist with this problem, however, was that the bioremediation of chlorinated solvents is an anaerobic process, in which oxygen is in fact toxic and hydrogen is required as a chemical reductant (an electron donor rather than an electron acceptor). This sets the stage to explain how bioremediation is employed in a given setting, noting that further details will follow. For hydrocarbons, the operant mechanism is oxidation, and for chlorinated solvents, it is reduction, noting that these terms have specific chemical definitions. Oxidation involves the removal of electrons, which hopefully alter the contaminant for the better and end up on an electron acceptor such as oxygen. Reduction is the flip side to this, in that electrons are donated and facilitate chemical transformations. If this seems confusing, note that a negatively charged electron is what is being donated, so that it is, in effect, a reduction. Again, as this historical accounting unfolds, there is a clear and present role for molecular diagnostics in defining and guiding the remediation engineering that is employed to solve the problems.

The end result at a point in the 1990s, almost two decades after the inception of bioremediation, was that engineers adept at moving soil and installing subsurface treatment systems now had a partnership with wastewater engineers. More importantly, this dimension was added because this body of scientific and technical knowledge was geared to addressing contaminated aqueous environments. After all, when a volume of soil is impacted, there are other considerations, such as the impact on the soil-water matrix as well as the aquifer itself.

On that basis, we can turn to the engineering aspects, meaning that by understanding the processes, we can devise the means to provide what is needed, thus forming the engineering component of the remediation. In this specific instance, remembering that this is just an early founding set of operations, delivering the limiting factor—oxygen—is the task. In this initial context of impacted soils, the challenge then becomes how to deliver the goods. The form this takes is to either mechanically turn a biopile to provide oxygen or to plumb it—meaning to install an array of pipes to provide the aeration. Note that if mineral nutrients or even bioaugmentation is considered appropriate, this is provided at the outset of forming a biopile and depending on how the biopile is managed, these features can be provided during the operation.

These are referred to as *ex situ methods*, but what if you cannot dig up the soil conveniently to treat it or are constrained in other ways? In addition, we must consider what happens when we are faced with contamination in the soil-water matrix as well as the aquifer itself. This calls for other operational approaches that in some cases, apply the principles of mass transfer and are typically serviced by the application of injected or sparged gases. The principles are straightforward, in that in an unsaturated but otherwise oxygenated soil matrix, forcing air into the system can drive out the volatile fraction of contaminants that may be present. This process is aptly called *soil vapor extraction*, but it does not preclude the action of microbes, which have a parallel, if less significant, role, driving a collateral metabolic process.

Then, as one takes the processes deeper into the subsurface where pore water and open aquifers are encountered, the forced air, or even pure oxygen if desired, facilitates aerobic metabolism—as pioneered by Dick Raymond. This air sparging process is designed to address dissolved phase contamination, depending on the solubility of the compounds of concern. This is just a quick sketch, because there are many variations on the theme, one of them dealing with a palpable mass that forms beyond the saturation threshold and sits on or in the soil-water interface. This mass can be extracted as a liquid or sequentially eaten away in place by mass transfer. To understand this, we reference Fick's Law, which teaches that concentration gradients for a compound will form, such that as a volatile fraction is removed, the source mass replenishes it to re-establish the gradient equilibrium. Also, when using terms such as *eaten*, we are of course referencing bioremediation, and where there is bioremediation, the use of diagnostics is not far behind.

Before we get into a further review of these options, we must stop and recognize that where petroleum hydrocarbons are concerned, there was very little use for molecular diagnostics, noting that in the early days they were not really available to any extent. Said another way, the methods were available, but they took time to be integrated. Still, it was a beginning, and at this point, whenever you see bioremediation cited, you will now know there are tools and processes to guide you.

We are now ready, in the rest of this chapter, to discuss the advent of environmental molecular diagnostics (EMD), which like bioremediation itself, is a compilation of things borrowed from other fields. Within the subject of EMD, there are advanced chemical techniques, such as compound-specific isotope analysis (CSIA), but in this chapter, we are restricting our discussion to what are termed *molecular biological tools* (MBTs). MBTs rely on the diagnostics of nucleic acids (DNA and RNA) and occasionally the analysis of proteins and other biomolecules such as phospholipid fatty acids (PLFAs). In essence, it is like being Dr. Doolittle, who talked to the animals; only we are talking to the bacteria, and through the diagnostics, we are asking them who they are and what they do for a living—as in what enzymes are produced.

The thing of note here is that the MBT techniques, at least in the nucleic and protein realms, have come about from billions of dollars invested in diagnostics by medical and defense interests. We have simply accessed the wisdom of the science and the methods and used the mechanics offered to us by the engineers in the form of biopile, landfarming, direct injection, and extraction operations. One way to appreciate this in context is to consider how a physician (hence the reference to being an environmental physician) uses the science to diagnose a patient for an infectious disease or an aberrant enzyme profile, or how, as first responder in a bioterrorism event, he or she sorts through a scene to assess the status. Using MBTs upfront and along the way, we are poised to make remedial operations better, faster, and cheaper.

Remember that the MBTs are designed to identify the presence of key microbes and their enzymes, and a full literature has evolved whereby conditions can be diagnosed to determine the best course of action. Further, the role of diagnostics can now be connected to how we deal with other contaminants and the collateral impacts of other physical processes. All this being said, the problem is bigger than petroleum hydrocarbons and chlorinated solvents. In addition, there are uses for the MBTs in other circumstances, as will be presented.

First, on the contaminant side, many other problems in terms of compounds of concern manifested over time, and so a series of other chemical insults joined the repertoire. There were gasoline oxygenates, such as methyl tertiary butyl ether (MTBE), that were used to oxygenate gasoline so that it would improve combustion efficiency and lower toxic air emissions. These predictably got out into the environment and proved to be somewhat resistant to bioremediation, at least in early studies, although the conventional wisdom of that has changed with greater insight. Now, it is crucial to point out that the use of molecular diagnostics was a key part of our understanding of mechanisms and remediation strategies in that specific problem.

We cannot cover everything that has served as a pollutant in the last 40 years or so, but two other kinds of chemicals are significant. One is the metals, and the other is what we call *emerging contaminants*. Some of the more recent ones, in fact, are presenting significant concern, as will be discussed.

Regarding metals, the fundamental principle at stake here is that metals have variable solubilities governed by pH and the redox state—whether conditions are oxic or reductive. The way this is played by engineers is that one might want to get the offending metals into solution to flush them or get them to precipitate on the matrix; out of sight, out of mind. So, where do diagnostics fit in? The answer is that the microbial ecology mediates these reactions. As an example, if we are interested in metal reduction, we can identify metal-reducing microbes, and the diagnostic testing can help us determine whether they are present and the size of the populations.

Then, most recently, we have been made aware of a series of emerging contaminants such as 1,4-dioxane (1,4-D) and PFASs, both of which are, in so many words, site re-openers or wholly new serious matters. The more insidious contaminant of the two is a family of compounds called *per- and polyfluoroalkyl substances* (PFASs). These were formally known as *perfluorinated compounds* (PFCs), but the definition has expanded due to the polyfluorinated species. As far as bioremediation is concerned, we may have met our match, because we did such a good job of making these molecules inert that we seem to be stuck with minimally or non-biodegradable materials. As far as MBT applications, it may be a long time, if ever, before any microbiology can impact these molecules, but this is not the case for 1,4-D.

1,4-D is an ether, like MTBE, and in fact, recently, there has been great progress in using enzyme diagnostics to determine the degree to which a contaminated environment may be able to respond to bioremediation. Of course, there are microbes behind the secretion of the enzyme markers, and at the moment, we tend to focus on the enzymatic signatures. We can expect to hear a lot more about the use of MBTs in dealing with this wide-spread problem.

Then, it is important to understand that there is a role for MBTs in treatment modalities where bioremediation may be termed *stoichiometrically challenged*. This means that there is a point after which the routine treatment formulae are no longer applicable. For example, when a mass of hydrocarbon is extensive, one can calculate the oxygen demand for having the microbes turn organic carbon into carbon dioxide and water—this being the stoichiometry of the reaction. Therefore, as noted, when we exceed practical limits for providing oxygen in water due to solubility limits, it is time for the "big guns"; one example of this is *in situ chemical oxidation* (ISCO). With ISCO, one uses reagents such as hydrogen peroxide, permanganates, or persulfates—all powerful oxidants. While this is not bioremediation, bioremediation may be used as a polishing agent once the "big rocks are broken down into little rocks," and therein lies the point.

The harsh nature of ISCO is not friendly to living things and can have a devastating effect on the microbial ecology. That said, the value and usefulness of MBTs are apparent. The fact is that it is pretty difficult, if not impossible, to sterilize an aquifer, and so it becomes important to understand that when the populations grow back, they can be fed and nurtured so that they can polish the site. The same is true of popular thermal treatments, whereby the subsurface is heated up to drive off the contaminants. Again, this is a harsh treatment, but microbes can even thrive at the edges of a thermal treatment, where temperatures are not lethal but favor growth. In these circumstances, MBTs can serve to monitor and restore conditions favorable to bioremediation.

This has been a sketch to serve as a basis for diving into the deeper waters to follow. In completing a study of this, the reader will have appreciated the particulars of the processes and how diagnostics illuminate our knowledge and awareness of how we can achieve environmental clean-up better, faster, and cheaper.

10.2 DIAGNOSTIC TOOLS AND PROCESSES

10.2.1 Introduction

As presented in part in the Preamble, bioremediation is a restorative process that uses microorganisms to reduce, eliminate, contain, or transform contaminants in the environment. Bioremediation is not new, as it has been practiced for over 40 years, starting with petroleum hydrocarbons, but in the last 20 years, there has been a significant increase in the types of contaminants it has addressed. A partial list of this expansion includes chlorinated solvents, aromatic hydrocarbons, polychlorinated biphenyls (PCBs), heterocyclic organic compounds such as 1,4-dioxane, and metals, including radioactive constituents. Perhaps the greatest advantage of bioremediation is that it can be executed *in situ* (in place), where key reagents are applied to the subsurface by a variety of means. These operations are usually much less expensive and disruptive to the surrounding environment than gross engineering methods such as "pump and treat" or excavation and disposal.

The successful use of bioremediation depends on the presence of microorganisms in appropriate numbers under the correct geochemical conditions that are capable of supporting them. The prokaryotic organisms, as represented by the archaea and bacteria, are usually the "heavy lifters" in bioremediation; however, eukaryotes such as fungi and algae can also degrade contaminants. For the purposes of this chapter, biotransformation is defined as any alteration of the molecular or atomic structure of a contaminant by microorganisms and as such, is linked to the process of biodegradation, which is the decomposition of an organic substance into smaller organic or inorganic components. The decomposition reactions may proceed all the way to mineralization (the complete biodegradation of an organic contaminant into inorganic constituents such as carbon dioxide or methane, mineral salts, and water).

Metals and radionuclide contaminants cannot be degraded but can be biotransformed from one state to another through adding or removing electrons in processes called *reduction* (adding a negatively charged electron) or *oxidation* (which constitutes a removal of electrons). The oxidation and reduction process are central to any bioremediation process, but in the case of metals, this action can increase or decrease the solubility of the metal in the environment. In practical terms, increasing the solubility can allow the contaminants to be flushed from the system, whereas precipitation reactions sequester (immobilize) the metals.

The focus of this chapter is on below-ground bioremediation of contaminants, where we will explore the tools that allow us to understand and manage the agents of environmental repair—the microorganisms.

10.2.2 Microorganisms

Various categories of bacteria (e.g., sulfate reducers, dehalogenators, etc.) capable of biotransformation of contaminants in soils and sediments are referred to collectively as *microorganisms*. Microorganisms can be thought of as small, self-feeding, and self-replicating life forms that carry out chemical transformations in the process of metabolism (Madigan et al. 1996). Just as humans require water, food, and oxygen to sustain life, microorganisms also require food (in the form of organic carbon), energy, and a way to respire (breathe). Microorganisms have been on the earth for billions of years, and due to their sheer numbers and diversity, have evolved to be capable of biotransforming a wide range of contaminants. An important feature of microbes, especially important in bioremediation, is whether they are aerobes (oxygen requiring) or anaerobes (requiring an oxygen-free environment). This section describes microbial structure, physiology, and function as they relate to general contaminant biotransformation and degradation.

Madigan et al. (1996) describe a single-cell microorganism as a complex chemical system. There are five major characteristics that set living cells apart from nonliving chemical systems:

1. *Self-feeding or nutrition*: Microorganisms take up chemicals, including contaminants, from the environment and transform them from one form to another. During these biotransformations, there is a release of energy and the excretion of cellular waste products.
2. *Self-replication and growth*: Microorganisms are capable of directing their own existence. As a result of metabolic nutritional processes, microorganisms can grow and replicate.
3. *Differentiation*: Microorganisms can undergo changes in form and function. These changes are usually a direct result of environmental pressures.
4. *Communication*: Microorganisms respond to chemical and physical stimuli in their environment and can interact and communicate with other microorganisms through chemical

signaling. Many microorganisms are also capable of movement and respond to various stimuli (e.g., carbon and energy).
5. *Evolution*: Microorganisms are capable of hereditary changes that can influence their overall fitness.

The following topics on the cell and the fine points of metabolism and cellular energetics are designed to set the stage for a discussion of the important diagnostic tools we use to manage microorganisms as they execute their ability to perform bioremedial activities.

10.2.3 The Cell

Microorganisms convert energy from one form to another by breaking down large molecules into smaller ones and building up larger molecules from smaller ones. *Metabolism* is the term used to describe the chemical transformations that occur within the cell to maintain life. Figure 10.1 shows a representative cell and the associated cellular components.

- Cell structure is maintained via the cell wall and membrane lipids. It is through the cell membrane that all nutrients and other substances pass in and out of the cell.
- Inside the cell is a mixture of substances, which are effectively in solution and constitute the cytoplasm. This "soup" is the ideal medium for the functions of the cell to be carried out.
- The cell's genetic coding (deoxyribonucleic acid [DNA]) is held by the chromosome and is how the cell stores information, for example, for replication and the coding of enzymes.
- Ribosomes are a ribonucleic acid (RNA) and protein complex that carry out the process of *translation*, where genetic information directs the synthesis of enzymes.

10.2.4 Cellular Metabolism

Just like humans, single-cell microorganisms need to eat and breathe to stay alive. Unlike humans, microbial cells have a much more diverse capacity regarding the types of food (carbon and energy) they can eat and the types of "breathing medium" (respiration) they can use. Figure 10.2 shows the basics of cellular requirements for metabolism for an aerobic (oxygen-requiring) microbe.

- *Electron donor (carbon and energy)*: This is the food "carbon" the cell needs for growth and also the energy (via an electron donor) the cell needs to survive. Depending on the type of cell, the electron donor can be a range of compounds. Heterotrophic microorganisms derive their nutritional requirements from complex organic substances. In the case of

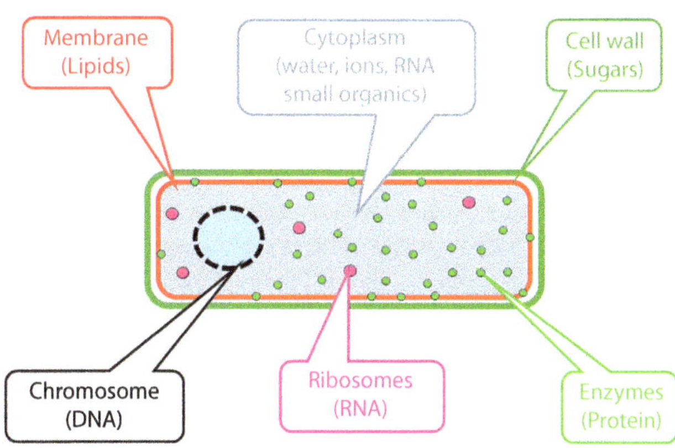

FIGURE 10.1 Cellular components. (From Michael R. Hyman, NC State University.)

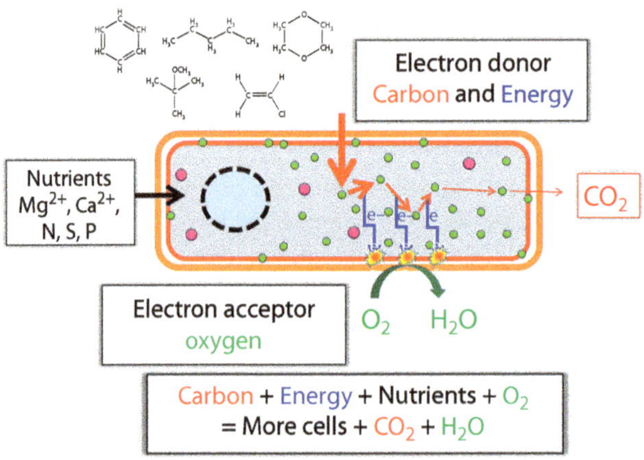

FIGURE 10.2 Aerobic cellular metabolism. (From Michael R. Hyman, NC State University.)

aerobic heterotrophic organic biotransformation, organic contaminant compounds (e.g., contaminants in the form of monoaromatic hydrocarbons) can provide the carbon for cell growth and the energy. In the course of cellular action, the contaminant molecule is either biotransformed into a metabolic derivative compound or completely biodegraded and converted to CO_2 (red lines denote the conversion pathway through enzymes [green dots] in Figure 10.2). It is important to note that in most cases, these biotransformation processes occur in aqueous solution; therefore, the solubility of the electron donor (contaminant) will strongly influence the availability of the compound to be degraded and the rate at which it is degraded.

- For the cell to process the organic contaminant for carbon and energy and convert it to CO_2, a series of steps are required. Depending on the particular cellular mechanism that is used, electrons originating from the contaminant are passed via enzymatic pathways through the cell and are then discharged to a terminal electron acceptor (blue lines denote electron transfer, and the green curved arrow shows the terminal electron acceptor). The terminal electron acceptors that microorganisms can use vary widely and include oxygen, nitrate, manganese (IV), iron (III), sulfate, and carbon dioxide.
- Besides an electron donor and electron acceptor, the cell also needs trace nutrients, which can be inorganic (certain metals) or organic (e.g., vitamins), to survive and grow.
- The sum of the process, carbon plus energy plus nutrients plus an electron acceptor, equals more cells and the biotransformed products of the original organic contaminant.

10.2.5 Terminal Electron-Accepting Processes—Cellular Energetics in Detail

Unlike humans, who can only use oxygen for respiration, microorganisms have a very wide variety of respiratory capabilities for the transfer of electrons. Microorganisms can gain most energy if the electron acceptor has a large positive reduction potential, which leads to a large net differential between the primary donor and the terminal electron acceptor. Figure 10.3 shows microbial terminal electron acceptors and their relative energy potential. Oxygen has the highest energetic yield of any terminal electron acceptor, followed by nitrate, manganese, iron, sulfate, and finally, carbon dioxide (methanogenesis). The use of oxygen as a terminal electron acceptor is defined as *aerobic respiration*, and the use of terminal electron acceptors other than oxygen is defined as *anaerobic respiration*. The use of carbon dioxide as a terminal electron acceptor, which results in the production of methane, is defined as *methanogenesis*.

There is another microbial energy producing process (not shown in the diagram in Figure 10.3) called *fermentation*, in which organic molecules act as both electron donor and acceptor. In this

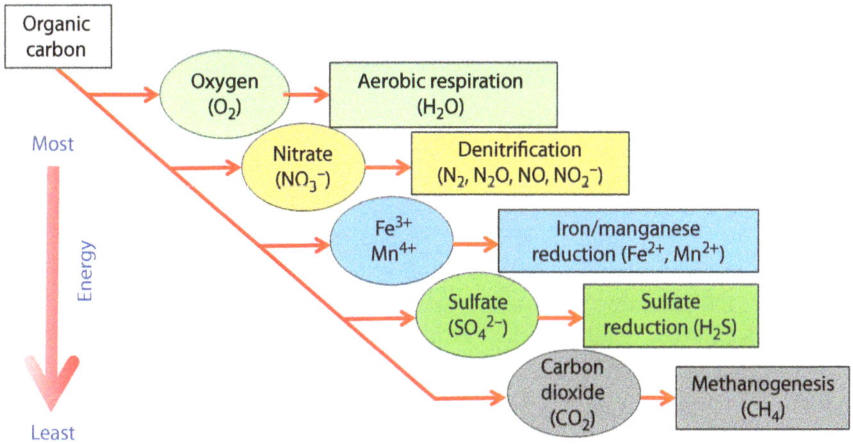

FIGURE 10.3 Microbial terminal electron acceptors and processes.

process, fermentative bacteria synthesize hydrolytic enzymes that degrade complex polymers to soluble monomers. The soluble products are then fermented to H_2, CO_2, simple alcohols, fatty acids, and acetate. Typically, higher-energy processes tend to occur preferentially; however, as will be discussed later, it is an over-simplification to state that they occur sequentially in order of decreasing energy yield.

Other factors besides energetic yield can influence the use of various terminal electron acceptors in the subsurface by microorganisms. For example, nitrate and sulfate are much more soluble in water than oxygen, whereas manganese (IV) and iron (III) are essentially insoluble in water and exist as solids. In areas where oxygen is available, some microorganisms may still prefer nitrate, especially if the concentrations are high. The availability and type of acceptor may override energetic yield, particularly when there is an oversupply of organic carbon. In such cases, there can be several simultaneous terminal electron-accepting processes occurring. So, instead of a definitive redox ladder (i.e., aerobic processes occur first, then nitrate reduction, then iron/manganese reduction, etc.), it may be better to think of these processes as a web. Additionally, the type of subsurface matrix has an influence on terminal electron-accepting processes via clay, mineral, and organic matter content.

Certain types of microorganisms specialize in different terminal electron-accepting processes and can be thought of in human terms as microbial guilds. For the purposes of this discussion, the general guilds can be defined as:

- *Aerobes*: Microorganisms that use oxygen as a terminal electron acceptor.
- *Nitrate reducers*: Microorganisms that use nitrate as a terminal electron acceptor.
- *Metal reducers*: Microorganisms that can use solid manganese (IV) or iron (III) as a terminal electron acceptor.
- *Sulfate reducers*: Microorganisms that can use sulfate as a terminal electron acceptor.
- *Fermenters*: Microorganisms that can use organic molecules as both electron donors and acceptors.
- *Methanogens*: Microorganisms that use CO_2 as a terminal electron acceptor and produce methane.

10.2.6 Microbial Ecology and Biotransformation

Contaminant biotransformation and degradation in the subsurface are directly dependent on biogeochemical carbon and energy cycles that are facilitated by microorganisms. There are multiple chemical reactions possible within the subsurface, including ion association, ion exchange, complexation, multivalent ion hydrolysis, oxidation, reduction, degradation of organics, crystallization, sorption, and solubilization/dissolution. From the perspective of a microorganism, subsurface soils and sediments are a construction of heterogeneous solids, liquids, and gases. The solid portion contains both inorganic and organic material, with the inorganic material ranging in size from clays and silts to sand, gravel, and rocks. The inorganic portion includes minerals such as quartz, feldspars, and montmorillonite with specific surface areas (both external and internal) of 10–1000 m^2/g. Natural organic material is also present but usually at very low amounts relative to inorganic constituents. Subsurface soil and sediment voids may be occupied by water or gases, depending on the depth of the water table (a geologic layer in which fractures and pores are filled with water). Water may also exist as films on particle surfaces or within micropores of subsurface particles.

Microorganisms can live at the particle-fluid interface of subsurface material or within a pore that is filled with liquid. Microorganisms may be transported in the subsurface with groundwater flow but could have self-mobility that would allow movement through fluids. Most microorganisms in subsurface environments are believed to be attached to mineral particle surfaces. To gain an understanding of how to employ bioremediation, it is useful to know how the subsurface ecosystem functions. There are many components to the system; however, the four primary components are carbon and energy, microbial community, geochemical, and contaminant environment.

With the background just provided, we are now in a position to examine the diagnostic tools we employ to manage bioremediation.

10.2.7 Tools to Characterize and Monitor the Microbial Community

As discussed, cells contain DNA, RNA, and proteins/enzymes. The characterization and quantification of these cellular constituents can provide information as to the capabilities and activity of subsurface microorganisms to biotransform or degrade a contaminant of interest. The detection of a gene (DNA) or a transcript (mRNA) is a useful surrogate measurement for the presence of a specific enzyme or activity (Figure 10.4).

10.2.7.1 Quantitative Polymerase Chain Reaction (qPCR)

One of the tools that are used to characterize and quantify the ability of the subsurface microorganisms to biotransform or degrade a contaminant is qPCR. This method targets specific bacteria or functional genes in biodegradation pathways. As an example of the use of qPCR, a factory that used chlorinated solvents for metal cleaning discovered that there had been a leak of trichloroethylene (TCE) waste over several years that had resulted in groundwater contamination. As part of the characterization effort, it was decided to use qPCR testing to see whether there were bacteria in the groundwater that were capable of degrading the TCE. The environmental professionals responsible for the characterization of the site decided to test for several bacteria known to

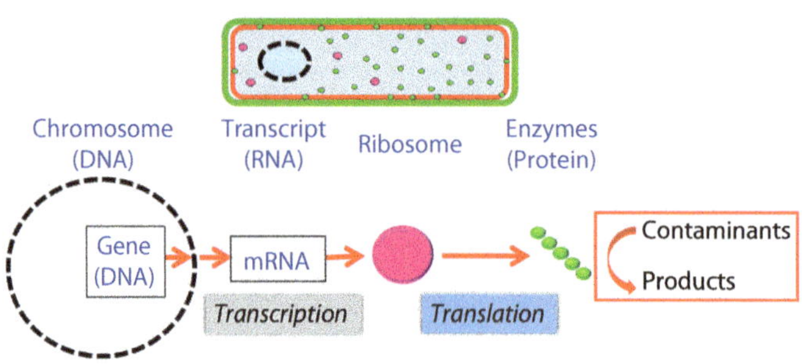

FIGURE 10.4 Protein biosynthesis and molecular targets. (From Michael R. Hyman, NC State University.)

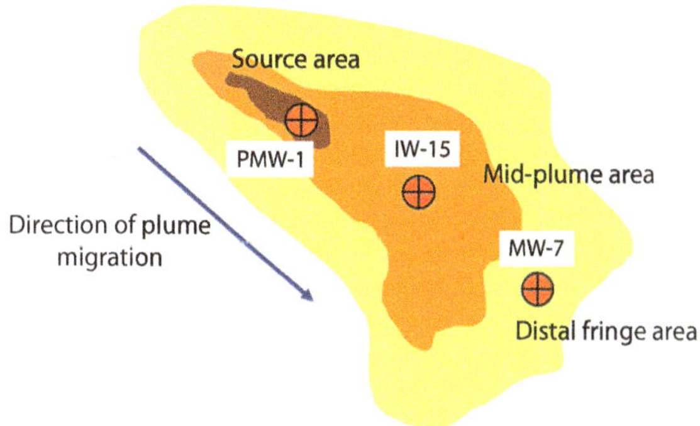

FIGURE 10.5 Well sampling plume map.

degrade TCE and for functional genes specific to TCE degradation. The qPCR targets included the bacteria *Dehalococcoides* spp., *Dehalobacter* spp., *Dehalogenimonas* spp., and *Desulfitobacterium* spp. Both *Dehalococcoides* spp. and *Dehalogenimonas* spp. are known to completely degrade TCE to ethene under anaerobic conditions, while *Dehalobacter* spp. and *Desulfitobacterium* spp. can degrade TCE to dichloroethylene (DCE). The functional genes targeted included trichloroethylene reductase and two vinyl chloride reductases. The presence of reductase genes would show that the capability for complete reductive dechlorination was possible at the site.

The remediation professionals chose three groundwater monitoring wells that represented different areas of the plume (Figure 10.5). Well PMW-1 was in the source area of the TCE spill, while well IW-15 was in the mid-plume downgradient of the source area. The last monitoring well, MW-7, was at the distal edge or fringe of the plume. The samples were collected by using a pump to collect 1 L of groundwater. To make sure that formation water was collected, the monitoring well was purged prior to taking the sample. The water samples were then sent to the laboratory on ice via overnight delivery. The short hold time ensured that changes in the results due to shipping would be kept to a minimum. Once the samples arrived at the laboratory, they were filtered, and the microbial DNA was extracted from the filter and purified. There were some quality control steps to make sure the DNA was of good quality and not contaminated. The DNA was then run on the qPCR instrument. Once that was complete, there were more quality checks, and a report was produced.

The qPCR report (Figure 10.6) showed there were 4.13×10^6 *Dehalococcoides* spp. gene copies per liter of water in the source monitoring well PMW-1. It is important to note that the qPCR test measures gene copies and not cells. The number of 16S genes in *Dehalococcoides* spp. is one per cell, so there can be a direct conversion to cells. The other results from the source area monitoring well showed the presence of large numbers of *Dehalogenimonas* spp. and relatively lower levels of both *Dehalobacter* spp. and *Desulfitobacterium* spp. The functional gene results showed high levels of both vinyl chloride reductase genes and the trichloroethylene reductase gene. The total eubacteria result was 6.50×10^9. The measurement of all microbial 16S genes in the sample provides an index of the bacterial biomass. Sample IW-15 in the mid-plume area showed a drop in *Dehalococcoides* spp. and *Dehalogenimonas* gene copies of about one order of magnitude. This pattern was repeated for most of the gene targets tested except for the total eubacteria, which was an order of magnitude higher than in the source area. The distal fringe area monitoring well showed low levels of *Dehalococcoides* spp. gene copies (1.57×10^2), while the *Dehalogenimonas* spp. level was a bit higher at 1.94×10^4, but well below that of the source and mid-plume wells. There was also a low level of vinyl chloride reductase genes detected in the fringe area well, and two of the functional genes (bvcA and tceA) were not detected. The biomass index as measured by total eubacteria was an order of magnitude lower than in the source area and two orders of magnitude lower than in the mid-plume sample.

		Matrix: Water	
Sample ID:	IW-15		
Description:		Gene copies per liter of water	
Dehalococcoides spp:	1.00E+05	Dehalobacter spp:	6.92E+03
Vinyl chloride reductase (vcrA):	3.40E+05	Dehalogenimonas spp:	4.03E+06
Trichloroethene reductase (tceA):	3.71E+03	Desulfitobacterium spp:	3.74E+04
BAV1 vinyl chloride reductase (bvcA):	2.61E+03	Total eubacteria:	5.67E+10
		Matrix: Water	
Sample ID:	PMW-1		
Description:		Gene copies per liter of water	
Dehalococcoides spp:	4.13E+06	Dehalobacter spp:	4.67E+03
Vinyl chloride reductase (vcrA):	1.28E+07	Dehalogenimonas spp:	1.40E+07
Trichloroethene reductase (tceA):	3.94E+04	Desulfitobacterium spp:	3.84E+03
BAV1 vinyl chloride reductase (bvcA):	2.21E+05	Total eubacteria:	6.50E+09
		Matrix: Water:	
Sample ID:	MW-7		
Description:		Gene copies per liter of water	
Dehalococcoides spp:	1.57E+02	Dehalobacter spp:	< 1
Vinyl chloride reductase (vcrA):	1.85E+02	Dehalogenimonas spp:	1.94E+04
Trichloroethene reductase (tceA):	< 1	Desulfitobacterium spp:	< 1
BAV1 vinyl chloride reductase (bvcA):	< 1	Total eubacteria:	2.51E+08

QA/QC	NIC	UNC	IPC	UPC
	Pass	Pass	Pass	Pass

FIGURE 10.6 qPCR sampling results (gene copies per liter).

The qPCR results from the site provided answers to several questions.

- Were contaminant-degrading microorganisms present?
  - Yes, qPCR showed that bacteria capable of completely degrading TCE were present at all three wells tested. The presence of functional gene targets for the specific reductases verified the capability for complete reductive dechlorination at the source and mid-plume areas.
- What were the relative numbers and where?
  - The largest numbers of degrading bacteria were in the source area, followed by the mid-plume area and then much lower levels in the distal fringe area.
- Is monitored natural attenuation (MNA) feasible?
  - Given the numbers of dechlorinating bacteria identified at the site, if the rate of degradation is acceptable, this could be a candidate for MNA.
- Is an amendment (e.g., carbon source) necessary?
  - To answer this question fully, there would need to be a characterization of the organic carbon levels using total organic carbon and/or volatile fatty acid analysis. If those levels were sufficient due to natural organic matter or a co-contaminant, no amendment would be necessary in the source or mid-plume areas. The distal fringe area would need further assessment.
- Is bioaugmentation necessary?

Given the levels of dechlorinating bacteria in the source and mid-plume areas, there would be no urgent need to bioaugment. The distal fringe area has much lower levels of dechlorinating bacteria and may be a candidate for bioaugmentation.

10.2.7.2 16S Amplicon Genetic Sequencing

Another molecular tool that is used to characterize the ability of subsurface microorganisms to biotransform or degrade a contaminant is 16S amplicon genetic sequencing. This method identifies (to the extent possible) all bacteria in a sample. It is more of a general population approach determining the phylogenetic position of the unknown microorganism among known microorganisms. The identification is based on the gene in the cell's DNA that encodes for 16S ribosomal RNA. The 16S target has proved to have excellent properties for making comparisons and is considered a standard for phylogeny.

The groundwater under a specialty chemical manufacturing plant was discovered to be contaminated with chlorobenzenes. The source area was shown to be highly contaminated; however, along the groundwater flow path the contamination levels dropped rapidly. The stakeholders wanted to know whether the reduction of contamination mass was due to dilution and dispersion alone or whether contaminant degradation was taking place along the flow path. One of the molecular biological tools used to characterize the bacterial community at the site was 16S genetic sequencing. The 16S data were meant to investigate which bacteria were present and ascribe a putative function.

Table 10.1 shows an example of 16S results and the associated putative functions of the identified bacterial genus. The highlighted rows denote bacteria with known putative capabilities to degrade the contaminants of interest.

10.2.8 Biological Degradation Pathways and Biomarker Analysis

The biological pathways and biomarker analysis identified putative biological degradation pathways for the primary contaminants based on the relative presence or absence of 16S biomarkers for bacteria that have a demonstrated capability to produce enzymes involved in degradation. The mechanisms can vary based on the contaminant. For the identified project, there was a specific focus on benzene, chlorobenzene, and dichlorobenzene(s). There are multiple degradation pathways for chlorinated benzenes, including direct aerobic metabolism, aerobic cometabolism, and anaerobic dechlorination.

Aerobic Cometabolism: In this type of reaction, the primary contaminant is simultaneously degraded along with the use/degradation of a second compound. The presence of the second compound is required for the degradation of the primary contaminant. There are many reported examples of cometabolism of chlorinated benzenes and co-substrates. Table 10.2 provides examples of bacteria capable of the cometabolism of chlorinated benzenes (CBs).

One particular enzyme involved in the cometabolism of CBs is methane monooxygenase (MMO). Methanotrophic bacteria can produce particulate methane monooxygenase (pMMO) or soluble methane monooxygenase (sMMO).

Aerobic Direct Metabolism: The degradation of the chlorinated benzenes is accomplished via aerobic growth of bacteria using the contaminant as a sole carbon and energy source. Table 10.3 provides multiple examples and references for bacteria capable of the direct metabolism of chlorinated benzene(s).

Anaerobic Dechlorination: During the dechlorination of a compound, where the compound is used as an electron acceptor, the bacteria may or may not gain energy by reduction of the compound. This reaction removes one chloride atom from the compound and replaces it with a proton. Tables 10.4 and 10.5 list examples of anaerobic reduction of CBs and the bacteria involved in the process.

There are also aerobic and anaerobic mechanisms for benzene degradation. The aerobic degradation of benzene in contaminated groundwater is primarily controlled by the abundance of oxygen. The degradation of benzene under aerobic conditions is carried out by ubiquitous subsurface bacteria. The anaerobic degradation of benzene has been shown to occur under varying conditions (nitrate, iron, and sulfate reduction as well as methanogenesis) and is recognized as a relevant environmental process. A minireview of anaerobic benzene degradation by bacteria (Vogt et al. 2011) lists some recognized genus-level phylotypes (Table 10.6).

TABLE 10.1
Example Genus-Level 16S Sample Profile with Putative Functions

		Monitoring Well LLL		
Genus (Closest Match)	**% of Profile**	**Metabolism**	**Gram**	**Putative COC Capability**
Sulfuritalea	6.33	Facultative/SOX (Sulfur oxidizing)	Negative	
Chlamydia	5.39		Negative	
Nitrosovibrio	4.45	Aerobic	Negative	
Streptomyces	4.45	Aerobic	Positive	
Candidatus Protochlamydia	3.79	Aerobic		
Methylobacter	3.59	Aerobic/MOB (Methane oxidizing bacteria)	Negative	Cometabolic potential
Eubacterium	3.27	Anaerobic	Variable	
Geobacter	1.79	Anaerobic/MRB (Metal reducing bacteria)	Negative	Produces cofactor
Methylocystis	1.70	Aerobic/MOB		Cometabolic potential
Verrucomicrobium	1.50	Anaerobic/fermenter	Negative	
Aquicella	1.49	Aerobic	Negative	
Arthrobacter	1.48	Aerobic	Positive	
Nitrospira	1.47	Aerobic/NOB (Nitrite oxidizing bacteria)	Negative	
Candidatus Metachlamydia	1.46	Aerobic	Negative	
Candidatus Rhabdochlamydia	1.25	Aerobic	Negative	
Opitutus	1.23	Anaerobic	Negative	
Dehalococcoides	1.21	Anaerobic	Positive	Reductive dechlorination
Pelotomaculum	1.19	Anaerobic	Positive	Anaerobic benzene
Clostridium	1.16	Anaerobic/fermenter	Positive	Produces cofactors
Salinibacter	1.16			
Dechloromonas	1.12	Anaerobic/NRB (Nitrate reducing bacteria)	Negative	Anaerobic benzene
Candidatus Babela	1.11			
Planktothricoides	1.07			
Prochlorococcus	1.02	Aerobic	Negative	
Desulfomonile	0.99	Anaerobic	Negative	Reductive dechlorination
Anaeromyxobacter	0.93	Anaerobic/MRB	Negative	
Acidobacterium	0.91	Aerobic	Negative	

TABLE 10.2
Example Bacterial Genera with Members Capable of Chlorinated Benzene Cometabolism

Genus	Contaminant	Cometabolite	Reference
Pseudomonas	1,2-DCB, 1,3-DCB	CB	Haigler et al. (1992)
Pseudomonas	1,2-DCB, 1,2,4-TCB	CB	Brunsbach and Reineke (1994)
Methylosinus	1,2,3-TCB	Methane/formate	Sullivan and Chase (1996)
Methylocystis	CB	Methane	Jechorek et al. (2003)
Mycobacterium	CB	Propane	Burback and Perry (1993)
Pseudomonas	CB	Benzene	Bestetti et al. (1992)
Pseudomonas	1,2-DCB	Benzene	Ballschmiter and Scholz (1981)

TABLE 10.3
Aerobic Bacterial Strains That Use Chlorinated Benzenes (CB) as a Sole Source of Carbon and Energy

Bacterial Strain	Congener	References
Burkholderia sp. strain PS12	CB	Sander et al. (1991)
Burkholderia sp. strain PS14	CB	Sander et al. (1991)
Escherichia hermanii	CB	Kiernicka et al. (1999)
Hybrid strain WR1313	CB	Oltmanns et al. (1988)
Pseudomonas aeruginosa RHO1	CB	Brunsbach and Reineke (1994)
Pseudomonas putida GJ31	CB	Oldenhuis et al. (1989)
Pseudomonas sp. strain JS100	CB	Haigler et al. (1988)
Pseudomonas sp. strain JS150	CB	Haigler et al. (1992)
Pseudomonas sp. strain JS6	CB	Pettigrew et al. (1991)
Ralstonia sp. strain JS705	CB	van der Meer et al. (1998)
Rhodococcus phenolicus	CB	Rehfuss and Urban (2005)
Rhodococcus sp.	CB	Vogt et al. (2004a)
Pseudomonas sp.	CB	Vogt et al. (2004a)
Xanthobacter sp.	CB	Vogt et al. (2004a)
Paenibacillus sp.	CB	Vogt et al. (2004a)
Kocuria sp.	CB	Vogt et al. (2004a)
Stenotrophomonas sp.	CB	Vogt et al. (2004a)
Unidentified strain 1469	CB	Nishino et al. (1992)
Unidentified strain 1474	CB	Nishino et al. (1994)
Unidentified strain WR1306	CB	Reineke and Knackmuss (1984)
Planococcus sp. strain ZD22	CB	Li et al. (2006)
Acidovorax facilis 13517	CB	Vogt et al. (2004a,b)
Cellulomonas turbata B529	CB	Vogt et al. (2004b)
Pseudomonas veronii 13547	CB	Vogt et al. (2004b)
Pseudomonas veronii B549	CB	Vogt et al. (2004b)
Paenibacillus polymyxa B550	CB	Vogt et al. (2004b)
Burkholderia sp. strain PS12	1,2-DCB	Sander et al. (1991)
Burkholderia sp. strain PS14	1,2-DCB	Sander et al. (1991); Rapp and Timmis (1999)
Pseudomonas sp. strain GJ60	1,2-DCB	Oldenhuis et al. (1989)
Pseudomonas sp. strain P51	1,3-DCB	Van Der Meer et al. (1987)
Alcaligenes sp. strain OBB65	1,3-DCB	Debont et al. (1986)
Burkholderia sp. strain PS12	1,3-DCB	Sander et al. (1991)
Burkholderia sp. strain PS14	1,3-DCB	Sander et al. (1991); Rapp and Timmis (1999)
Pseudomonas sp. strain JS100	1,2-DCB	Haigler et al. (1988)
Pseudomonas sp. strain P5	1,2-DCB	Van Der Meer et al. (1987)
Acidovorax avenae	1,2-DCB	Monferran et al. (2005)
Alcaligenes sp. strain R3	1,4-DCB	Oltmanns et al. (1988)
Burkholderia sp. strain PS12	1,4-DCB	Sander et al. (1991)
Burkholderia sp. strain PS14	1,4-DCB	Sander et al. (1991); Rapp and Timmis (1999)
Hybrid strain WR1323	1,4-DCB	Oltmanns et al. (1988)
Pseudomonas aeruginosa RHO1	1,4-DCB	Oltmanns et al. (1988); Brunsbach and Reineke (1994)
Pseudomonas sp. B1	1,4-DCB	Oltmanns et al. (1988)
Pseudomonas sp. strain JS150	1,4-DCB	Haigler et al. (1992)
Pseudomonas sp. strain JS6	1,4-DCB	Spain and Nishino (1987)

(Continued)

TABLE 10.3 (CONTINUED)
Aerobic Bacterial Strains That Use Chlorinated Benzenes (CB) as a Sole Source of Carbon and Energy

Bacterial Strain	Congener	References
Pseudomonas sp. strain P51	1,4-DCB	Van Der Meer et al. (1987)
Sphingomonas (Alcaligenes) A175	1,4-DCB	Schraa et al. (1986)
Unidentified strain 1474	1,4-DCB	Nishino et al. (1994)
Xanthobacter flavus 14p1	1,4-DCB	Sommer and Gorisch (1997)
Rhodococcus phenolicus	1,4-DCB	Rehfuss and Urban (2005)
Burkholderia sp. strain PS12	1,2,4-TCB	Sander et al. (1991)
Burkholderia sp. strain PS14	1,2,4-TCB	Sander et al. (1991); Rapp and Timmis (1999)
Pseudomonas sp. strain P51	1,2,4-TCB	Van Der Meer et al. (1987)
Pseudomonas chlororaphis RW71	1,2,3,4-TeCB	Potrawfke et al. (1998)
Burkholderia (Pseudomonas) PS12	1,2,4,5-TeCB	Beil et al. (1998)
Burkholderia sp. strain PS14	1,2,4,5-TeCB	Sander et al. (1991); Rapp and Timmis (1999)

TABLE 10.4
Examples of Anaerobic Reductive Dechlorination of CBs

Congener	Reference
Trichlorobenzenes	Holliger et al. (1992)
Trichlorobenzenes	Jayachandran et al. (2004)
Trichlorobenzenes	Adrian et al. (2000a,b)
Dichlorobenzenes	Fung et al. (2009)
Dichlorobenzenes	Bosma et al. (1988)
Monochlorinated benzene	Fung et al. (2009)
Monochlorinated benzene	Nowak et al. (1996)
Monochlorinated benzene	Masunaga et al. (1996)
Monochlorinated benzene	Liang et al. (2013)
Benzene	Liang et al. (2013)

TABLE 10.5
Bacteria Identified in CB Anaerobic Dechlorination

Bacterial Genus	Reference
Dehalococcoides sp.	Jayachandran et al. (2003)
Dehalococcoides sp.	Adrian et al. (2000a,b)
Dehalococcoides sp.	Fennell et al. (2004)
Dehalococcoides sp.	Zhou et al. (2015)
*Geobacter sp.	Zhou et al. (2015)
Dehalococcoides sp.	Vandermeeren et al. (2014)
Desulfitobacterium sp.	Vandermeeren et al. (2014)
Dehalobacter sp.	Vandermeeren et al. (2014)
Sedimentibacter sp.	Vandermeeren et al. (2014)
Dehalobacter sp.	Nelson et al. (2014)

TABLE 10.6
Example Anaerobic Benzene-Degrading Bacteria

Bacterial Genus	Electron Acceptor	Reference
Azoarcus	NO_3^-	
Dechloromonas	NO_3^-	
Peletomaculum	NO_3^-	
Chlorobi	NO_3^-	
Geobacter	Fe(III)	
Peletomaculum-related	SO_4^-	Vogt et al. (2011)
Pelobacter-related	SO_4^-	
Desulfobacterium	HCO_3^- or SO_4^-	
Desulfotomaculum	HCO_3^- or SO_4^-	
Spirochaetes	HCO_3^- or SO_4^-	
Desulfosporosinus	HCO_3^- or SO_4^-	

RESULTS FROM THE 16S ANALYSIS

Aerobic cometabolism of CBs: Table 10.7 reports methane-oxidizing bacteria (methanotrophs) and Mycobacteria (known to degrade CBs via propane cometabolism). Methane-oxidizing bacteria have the capability to produce soluble or particulate methane monooxygenase (MMO) and are of interest due to the high cometabolic activity of this enzyme with contaminants such as chlorobenzenes.

Aerobic direct metabolism of CBs: Table 10.8 is the sum of the bacteria identified (genus level) that may be able to carry out aerobic degradation of CBs. It is important to note that distinguishing between cometabolism and direct aerobic degradation is difficult, and some bacterial genera have members that can perform either function.

Reductive dechlorination of CBs: There were several genus-level detections of bacteria known to carry out various reductive dechlorination reactions (Table 10.9), including *Dehalobacter*, *Dehalobacterium*, *Dehalococcoides*, *Dehalogenimonas*, and *Desulfitobacterium*.

TABLE 10.7
Bacterial Putative Function of Aerobic Cometabolism (Estimated Percentage of Profile)

		Source Area Samples				
Sample ID	BBB	CCC	EEE	AAA		
% of profile	1.24	9.01	0.48	0.58		
Mid-Plume Area Samples						
Sample ID	JJJ	HHH	III	FFF	GGG	
% of profile	25.03	0.27	4.37	0.18	0.25	
Distal Area Samples						
Sample ID	LLL	QQQ	MMM	RRR	OOO	PPP
% of profile	9.69	15.01	5.85	3.58	26.90	3.33

TABLE 10.8
Bacterial Putative Function of Aerobic Direct Metabolism (Estimated Percentage of Profile)

	Source Area Samples					
Sample ID	BBB	CCC	EEE	AAA		
% of profile	0.93	1.62	0.38	10.78		
Mid-Plume Area Samples						
Sample ID	JJJ	HHH	III	FFF	GGG	
% of profile	0.72	3.29	6.01	0.64	16.06	
Distal Area Samples						
Sample ID	LLL	QQQ	MMM	RRR	OOO	PPP
% of profile	1.50	1.72	1.53	3.82	0.63	0.81

TABLE 10.9
Bacterial Putative Function of Reductive Dechlorination (Estimated Percentage of Profile)

	Source Area Samples					
Sample ID	BBB	CCC	EEE	AAA		
% of profile	4.29	3.00	1.28	4.65		
Mid-Plume Area Samples						
Sample ID	JJJ	HHH	III	FFF	GGG	
% of profile	1.31	0.17	1.91	1.26	0.47	
Distal Area Samples						
Sample ID	LLL	QQQ	MMM	RRR	OOO	PPP
% of profile	2.33	1.26	1.10	4.74	1.93	1.40

16S analysis data showed that there were bacteria capable of chlorobenzene and benzene degradation throughout the site. The specific putative capabilities of the microbes site-wide to degrade the contaminants are shown in Figure 10.7. The results from the 16S analysis do not stand alone, but when combined with contaminant, geochemical, and isotopic data, they can provide a picture of the possible contaminant degradation and metabolic activity in the subsurface.

PRACTICE AND EXERCISES:

Question 1: What has happened with bioremediation in the last 20 years that makes it more applicable to remediation?

Question 2: Why might bioremediation be less costly than other more active remediation technologies?

Question 3: What is biotransformation?

Question 4: Name one characteristic of a living cell.

Question 5: Where is the cell's genetic coding held?

Question 6: Name two of the general bacterial "guilds" as described by their terminal electron acceptors.

Question 7: Describe two questions that qPCR analysis can answer regarding the suitability of bioremediation.

Question 8: What is the difference between 16S rRNA analysis and functional gene analysis?

Question 9: What is meant by "putative" function?

Question 10: A site is contaminated with gasoline (benzene). What would be your approach to characterize the microbial community and discover whether there were benzene degraders at the site?

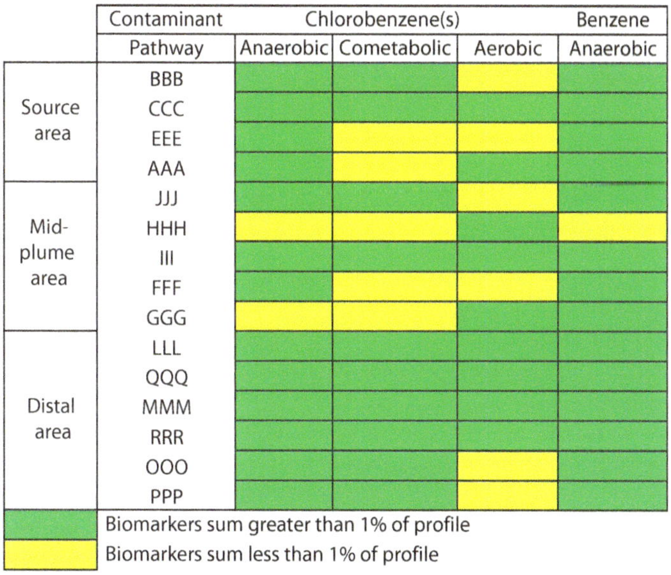

FIGURE 10.7 Putative metabolic capabilities of native bacteria.

ANSWERS:

Answer 1: Answer should include the discovery that many that are amenable to bioremediation and biotransformation.

Answer 2: Answer should describe that bioremediation can be carried out *in situ* and without the need for active mechanized approaches.

Answer 3: Biotransformation is any alteration of the molecular or atomic structure of a contaminant by microorganisms.

Answer 4: Answer should include one of the following:

a. Self-feeding or nutrition. Microorganisms take up chemicals, including contaminants, from the environment and transform them from one form to another. During these biotransformations, there is a release of energy and the excretion of cellular waste products.
b. Self-replication and growth. Microorganisms are capable of directing their own existence. As a result of metabolic nutritional processes, microorganisms can grow and replicate.
c. Differentiation. Microorganisms can undergo changes in form and function. These changes are usually a direct result of environmental pressures.
d. Communication. Microorganisms respond to chemical and physical stimuli in their environment and can interact and communicate with other microorganisms through chemical signaling. Many microorganisms are also capable of movement and respond to various stimuli (e.g., carbon and energy).
e. Evolution. Microorganisms are capable of hereditary changes that can influence their overall fitness.

Answer 5: Answer would include DNA and chromosome.

Answer 6: Answer would include two of the following:

a. Aerobes
b. Nitrate reducers
c. Metal reducers
d. Sulfate reducers

e. Fermenters
f. Methanogens

Answer 7: Answer would include two of the following:

a. Were contaminant-degrading microorganisms present?
b. What were the relative numbers and where?
c. Is monitored natural attenuation (MNA) feasible?
d. Given the numbers of dechlorinating bacteria identified at the site, if the rate of degradation is acceptable, this could be a candidate for MNA.
e. Is an amendment (e.g., carbon source) necessary?
f. Is bioaugmentation necessary?

Answer 8: Answer would include that 16S analysis is based on phylogeny or what type of bacteria is identified, and functional gene analysis directly identifies genes involved in biotransformation and biodegradation.

Answer 9: Answers would include that when we measure DNA, we are measuring genes and not cells. *Putative* is defined as *generally considered or reputed to be*. In the case of monitoring, for example, bacteria by 16S, we may identify bacteria that are generally considered or reputed to be capable of biotransforming a contaminant of interest. This is why monitoring DNA is only considered to be one line of evidence and must be accompanied by geochemical and contaminant analysis.

Answer 10: The answer would include detecting bacteria via 16S analysis that are known to carry out benzene degradation under the site geochemical conditions or by detecting the functional genes that code for the enzymes involved in benzene degradation.

REFERENCES

Adrian, L., Szewzyk, U., and Görisch H., 2000a. Bacterial growth based on reductive dechlorination of trichlorobenzenes. *Biodegradation* 11, 73–81.

Adrian, L., Szewzyk, U., and Görisch H., 2000b. Bacterial dehalorespiration with chlorinated benzenes. *Nature* 408, 580–583.

Ballschmiter, K. and Scholz, C., 1981. Microbial-degradation of chlorinated arenes .7. initial steps in the degradation of chlorobenzene derivatives by *Pseudomonas putida*. *Angew. Chem.-Int. Edit. Engl.* 20, 955–956.

Beil, S., Mason, J.R., Timmins, K.N., and Pieper, D.H., 1998. Identification of chlorobenzene dioxygenase sequence elements involved in dechlorination of 1,2,4,5-tetrachlorobenzene. *J. Bacteriol.* 180, 5520–5528.

Bestetti, G., Galli, E., Leoni, B., and Pelizzoni, F., 1992. Regioselective hydroxylation of chlorobenzene and chlorophenols by a *Pseudomonas putida*. *Appl. Microbiol. Biotechnol.* 37, 260–263.

Bosma, T., van der Meer, J.R., Schraa, G., Tros, M.E., and Zehnder, A.J.B., 1988. Reductive dechlorination of all trichlorobenzene and dichlorobenzene isomers. *FEMS Microbiol. Ecol.* 53, 223–229.

Brunsbach, F., and Reineke, W., 1994. Degradation of chlorobenzenes in soil slurry by a specialized organism. *Appl. Microbiol. Biotechnol.* 42, 415–420.

Burback, B. and Perry, J., 1993. Biodegradation and biotransformation of groundwater pollutant mixtures by *Mycobacterium vaccae*. *Appl. Environ. Microbiol.* 59, 1025–1029.

Debont, J., Vorage, M.J., Hartmans, S., and van den Tweel, W.J., 1986. Microbial degradation of 1,3-dichlorobenzene. *Appl. Environ. Microbiol.* 52, 677–680.

Fennell, D., Nijenhuis, I., Wilson, S.F., Zinder, S.H., and Häggblom, M.M., 2004. *Dehalococcoides ethenogenes* strain 195 reductively dechlorinates diverse chlorinated aromatic pollutants. *Environ. Sci. Technol.* 38, 2075–2081.

Fung, J., Weisenstein, B.P., Mack, E.E., Vidumsky, J.E., Ei, T.A., and Zinder, S.H., 2009. Reductive dehalogenation of dichlorobenzenes and monochlorobenzene to benzene in microcosms. *Environ. Sci. Technol.* 43(7), 2302–2307.

Haigler, B., Nishino, S.F., and Spain, J.C. et al., 1988. Degradation of 1,2-dichlorobenzene by a *Pseudomonas* sp. *Appl. Environ. Microbiol.* 54, 294–301.

Haigler, B., Pettigrew, C.A., and Spain, J.C., et al., 1992. Biodegradation of mixtures of substituted benzenes by *Pseudomonas* sp strain-JS150. *Appl. Environ. Microbiol.* 58, 2237–2244.

Holliger, C., Schraa, G., Stams, A.J., and Zehnder, A.J., 1992. Enrichment and properties of an anaerobic mixed culture reductively dechlorinating 1,2,3-trichlorobenzene to 1,3-dichlorobenzene. *Appl. Environ. Microbiol.* 58, 1636–1644.

Jayachandran, G., Gorisch, H., and Adrian, L., 2003. Dehalorespiration with hexachlorobenzene and pentachlorobenzene by *Dehalococcoides* sp strain CBDB1. *Arch. Microbiol.* 180, 411–416.

Jayachandran, G., Gorisch, H., and Adrian, L., 2004. Studies on hydrogenase activity and chlorobenzene respiration in Dehalococcoides sp. strain CBDB1. *Arch. Microbiol.* 182, 498 504.

Jechorek, M., Wendlandt, K.D., and Beck, M., 2003. Cometabolic degradation of chlorinated aromatic compounds. *J. Biotechnol.* 102, 93–98.

Kiernicka, J., Seignez, C., and Peringer, P., 1999. *Escherichia hermanii* – a new bacterial strain for chlorobenzene degradation. *Lett. Appl. Microbiol.* 28, 27–30.

Li, H., Liu, Y.H., Luo, N., Zhang, X.Y., Luan, T.G., Hu, J.M., Wang, Z.Y. et al., 2006. Biodegradation of benzene and its derivatives by a psychrotolerant and moderately haloalkaliphilic *Planococcus* sp strain ZD22. *Res. Microbiol.* 157, 629–636.

Liang, X., Devine, C.E., Nelson, J., Sherwood-Lollar, B., Zinder, S., and Edwards, E.A., 2013. Anaerobic conversion of chlorobenzene and benzene to CH_4 and CO_2 in bioaugmented microcosms. *Environ. Sci. Technol.* 47, 2378–2385.

Madigan, M., Martinko, J.M., and Parker, J.M., 1996. *Brock's Biology of Microorganisms* 8th edn. Prentice Hall College Div, Upper Saddle River, NJ. ISBN-10: 0135208750.

Masunaga, S., Susarla, S., and Yonezawa, Y., 1996. Dechlorination of chlorobenzenes in anaerobic estuarine sediment. *Water Sci. Technol.* 33, 173–180.

Monferrán, M.V., Echenique, J.R., and Wunderlin, D.A., 2005. Degradation of chlorobenzenes by a strain of *Acidovorax avenae* isolated from a polluted aquifer. *Chemosphere* 61, 98–106.

Nelson, J., Jiang, J., and Zinder, S., 2014. Dehalogenation of chlorobenzenes, dichlorotoluenes, and tetrachloroethene by three Dehalobacter spp. *Environ. Sci. Technol.* 48(7), 3776–3782.

Nishino, S., Spain, J., Belcher, L., and Litchfield, C., 1992. Chlorobenzene degradation by bacteria isolated from contaminated groundwater. *Appl. Environ. Microbiol.* 58, 1719–1726.

Nishino, S., Spain, J., and Pettigrew, C., 1994. Biodegradation of chlorobenzene by indigenous bacteria. *Environ. Toxicol. Chem.* 13, 871–877.

Nowak, J., Kirsch, N., Hegemann, W., and Stan, H., 1996. Total reductive dechlorination of chlorobenzenes to benzene by a methanogenic mixed culture enriched from Saale river sediment. *Appl. Microbiol. Biotechnol.* 45, 700–709.

Oldenhuis, R., Kuijk, L., Lammers, A., Janssen, D., and Witholt, B., 1989. Degradation of chlorinated and non-chlorinated aromatic solvents in soil suspensions by pure bacterial cultures. *Appl. Microbiol. Biotechnol.* 30, 211–217.

Oltmanns, R., Rast, H., and Reineke, W., 1988. Degradation of 1,4-dichlorobenzene by enriched and constructed bacteria. *Appl. Microbiol. Biotechnol.* 28, 609–616.

Pettigrew, C., Haigler, B., and Spain, J.C., 1991. Simultaneous biodegradation of chlorobenzene and toluene by a *Pseudomonas* strain. *Appl. Environ. Microbiol.* 57, 157–162.

Potrawfke, T., Timmis, K., and Wittich, R-M., 1998. Degradation of 1,2,3,4- tetrachlorobenzene by *Pseudomonas chlororaphis* RW71. *Appl. Environ. Microbiol.* 64, 3798–3806.

Rapp, P. and Timmis, K.N., 1999. Degradation of chlorobenzenes at nanomolar concentrations by *Burkholderia* sp strain PS14 in liquid cultures and in soil. *Appl. Environ. Microbiol.* 65, 2547–2552.

Rehfuss, M. and Urban, J., 2005. *Rhodococcus phenolicus* sp nov., a novel bioprocessor isolated actinomycete with the ability to degrade chlorobenzene, dichlorobenzene and phenol as sole carbon sources. *Syst. Appl. Microbiol.* 28, 695–701.

Reineke, W. and Knackmuss, H.J., 1984. Microbial-metabolism of haloaromatics – isolation and properties of a chlorobenzene-degrading bacterium. *Appl. Environ. Microbiol.* 47, 395–402.

Sander, P., Wittich, R-M., Fortnagel, P., Wilkes, H., and Francke, W., 1991. Degradation of 1,2,4-trichlorobenzene and 1,2,4,5-tetrachlorobenzene by *Pseudomonas* strains. *Appl. Environ. Microbiol.* 57, 1430–1440.

Schraa, G., Boone, M., Jetten, M., van Neerven, A., Colberg, P., and Zehnder, A., 1986. Degradation of 1,4-dichlorobenzene by *Alcaligenes* sp strain A175. *Appl. Environ. Microbiol.* 52, 1374–1381.

Sommer, C. and Gorisch, H., 1997. Enzymology of the degradation of (di)chlorobenzenes by *Xanthobacter flavus* 14p1. *Arch. Microbiol.* 167, 384–391.

Spain, J. and Nishino, S., 1987. Degradation of 1,4-dichlorobenzene by a *Pseudomonas* sp. *Appl. Environ. Microbiol.* 53, 1010–1019.

Sullivan, J. and Chase, H., 1996. 1,2,3-Trichlorobenzene transformation by *Methylosinus trichosporium* OB3b expressing soluble methane monooxygenase. *Appl. Microbiol. Biotechnol.* 45, 427–433.

Van Der Meer, J., Zehnder, A., and de Vos, M., 1987. Degradation of low concentrations of dichlorobenzenes and 1 2 4 trichlorobenzene by *Pseudomonas* sp strain p51 in nonsterile soil columns. *FEMS Microbiol. Ecol.* 45, 333–342.

Van der Meer, J., Werlen, C., Nishino, S., and Spain, J., 1998. Evolution of a pathway for chlorobenzene metabolism leads to natural attenuation in contaminated groundwater. *Appl. Environ. Microbiol.* 64, 4185–4193.

Vandermeeren, P., Herrmann, S., Cichocka, D., and Springael, D., 2014. Diversity of dechlorination pathways and organohalide respiring bacteria in chlorobenzene dechlorinating enrichment cultures originating from river sludge. *Biodegradation.* 25(5), 757–776.

Vogt, C., Alfreider, A., Lorbeer, H., Hoffman, D., Wuensche, L., and Babel, W., 2004a. Bioremediation of chlorobenzene-contaminated ground water in an in situ reactor mediated by hydrogen peroxide. *J. Contam. Hydrol.* 68, 121–141.

Vogt, C., Simon, D., and Alfreider, A., 2004b. Microbial degradation of chlorobenzene under oxygen-limited conditions leads to accumulation of 3-chlorocatechol. *Environ. Toxicol. Chem.* 23, 265–270.

Vogt, C., Kleinsteuber, S., and Richnow, H.H., 2011. Anaerobic benzene degradation by bacteria. *Microb. Biotechnol.* 4(6), 710–724.

Zhou, X., Zhang, C., Zhang, D., Awata, T., Xiao, Z., Yang, Q., and Katayama, A., 2015. Polyphasic characterization of an anaerobic hexachlorobenzene-dechlorinating microbial consortium with a wide dechlorination spectrum for chlorobenzenes. *J. Biosci. Bioeng.* 120(1), 62–68.

11 Environmental Statistics Methods

Ning-Wu Chang, Jun Lu, and Yue Rong

CONTENTS

Environmental projects typically involve the remediation of contaminants in heterogeneous site conditions, such as various geological settings. Also, contaminant-specific physical and chemical properties will affect contaminant behavior in heterogeneous site conditions. Statistics is a tool used to address various issues in a specific environmental project. This chapter is not a tutorial on basic mathematical statistics; rather, it addresses certain practical applications in environmental projects. The basic statistics and how to derive the statistics theories can be found in other statistics text books.

11.1 INTRODUCTION: HOW STATISTICS IS USED IN ENVIRONMENTAL PROJECTS

Statistical analysis is required under the Resource Conservation and Recovery Act (RCRA) program in several areas. To monitor whether a release has occurred at a regulated facility, a background or upgradient condition has to be established using statistical analysis. Once a release has occurred, the evaluation of whether the release exceeds the compliance levels will also require using statistical analysis.

For corrective action under the RCRA program or site remediation under the Comprehensive Environmental Response Compensation and Liability Act (CERCLA) program, statistical analysis can be used throughout the remediation process. During the investigation phase, similar background and compliance exceedance analysis can be conducted. A statistical trend analysis can be conducted to evaluate the behavior of contaminants at the characterization stage and to evaluate the progress of remediation toward compliance. Statistics also plays an important role in human health risk assessment, where the upper-end concentrations (95% upper confidence level [UCL]) usually are used to assess the risk at a site. To improve remedial efforts, statistical analysis also provides another line of evidence to identify remedial optimization opportunities.

In March 2009, the United States Environmental Protection Agency (EPA) published a guidance document, "Statistical analysis of groundwater monitoring data at RCRA facilities" (also known as the *Unified Guidance*), which was developed for the statistical analysis of groundwater data at RCRA facilities. Although this document is for RCRA facilities, the statistical concepts are also applicable for other environmental programs. EPA also published a series of data quality management guidance, the EPA QA/G series, which also includes guidance in statistics (Data Quality Assessment: Statistical Methods for Practitioners [EPA QA/G-9S 2006; also see U.S. EPA, 2006a,b,c,d). Both the Unified Guidance and data quality management guidance incorporate the concepts of a systematic planning approach in site remediation. The Interstate Technology & Regulatory Council (ITRC) also published two web-based guidance documents, *Groundwater Statistics and Monitoring Compliance, Statistical Tools for the Project Life Cycle* in 2013 and *Geospatial Analysis for Optimization at Environmental Sites* in 2016, to provide guidance on how statistics can be used in environmental projects.

In this chapter, the following statistical topics related to site remediation are discussed:

- How to handle nondetect value
- How to evaluate the background concentrations
- How to evaluate compliance
- How to use trend analysis
- How to use geospatial analysis

11.2 HOW TO HANDLE NONDETECT VALUE

The levels or concentrations of contaminant at an environmental site will be reported either through field instrument or fixed laboratory analysis. However, it is technically impossible for any analytical method to confirm the complete absence of contaminants. Therefore, laboratories will report a nonzero value as a *detection limit*, which is the lowest concentration that can be reliably detected for a given analytical method. This detection limit is then used to evaluate data. Results below the limit are usually referred to as *nondetect*. The presence of nondetects is unique in environmental data evaluation. A contaminant may be present below that lowest concentration (detection limit). Therefore, inappropriate handling of this nondetect value in data analysis could affect the overall representativeness of the data set.

In statistical analysis, the nondetect should always be treated as an inequality rather than a fixed numerical value. The general approaches to handling nondetects include statistical approaches designed to handle nondetects, ranking-based nonparametric tests, and censored estimation technique to estimate sample statistics. The most commonly used methods (ITRC, GSMC-1) include:

- *Nonparametric methods*: Nonparametric methods treat nondetects as inequalities, order the data values, and then replace the data values with the corresponding ranks.
- *Omission of nondetects*: Although omitting the nondetects from the data set can introduce bias, the omission of non-detects may be a reasonable approach under certain conditions (for example, when only detection is of interest).
- *Substitution method*: Numerical values are assigned to the nondetects. In the past, half of the detection limit or reporting limit has often been used.
- *Kaplan-Meier method*: This method is a nonparametric method that counts the number of data points below each detected concentration and uses that information to generate the probability distribution function for the data set.
- *Robust regression on order statistics (ROS)*: This method plots the detected concentrations and determines a regression line through the data set. The regression line then extrapolates and generates estimates for the censored data of nondetects.
- *Maximum likelihood estimation (MLE)*: This is a parametric method and must assume a distribution type of the data set. The best fit parameters for that distribution will be determined through simulation and then used to calculate the other statistics parameters for that data set.
- *Cohen's method*: This is an adjustment method that assumes a normal distribution. By adjustment of the mean and standard deviation for a given data set, these statistical values are subsequently used to extrapolate to nondetect estimates.

TABLE 11.1
Summary of Results from Various Nondetects Methods

Nondetect Method	Estimated Mean of the Data Set
Omission of nondetect	89.4
Kaplan-Meier	43.64
ROS	40.64
Substitution method (using one-half reporting limit to replace nondetect)	42.14

The assumptions, limitations, strengths, and weaknesses of these methods are further discussed in the ITRC GSMC-1 guidance document.

An example of how nondetect would affect the data evaluation is illustrated here using the ProUCL software (U.S. EPA ProUCL).

There are 11 sampling locations with measured values of (the data set) 67, <5.5, 13, 82, 220, <5.5, <5.5, <5.5, 65, <5.5, and <5.5. There are six locations with measured results below the reporting limit of 5.5. The results using various methods are shown in Table 11.1.

Because of the high percentage of nondetects (6 out of 11 data points) in this data set, omitting the nondetects and substituting typically half of the reporting limit will not be appropriate and will tend to skew the results. Similarly, ROS may not be appropriate for data sets with more than 50% of nondetects, as it creates multiple tied observations at the reporting limit. In this particular example, the appropriate approach will be to use the Kaplan-Meier method.

11.3 HOW TO EVALUATE THE BACKGROUND CONCENTRATIONS

The background concentration can be defined as the concentration of a chemical that is not influenced by the releases from a site and is usually described as *naturally occurring* and/or *anthropogenic* (EPA, 2002). At a suspected or confirmed contaminated site, background concentration is determined on a site-specific basis using samples from nearby areas where a discharge or release did not occur and the soil and/or groundwater conditions resemble as closely as possible those at the site of concern. In some cases, a regional background concentration may be established and used as a screening tool (Chernoff et al., 2005).

During remedial site investigation, one of the critical tasks is to compare the concentrations of chemicals of concern (COC) with the background concentration so that the scope of site-specific COC can be established and risk-based remediation levels for soil and groundwater determined. There are two different approaches for comparing site and background concentrations: nonstatistical and statistical. Nonstatistical methods include (1) direct comparison of site concentrations with background; (2) comparison of 95% UCL between site and background; and (3) weight of evidence demonstration that contaminants are not site related (Florida Department of Environmental Protection, 2012).

In contrast to nonstatistical methods, statistical methods offer a stronger, more robust method, because they enable specifying the probabilities of making decision errors and extrapolating from a set of measurements to the entire site in a scientifically valid fashion. There are two types of statistical methods: parametric and nonparametric. Their applications are summarized in Table 11.2.

To illustrate the applications, an example is provided using the *t*-test to compare on-site lead concentrations with background concentrations (Gauthier, 2002). The data set is shown in Table 11.3.

As indicated in Table 11.2, an approximate normality is required for the data set for a *t*-test to be used. A chi-square (χ^2) goodness of fit test (not provided) demonstrated that the on-site data resemble a normal distribution, and the background data deviate from normal, but not significantly. The calculated *t*-statistic for the comparison of the two data sets is 5.77, and the tabular value for

TABLE 11.2
Summary of Methods for Background Concentrations

Method	Application	Comments
Parametric Tests • Student *t*-test • Behrens-Fisher Student *t*-test	Tests require approximate normality of the estimated means. Use if a larger number of data points are available ($n > 25$). For smaller data sets, examine data for normality or log normality in distribution	Statistically robust and used frequently in parametric data analysis
Nonparametric Tests • Wilcoxon Rank Sum Test (also called the Mann-Whitney Test) • Gehan Test	Use when data are not normally distributed, as rank-ordered tests make no assumption about distribution.	Statistically robust and used frequently in background estimation

Source: Modified from U.S. EPA, *Guidance for comparing background and chemical concentrations in soil for CERCLA sites*. Office of Emergency and Remedial Response. EPA 540-R-01-003, 2002.

TABLE 11.3
On-Site and Background Lead Concentrations

On-site	Background	On-site	Background
353	86	581	263
300	111	762	171
288	306	315	176
459	111	406	266
345	346	279	146
324	142	504	155
473	73	233	81
339	217	299	336
236	149	371	106
310	417	452	80
359	111	384	101
420	584	480	178
357	68	—	—

Student's *t*-distribution with significance level (alpha) of 0.05 and associated degrees of freedom of 48 is 2.01. Since the calculated value (5.77) is larger than the tabular value (2.01), the null hypothesis of the two means being equal must be rejected. This is a piece of strong evidence that the on-site lead concentrations are higher than background levels, suggesting that the facility has indeed affected the soils on-site. In addition, the *t*-test has allowed us to arrive at this conclusion with a quantitative estimate of the confidence in our conclusion.

11.4 HOW TO EVALUATE COMPLIANCE

Once a release has occurred and been confirmed, the next step will be to evaluate whether the release has caused contamination exceeding the regulatory threshold. If the regulatory threshold is exceeded, corrective action will be required. Similarly, a baseline risk assessment during the site investigation stage will be required to determine whether site remediation is needed. Once it is

TABLE 11.4
Statistics Summary for the Kaplan-Meier Method

Mean	**43.64**
Standard error of mean	21.14
95% UCL ($t_{0.95,\ 10} = 1.812$)	81.94
95% LCL ($t_{0.95,\ 10} = 1.812$)	5.33

confirmed that a corrective action or site remediation is needed, the progress of the remediation will be evaluated against the cleanup standards and to determine when the cleanup is complete.

Due to the heterogeneous nature of a remediation site, the environmental data may vary significantly from location to location at a site; therefore, it may be challenging to determine whether compliance is achieved using a single data point or point-to-point comparison with the threshold or standard. In most circumstances, using the statistical approach to demonstrate compliance will be acceptable.

However, certain statistical approaches at different stages of the remediation should be considered. For RCRA release detection, the program is designed to avoid false positive error (40CFR, 264.99 and USEPA, 2009) and to have high confidence that the true concentration has exceeded the threshold or standard. In this case, typically the 95% lower confidence level (LCL) will be used.

During site investigation, to evaluate the risk, the most probable maximum exposure concentrations usually are considered. In this case, typically the upper 95% upper confidence level (UCL) will be used to represent the most probable maximum exposure concentrations. Similarly, as remediation progresses, to achieve the cleanup goals, the false negative error should be avoided with high confidence that the true concentration will be below the threshold or standard. In this circumstances, typically the 95% UCL will also be used.

Use the example from Section 11.2, the Kaplan-Meier method has been selected as the appropriate method to handle the nondetects. The statistical parameters are calculated as shown in Table 11.4.

If the threshold or standard is equal to 10, then for release detection, the 95% LEL (5.33) is below this threshold or standard. Therefore, the data will indicate that the site is still in compliance due to a high percentage of nondetects and as a result, may not require immediate active corrective action. Further monitoring may be needed to confirm the high detection shown in the data set.

If the data set is to evaluate the progress of remediation, even though there is a high percentage of nondetects and detections is below the threshold or standard, the 95% UCL (81.94) exceeds the threshold or standard. Therefore, it will be concluded that the progress of remediation is not in compliance yet, and continued remediation may be required. However, an optimized or more focused remediation may be warranted using the statistical methods in the following sections as a tool.

11.5 HOW TO USE TREND ANALYSIS

Nonparametric trend analyses (such as the Mann-Kendall, seasonal Mann-Kendall, and Sen's Slope) and parametric trend analyses (such as linear regression) are routinely used to evaluate trends in concentrations of contaminants over time to confirm the conceptual site model or to evaluate the progress of remediation. The trend analysis will identify whether there is an increasing trend, a decreasing trend, or no trend and the significance of the trends. For decreasing trends during remediation, the trend can be used to predict the time required to meet remedial objectives. This information can be used to design and optimize the remedies or to demonstrate the expected time frame to meet remedial objectives.

An example is provided to illustrate how a trend analysis is conducted. The data set is shown in Table 11.5 and on Figure 11.1.

TABLE 11.5
Data for Trend Analysis (MTBE)

Date	Concentration	Date	Concentration	Date	Concentration
10/26/1995	7.7	10/9/2003	240	4/26/2011	110
4/22/1996	0.01	4/6/2004	180	10/19/2011	150
6/20/1996	19	10/4/2005	170	4/20/2012	260
10/23/1996	83	4/6/2006	170	10/12/2012	170
4/11/1997	143	10/5/2006	160	4/25/2013	280
11/17/1997	150	1/10/2007	200	10/17/2013	110
4/16/1998	130	4/9/2007	120	4/17/2014	160
10/12/1998	220	7/16/2007	180	10/23/2014	130
4/14/1999	620	10/3/2007	170	4/13/2015	91
10/11/1999	750	1/10/2008	81	4/13/2015	91
4/3/2000	1300	4/22/2008	200	10/19/2015	65
10/4/2000	1300	7/28/2008	120	10/19/2015	65
4/12/2001	360	11/11/2008	140	5/4/2016	61
10/10/2001	130	1/14/2009	190	5/4/2016	61
4/16/2002	170	7/16/2009	140	10/18/2016	42
10/11/2002	1000	4/21/2010	200	10/18/2016	42
4/1/2003	400	10/22/2010	210	5/4/2017	38

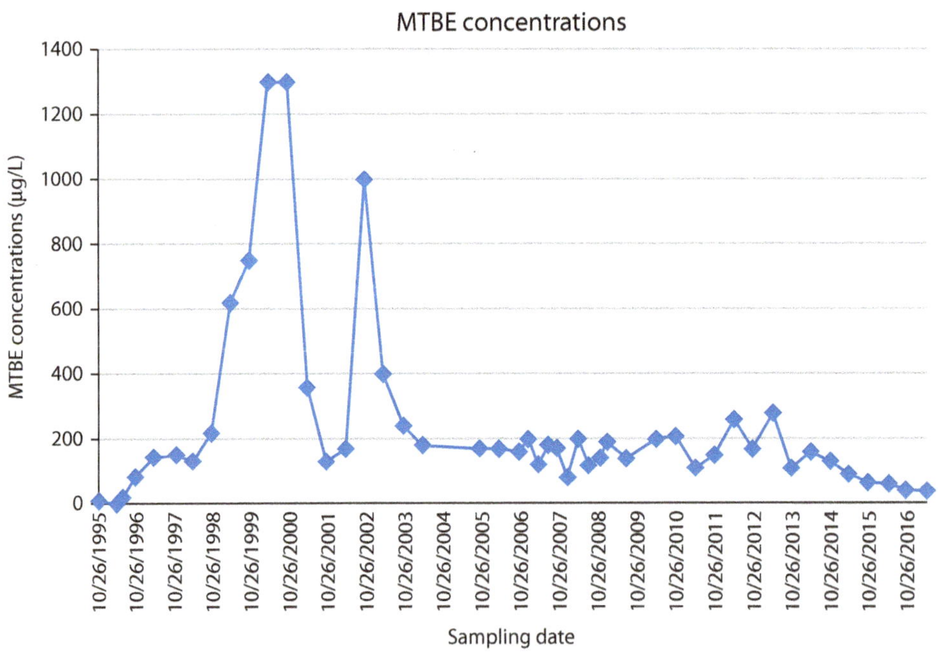

FIGURE 11.1 Plot from trend analysis data.

The data were collected from a monitoring well near an area where MTBE-containing light non-aqueous phase liquid (LNAPL) is an on-going MTBE source for groundwater. As can be observed, from the mid-1990s to 2000, MTBE concentration increased steadily; it then fluctuated in the next few years, followed by a generally stable level. From 2013 to the present, the MTBE concentration showed a steadily decreasing trend.

To confirm the visual observations, Mann-Kendall nonparametric trend analysis was conducted using the MAROS program, and the early trend (I: increasing trend), middle period trend (NT: no trend or fluctuating), and recent trend (D: decreasing trend) are shown in Figure 11.2.

For trend analysis, it is important to understand the history of the site and how the concentrations of contaminants of concern change over time. In this case, almost the entire life cycle of MTBE concentrations was observed. The conclusion may be drawn that the MTBE concentration will likely continue to decrease if there are no additional sources contributing to the plume. If monitored data are limited to a short period of time, greater caution should be taken in drawing conclusions regarding the trend of contaminants of concern in the plume.

11.6 HOW TO USE GEOSPATIAL ANALYSIS

Geospatial analysis is the process of compiling and analyzing data related in time or space (ITRC, GRO-1, 2016). When a contour map of groundwater elevation or concentration of contaminants has been generated, some level of geospatial analysis can be performed. To conduct the geospatial analysis, certain geospatial methods are applied. These methods incorporate the spatial and temporal dependence between data points, which is an important feature of almost all data collected for remediation projects, and estimate values between sampling points to generate the contour lines or depict changes over time. These methods can include statistical analyses involving spatial or temporal interpretations of environmental data as well as uncertainty evaluations of those data.

The geospatial methods are categorized in order of their complexity, ranging from simple to more complex to advanced (ITRC GRO-1, 2016). The simple methods, such as inverse distance weighting (IDW), natural neighbors, and Thiessen polygons methods, typically do not require statistical assumptions. The more complex methods, such as parametric and nonparametric regression, local linear regression, splines and kernel methods, make inferences about the statistical distribution of a sampled population. Advanced geospatial methods, such as geostatistical methods (Kriging) and conditional simulation, are generally more robust. Geospatial methods typically involve intensive calculation. Therefore, the geospatial analysis typically requires specialized software to perform the calculations.

The results of geospatial analysis can be used not only to generate contour maps but also to augment the data gap. These results can be used for any stage of the remediation project, such as estimating the contamination mass during characterization and estimating the remaining contamination during the progress of remediation to optimize the overall remediation efforts. It is particularly useful in complex remediation projects.

The fundamental concepts and a discussion of geospatial analysis methods are provided in the ITRC GRO-1 guidance document (2016) along with how to conduct geospatial analysis. It is important to recognize that geospatial analysis relies on relationships between data points, auto-correlation in statistical terms, and interpolation of data. This is different from the classical statistical approach, where the data set typically assumes random and independent observations.

The following examples illustrate how geospatial analysis can be used in remediation projects.

Example 1: This example illustrates how IDW can be used to estimate the average concentration in the area. The sampling location with representative area is shown in Figure 11.3. Use IDW to estimate each polygon concentration with the zonal statistics approach from ArcGIS. The results are shown in Table 11.6. The average concentration from the actual measured results is 98.4, which is slightly lower than the calculated average concentration of 99.6 using the IDW methods. Therefore, in this case, without using geospatial analysis, the estimated average concentration for exposure and the estimated total mass of contaminants may be lower than anticipated.

Example 2: Figure 11.4 illustrates the geospatial method–generated contour maps demonstrating the progress of a remediation. Using the proper interpolation of the monitoring results

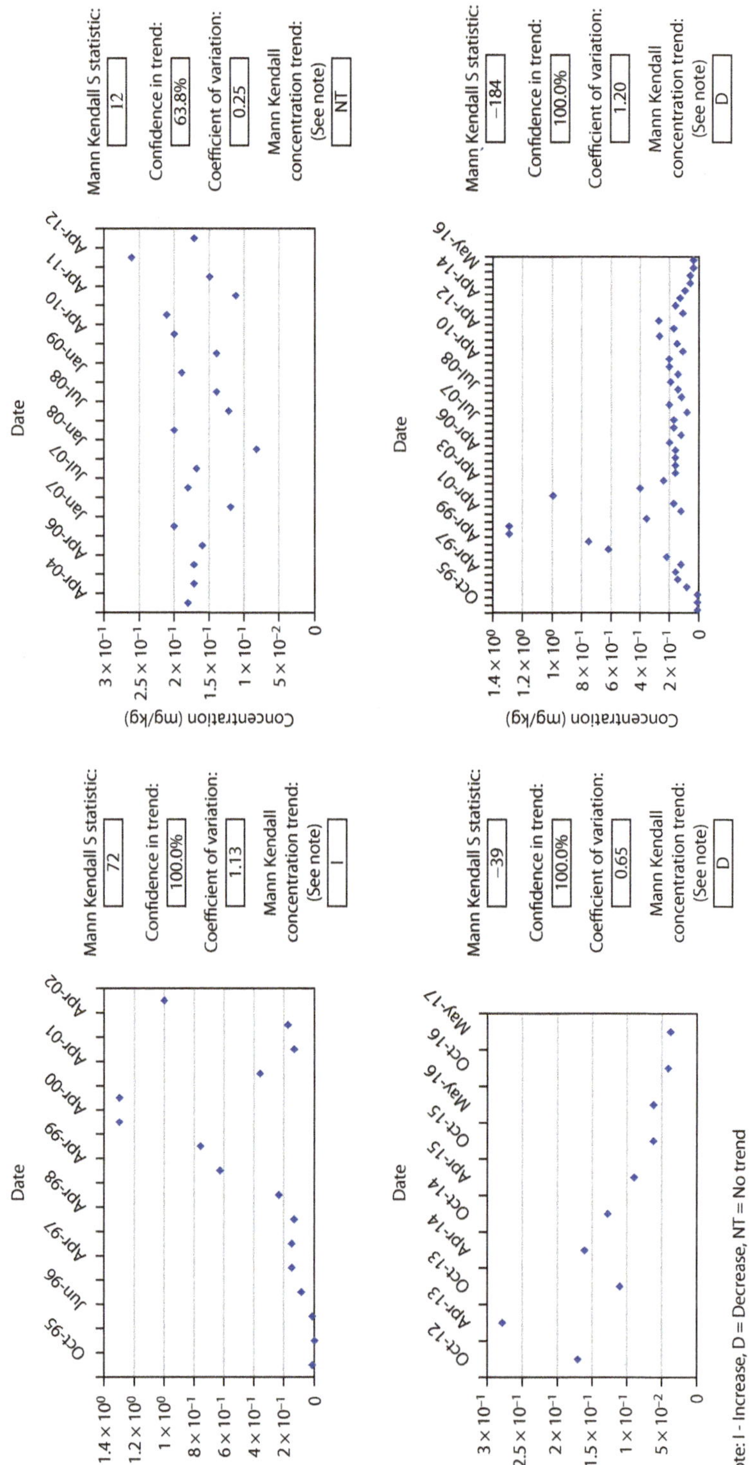

Note: I - Increase, D = Decrease, NT = No trend

FIGURE 11.2 Results of Mann-Kendall nonparametric trend analysis.

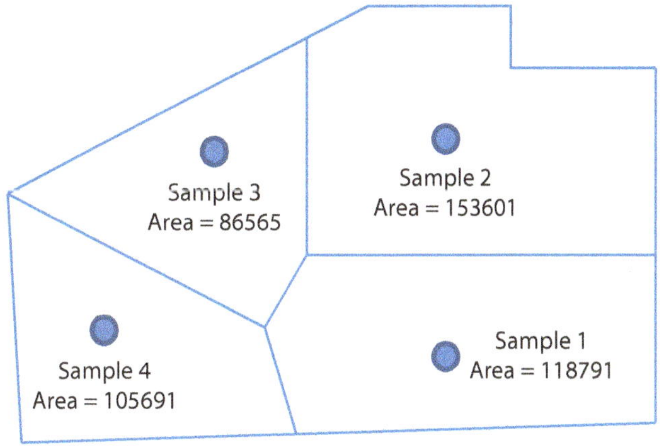

FIGURE 11.3 Example of IDW sampling locations in representative areas.

TABLE 11.6
IDW Data Analysis Results

Sample Location	Polygon Area	Actual Measured Concentration	IDW/Zonal Statistics within the Polygon			
			Minimum Concentration	Maximum Concentration	Mean Concentration	Standard Deviation
1	118,791	124	105.6	124	115	5.17
2	153,601	95.6	83.8	108.4	98.7	4.72
3	86,565	46.1	46.1	93.5	69.5	13.46
4	105,691	128	88.2	128	108.2	9.64

Estimated average concentration for Areas 1 through 4 = 99.6.

and visual illustration, it is clearly demonstrated that the remedial efforts had reduced the source area and overall extent of the plume. However, the hot spot remained, but with a smaller area. In addition, the remedial efforts may not affect an area to the west of the original hot spot. These visual illustrations will provide additional information to optimize the remediation by addressing the remaining hot spot and the area not being effectively remediated.

11.7 SUGGESTIONS

Statistics is a good tool in environmental site assessment and remediation. However, this tool must be used with caution. Before statistics is used, the concepts and methods of the statistics must be understood. Environmental data so often are detected in a wide range of concentrations due to heterogeneous site conditions, which is not like conventional data for classical statistics. Therefore, site-specific information on assessment and remediation should be learned first. Before one or more statistical methods are used, the raw data should be reviewed to check data key parameters such as the mean and standard deviation and the data distribution plot. If possible, more than one statistical method should be used for the environmental project. In addition, assumptions and uncertainties associated with statistical methods must be understood prior to interpreting the results. Statistical analysis results should always be interpreted on the basis of site-specific physical meanings. It will be prudent to use multiple lines of evidence for a final decision, not solely based on statistical methods.

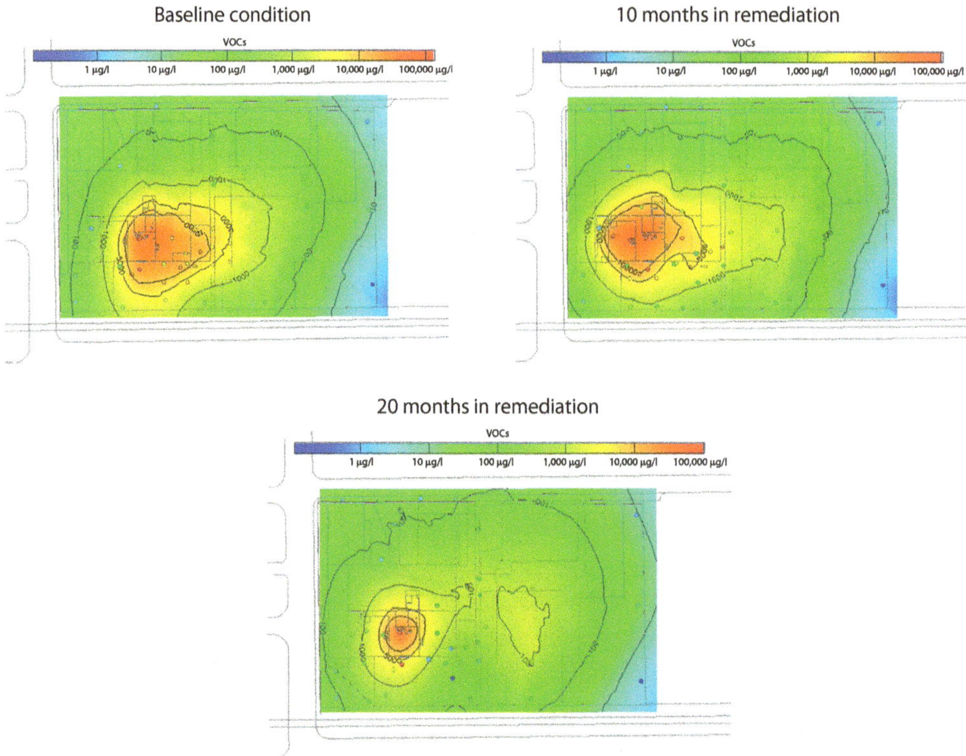

FIGURE 11.4 Geospatial method and contour maps.

PRACTICE AND EXERCISES

Question 1: Estimation of Percentile and Confidence Interval

Use data in Table E1 to calculate percentiles and confidence interval under the normal distribution and log-normal distribution, respectively. Explain why the results are different.

Answer 1:

Calculation of Percentiles

To understand data distribution, a percentile value usually needs to be calculated. Given the mean and standard deviation of the data distribution, percentiles of a normal distribution can be calculated using a normal distribution probability table as follows (Winkler and Hays, 1975):

$$x_p = \mu + Z_p \cdot \sigma \tag{11.1}$$

where:

- x_p is the pth percentile value of variable x
- μ and σ are the mean and standard deviation of the variable x distribution, respectively
- Z_p is the normal curve table value for pth percent of probability under the probability density curve

For example, $Z_{0.95} = 1.645$ is the 95 percentile value from the normal distribution table. For a standard normal distribution ($\mu = 0$ and $\sigma = 1$), Equation 11.1 gives the 95 percentile value equal to $x_{0.95} = 0 + 1.645 \times 1 = 1.645$.

TABLE E1
Case Study Data for Percentile and Confidence Interval

Sampling Date	Benzene (μg/L)	Log (Benzene)
March 1996	810	6.70
June 1996	1200	7.09
September 1996	200	5.30
December 1996	980	6.89
February 1997	340	5.83
May 1997	320	5.77
July 1997	3.8	1.34
October 1997	480	6.17
March 1998	180	5.19
June 1998	38	3.64
August 1998	55	4.01
December 1998	110	4.70
March 1999	35	3.56
May 1999	230	5.44
August 1999	130	4.87
December 1999	120	4.79
Mean	327	5.08
Standard deviation	362	1.465
Distribution type	Normal	Log-normal
50% ile (median)	327	161
95% ile	922	1788
Mean confidence interval at 95%	[150, 505]	[78, 329]

Unlike any other distribution curves, the log-normal distribution does not have a table for all percentile values. We usually "borrow" the normal probability distribution table for log-normal distribution calculations. If x is normally distributed with mean (μ) and standard deviation (σ), that is, $x \longrightarrow$ Normal (μ, σ), then $y=e^x$ is defined as being log-normally distributed. Since x is normally distributed, the logarithmic value of y: $\log(y)=x$ is normally distributed; that is, $\log(y) \longrightarrow$ Normal (μ, σ).

If y is known to be log-normally distributed, then the natural logarithm of y, $\log_e(y)$, is normally distributed. For the pth percentile, Equation 11.1 becomes

$$\log_e(y_p) = \mu + Z_p \cdot \sigma \tag{11.2}$$

Thus, the pth percentile of the log-normally distributed y is

$$y_p = \exp(\mu_y + Z_p \cdot \sigma_y) \tag{11.3}$$

where:

y_p	is the pth percentile value of the log-normally distributed variable y
μ_y and σ_y	are the logarithmic mean and logarithmic standard deviation of the log-normally distributed variable y, respectively

For example, soil concentration data of a log-normal distribution is transformed into logarithmic values that produce a mean of 2.724 and a standard deviation of 1.995. Thus, $\mu=2.724$, $\sigma=1.995$, $Z_{0.95}=1.645$. Equation 11.3 yields $y_{0.95}=\exp\ (\mu+Z_{0.95}\cdot\sigma)=\exp\ (2.724+1.645\times1.995)=406$. Therefore, the 95 percentile of this soil concentration distribution is equal to 406. By using Equation 11.3 backward, we can also determine the percentage under the probability curve with known y_p. For example, for the same case, given $y_p = 60 = \exp(2.724 + Z_p \cdot 1.995)$, $Z_p = (\ln 60 - 2.724)/1.995 = 0.69$. From the normal probability distribution table, $N(Z = 0.69) = 0.75$. Therefore, soil concentration at 60 is at 75 percentile of the log-normal distribution.

Calculation of Confidence Interval

To evaluate environmental sample data relative to a singular value of a regulatory standard, a confidence interval for the sample mean is often calculated to compare with the standard (U.S. EPA, 1989). Given the data distribution mean and standard deviation of a normal distribution, the confidence interval is calculated as (Walpole, 1974)

$$X_{(1-\alpha)} = \overline{X} \pm Z_p \cdot \left(\sigma / n^{1/2}\right) \tag{11.4}$$

where:

$X_{(1-\alpha)}$	is the confidence interval for the mean of variable x at level of $1-\alpha$;
$\overline{X}$	is the sample mean
n	is the number of samples
$p = 1 - \alpha/2$	for both sides

For example, we assume a normal distribution, $\overline{X} = 327$, $\sigma = 362$, $n = 16$, $Z_{0.975} = 1.96$, $X_{(0.95)} = [\overline{X} - Z_{0.975} \cdot (\sigma/n^{1/2}), \overline{X} + Z_{0.975} \cdot (\sigma/n^{1/2})] = [150, 505]$.

If data follow a log-normal distribution, Equation 11.4 becomes

$$Y_{(1-\alpha)} = \exp\left[\overline{Y} \pm Z_p \cdot \left(\sigma_Y / n^{1/2}\right)\right] \tag{11.5}$$

where:

$Y_{(1-\alpha)}$	is the confidence interval for the mean of log-normally distributed variable y at level of $1-\alpha$
$\overline{Y}$	is the sample logarithmic mean
σ_Y	is the sample logarithmic standard deviation

For the same example, we assume a log-normal distribution, $\overline{Y} = 5.08$, $\sigma_Y = 1.465$, $n = 16$, $Z_{0.975} = 1.96$, $Y_{(0.95)} = \exp[\overline{Y} - Z_{0.975} \cdot (\sigma_Y/n^{1/2}), \overline{Y} + Z_{0.975} \cdot (\sigma_Y/n^{1/2})] = [78, 329]$.

Both percentile and confidence interval are calculated based on an assumed distribution. Different distributions result in very different outcomes, as we have demonstrated. Therefore, before we calculate percentile and confidence interval, the data distribution must be determined. The distribution may require a large enough set of data points.

Question 2: Correlation Coefficient

Use data in Table E2 to calculate respective correlation coefficients for Column A and Column B. Explain why the correlation coefficients are so different because of only one data point (line 29).

Answer 2:

The correlation coefficient is a way to quantify the linear relationship between two random variables. The correlation coefficient is calculated as follows (Berthouex and Brown, 1994):

$$\rho_{x1,x2} = (1/n) \sum_{i=1}^{n} \left(x_{1i} - \overline{X}_1\right)\left(x_{2i} - \overline{X}_2\right) / \left(\sigma_{x1} \times \sigma_{x2}\right) \tag{11.6}$$

TABLE E2
MTBE Case Study Data for Correlation Coefficient

	A (MTBE)		B (MTBE)	
Case No.	Soil Concentration (mg/kg)	Groundwater Concentration (μg/L)	Soil Concentration (mg/kg)	Groundwater Concentration (μg/L)
1	15	43	15	43
2	74.1	82,000	74.1	82,000
3	15	18,700	15	18,700
4	6.4	59,000	6.4	59,000
5	120	9,906	120	9,906
6	58	470	58	470
7	0.53	130	0.53	130
8	710	3,100	710	3,100
9	0.87	2,200	0.87	2,200
10	200	22,000	200	22,000
11	7.3	1,200	7.3	1,200
12	3.1	690	3.1	690
13	3.5	12,000	3.5	12,000
14	41	110,000	41	110,000
15	22.6	14,000	22.6	14,000
16	210	50,000	210	50,000
17	0.24	2,000	0.24	2,000
18	32	38	32	38
19	315	45.2	315	45.2
20	57	24,000	57	24,000
21	170	5	170	5
22	160	290	160	290
23	2.3	4,000	2.3	4,000
24	1.5	410	1.5	410
25	11.2	24,000	11.2	24,000
26	90.3	180,000	90.3	180,000
27	2.7	5,800	2.7	5,800
28	10	61	10	61
29	860	2,100,000	—	—
Correlation coefficient	—	0.700	—	−0.03
Mean	110.3	94,003.0	83.6	22,360.3
SD	204.2	387,917.5	147.2	41,141.0
Max	860	2,100,000	710	180,000

MTBE: methyl tertiary butyl ether; SD: standard deviation.

where:

$\rho_{x1,x2}$	is the correlation coefficient between variables x_1 and x_2
$\overline{X}_1$ and $\overline{X}_2$	are the mean of x_1 and x_2, respectively
σ_{x1} and σ_{x2}	are the standard deviation of x_1 and x_2, respectively
n	is the number of data points

Therefore, for Column A,

$$\rho_{x1,x2} = \left[(15-110)\times(43-94{,}003)+\cdots+(860-110)\times(2{,}100{,}000-94{,}003)\right]/29(204\times 387{,}917)$$

For Column B,

$$\rho_{x1,x2} = \left[(15-110)\times(43-94{,}003)+\cdots+(10-84)\times(61-22{,}360)\right]/28(147\times 41{,}141).$$

The correlation coefficient ranges from −1 to +1, which implies a negative and a positive correlation, respectively. In general, $\rho_{x1,x2}=0$ means no correlation at all, and an increasing coefficient in absolute value means an increasing degree of correlation. Equation 11.6 indicates that the correlation coefficient is actually an index of the sum of two respective comparisons between a variable with its mean. If x_1 is greater than its mean ($\overline{X_1}$) and x_2 is less than its mean ($\overline{X_2}$), Equation 11.6 gives a negative value. If x_1 is greater than its mean and so is x_2, the result of Equation 11.6 is a positive value. Equation 11.6 also shows that the correlation coefficient has an inverse relationship with the standard deviation. A potential pitfall is to use inconsistent data over a wide range of values to calculate a correlation coefficient that can be skewed by a pair of large values of data.

These two sets of data are almost identical except for the last pair of data points (line 29). The values of the last data points are not only the maximum among all data but also much greater than other pairs of data in magnitude. If a pair of data points is far greater than others in value, only this pair of data is countable, and other data become trivia in Equation 11.6. This phenomenon is clearly demonstrated in Table E2. The correlation coefficient is 0.7 with the last pair of data and −0.03 without it. Apparently, we can draw completely different conclusions based on these two correlation coefficients, of which 0.7 indicates a moderately good correlation and −0.03 shows no correlation at all. Environmental data are usually widespread in magnitude, probably because pollutant concentration is tested not only in the source areas but also in the background and because of lack of knowledge on pollutant release timing, frequency, and quantities. Therefore, field data should be screened for magnitude of the data prior to computing a correlation coefficient. Usually, the standard deviation would indicate the degree of data spreading. In this case study, it is more realistic to use a correlation coefficient of −0.03 than 0.7 for environmental decisions, since the coefficient of 0.7 seems to result from only one pair of the maximum data points.

Question 3: Regression Analysis

The regression method can be used to predict concentration *Y* at a downgradient well (MW2) based on concentration *X* at the source well (MW1). Use data in Table E3 to fit a regression model by applying (a) a linear model and (b) a logarithmic model. Discuss which model is better to predict the downgradient well (MW1) concentration given the source well (MW2) concentration.

Answer 3:

Regression is a method to study the relationship among variables, most often to study two variables of which one depends on the other. This relationship is approximated by a function that is fitted by field data, and then the "fitted function" is used to predict the result of one variable given the other variable. For example, Equation 11.7 is a simple linear regression model (Walpole, 1974):

$$y = a + b\cdot x + e \qquad (11.7)$$

TABLE E3
MTBE Groundwater Concentrations over Time for Regression Analysis

	MTBE (μg/L)	
Sampling Date	**MW1**	**MW2**
June 1996	21,000	2,700
Sept 1996	37,000	5,100
December 1996	16,000	4,200
February 1997	59,000	5,200
May 1997	56,000	4,300
July 1997	48,000	5,600
October 1997	45,000	7,500
March 1998	26,000	8,200
June 1998	35,000	6,300
August 1998	35,000	12,000
December 1998	16,000	13,000
March 1999	13,000	15,000
May 1999	13,000	20,000
August 1999	7,600	19,000
December 1999	3,000	20,000
Mean	28,706.7	9,873.3
SD	17,803.8	6,139.9
Max	59,000	20,000
Min	3,000	2,700

MTBE: methyl tertiary butyl ether; SD: standard deviation; MW1, Monitoring Well No. 1; MW2, Monitoring Well No. 2.

where:

- y is a dependent variable
- x is an independent variable
- a is the linear intercept
- b is the linear slope
- e is the error term of measurement

The coefficients a and b are usually determined by the "least squares" method, which virtually minimizes the sum of deviations of all data points from the artificial "fitted" line to make the error term as small as possible. Given a data series of x_i and y_i ($i=1 \ldots n$), the algebraic solution of the least squares method is

$$a = \left(\overline{Y} - b\right)\overline{X} \tag{11.8}$$

$$b = \left[\sum_{i=1}^{n}\left(x_i - \overline{X}\right)\left(y_i - \overline{Y}\right)\right] \Big/ \left[\sum_{i=1}^{n}\left(x_i - \overline{X}\right)^2\right] \tag{11.9}$$

where:

- $\overline{X}$ is the mean of x
- $\overline{Y}$ is the mean of y

The degree of "goodness" of fitting the regression models can be measured by an index of R^2:

$$R^2 = 1 - \left[\sum_{i=1}^{n} (y_i - Y')^2 \Big/ \sum_{i=1}^{n} (y_i - \bar{Y})^2 \right] \qquad (11.10)$$

where:

R^2 is the coefficient of determination of "goodness to fit" (also called "*R* squares")

Y' is the predicted y from Equation 11.7

R^2 ranges between 0 and 1. R^2 near unity implies an excellent fit of the regression line. The larger the R^2 is, the better the fitting line is, and vice versa.

More complicated regression models such as multiple linear regression or nonlinear regression are also available. The solutions for fitting the coefficients in those models are usually built in spreadsheet computer programs now. R^2 is still used as a common indicator of the goodness to fit. Algebraically, the regression method can be used to predict changes of one variable when other variable(s) are known.

This case study has groundwater monitoring data of MTBE from 1996 to 1999 at the source well (MW1) and downgradient well (MW2), respectively (Table E3). The data show that MTBE concentration at the source well MW1 reached its peak (59,000 μg/L) in February 1997 and gradually decreased after that. On the other hand, MTBE concentration at the downgradient well MW2 gradually increased and reached its peak (20,000 μg/L) around May 1999. The spatial and temporal variation of MTBE concentrations seems to be reasonable. Apparently, the MTBE plume originated at the source, as detected at the source well, and the center mass of the plume has migrated toward downgradient, as indicated at both the source well and the downgradient well.

The purpose of the regression model is to find a relationship of MTBE concentrations between these two wells and then to use the concentration at the source well to predict the concentration at the downgradient well. Using the built-in regression function in the Excel spreadsheet program, based on data in Table E3, we generate two regression lines: a linear model ($R^2 = 0.48$) and a logarithmic model ($R^2 = 0.6$), as shown in Figures E1 and E2, respectively. The two models are expressed as $y = -0.24 \cdot x + 16{,}755$ (linear) and $y = -5763 \log_e(x) + 67{,}590$ (logarithmic), respectively.

The later monitoring data from February 2000 show 1400 μg/L MTBE at the source well MW1 and 13,000 μg/L MTBE at the downgradient well MW2, which is consistent with the trend of historical data. However, using the regression models, given $x = 1400$, the linear model yields $y = -0.24 \ldots 1{,}400 + 16{,}755 = 16.419$, and the logarithmic model yields $y = -5{,}763 \log_e(1400) + 67{,}590 = 25{,}841$. Comparing with the real field data $y = 13{,}000$, we can determine the percentage difference to be (16,419–13,000)/13,000 = 0.263 for the linear

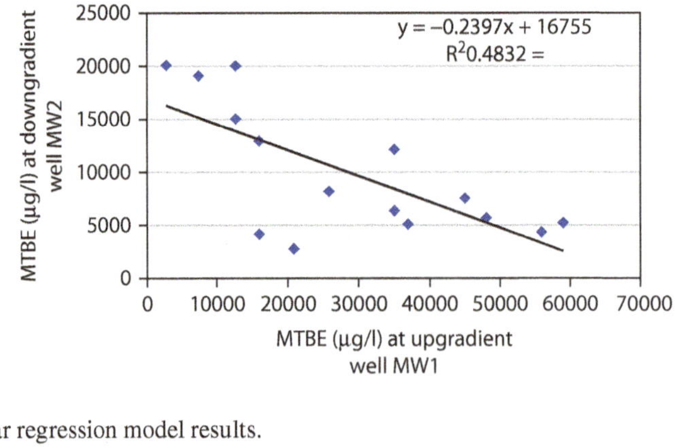

FIGURE E1 Linear regression model results.

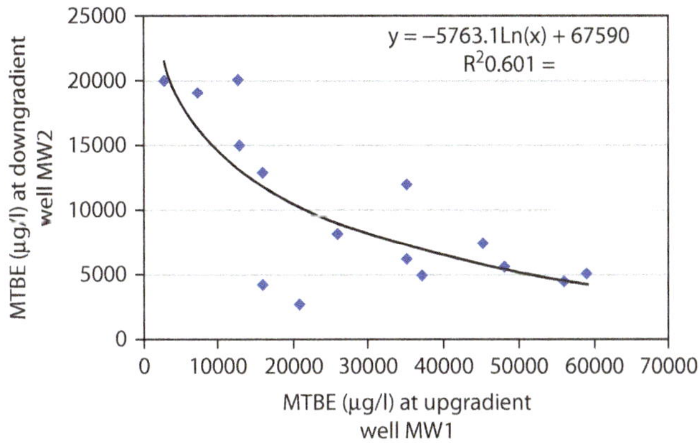

FIGURE E2 Logarithmic regression model for case study.

regression model and to be (25,841 – 13,000)/13,000 = 0.988 for the logarithmic regression model. Although the predictions of the regression models stay within an order of magnitude from the field observations, the accuracy of the prediction can be off from 26% to 99%. However, we should note that $y = 13{,}000$ is a random one-time observation. A series of observations may increase the accuracy of this value. Surprisingly, although the logarithmic model is the better fitted one ($R^2 = 0.6$) than the linear model ($R^2 = 0.48$) according to the R^2 values, the logarithmic regression model produces a prediction margin far more distant from the field observation than that produced by the linear regression model. Therefore, the fitting criterion R^2 may be used to evaluate the fitting condition based on the existing data set and is not really reliable in terms of prediction by the regression models.

If a regression model produces a prediction margin about 100% from a field observation, as in this case, this model may not be meaningful to model users. We recognize uncertainties associated with the field data at times and at locations. The uncertainties could produce a wide range of data sets, which consequently creates difficulties in fitting regression models. This might be the reason for the regression model's predictive inaccuracy. This case study reminds us that if there is a wide-range data set, we should apply the regression method very carefully. Otherwise, we simply carry the uncertainties over to the predictions, which may or may not be helpful to decision-making based on environmental data.

Question 4: Data Trend Analysis

Statistics practitioners often use statistical methods to analyze environmental monitoring data in a time sequence to see the trend of concentration changes over time. A popular method for groundwater monitoring data analysis is the Mann-Kendall method (Gilbert, 1987; Kendall, 1975). Let $x_1, x_2, \ldots x_j \ldots x_n$ be n data points, where x_j is a data monitoring value at time j. Then, the Mann-Kendall index is calculated by the method

$$S = \sum_{k=1}^{n-1} \sum_{j=k+1}^{n} \text{MK}\left(x_j - x_k\right) \tag{11.11}$$

where:

$\text{MK}(x_j - x_k) = 1$ if $(x_j - x_k) > 0$

$\text{MK}(x_j - x_k) = 0$ if $(x_j - x_k) = 0$

$\text{MK}(x_j - x_k) = -1$ if $(x_j - x_k) < 0$

$(k = 1, 2, \ldots n - 1)$ $(j = k + 1, k + 2, \ldots n)$

TABLE E4
TCE Groundwater Monitoring Data in Four Quarters

Unit (μg/L)	Case 1	Case 2
Quarter 1	1010	10,000
Quarter 2	995	5,600
Quarter 3	980	479
Quarter 4	969	34

TCE: trichloroethylene.

where:

S is the total score of the Mann-Kendall index

MK is the individual Mann-Kendall score at one comparison, where x_j is the data point at time j and x_k is the data point at time k

A very high positive value of S indicates an increasing trend, and a very low negative value indicates a decreasing trend. The method also takes many other statistical steps (probability, hypothesis, etc.) to further justify the significance of the trend. Nevertheless, the pitfalls in the fundamental Equation 11.11 of the Mann-Kendall method for data trend analysis will be discussed in the following.

Use data in Table E4 to do the Mann-Kendall trend analysis. Discuss the shortcomings and uncertainties of the method and explain why this method may not be helpful to show the trend of concentration changes.

Answer 4:

Using the Mann-Kendall method (Equation 11.11), the results clearly demonstrate that both Case 1 and Case 2 have the Mann-Kendall statistic S equal to −4; however, the reality is that Case 1 has very little decrease, while the decrease in Case 2 is very significant over a period of four quarters. Therefore, the Man-Kendall method is not able to distinguish Case 1 from Case 2.

The potential pitfall of this method is analyzing the data temporal trend without looking into the magnitude of the data values and the meanings of the values. Statistical methods like the Mann-Kendall method (comparing data pairs) may be good for a situation where a "Yes or No" determination is needed. For example, it is detection or no detection. But, this method does not take into account the magnitude of data values. In other words, the MK score is obtained regardless of the magnitude of values compared. For example, MK equals 1 if $x_j = 10{,}000$ and $x_k = 10$, whereas MK also equals 1 if $x_j = 10{,}000$ and $x_k = 9{,}999$. Apparently, despite both MKs being equal to 1, the former case shows a large decreasing trend from 10,000 to 10, while the latter case shows almost no decrease. The concentration magnitude matters in a situation where active remedial action is needed.

Question 5:

The method detection limit (MDL) is defined as the minimum concentration of a substance that can be measured and reported with 99% confidence that the analyte concentration is greater than zero.

Determination: take seven replicates from about one to five times the MDL (projected)

$$MDL = SD \times t_{0.99}$$

$$SD = \left\{ \sum_{i=1}^{n} (x_i - X_{avg})^2 / (n-1) \right\}^{1/2} \quad n = 7$$

Given the following concentrations from seven calibration spiking samples, calculate the MDL.

0.45	0.46	0.49	0.46	0.45	0.50	0.53 (unit: μg/L)

Answer 5:

Therefore, the mean (X_{avg}) = 0.48, standard deviation (SD) = 0.0304, table t-value, $t_{0.99}$ (n = 7) = 3.143.

Hence, the MDL is

$$MDL = SD \times t_{0.99} = 0.0304 \times 3.143 = 0.0955$$

Question 6: Field sampling results are shown in the following table. <0.4 means the detection limit is 0.4, and <0.4 = ND (nondetect) (sample locations and analytical result units are omitted to simplify the study).

Sample 1	<0.40
Sample 2	4.20
Sample 3	7.90
Sample 4	0.91
Sample 5	4.60
Sample 6	<0.40
Sample 7	<0.40
Sample 8	<0.40
Sample 9	5.60
Sample 10	4.90
Sample 11	<0.40
Sample 12	0.42
Sample 13	1.3
Sample 14	21
Sample 15	3.8
Sample 16	<0.40
Sample 17	2.5
Sample 18	<0.40
Sample 19	1.1
Sample 20	1.9
Sample 21	0.49

Consider the proper approach to handle the nondetects in the data set.

(You may use the free ProUCL software, downloaded from www.epa.gov/land-research/proucl-software, to conduct the calculation.)

1. What is the proper mean? Explain why.
2. What is the proper upper confidence level? Explain why.
3. If the remediation goal is 4.5, does the data set demonstrate completion of remediation?
4. If the background level is 0.5, does the data set demonstrate a release?

Answer 6:

1. Based on ProUCL output: Since the Kaplan-Meier (KM) and Gamma ROS methods are recommended due to the relatively high percentage of nondetects (7 out of 21 samples) in the data set, to calculate UCL, the mean derived from these two methods (3.02 and 2.89, respectively) will be used.

2. Based on ProUCL output: With the same rationale for using the Kaplan-Meier (KM) and Gamma ROS methods as described in Answer (1), the recommended UCLs are calculated using the mean and standard deviation from each method, with the critical value for 95% confidence level at degrees of freedom of 20:
   95% KM (BCA) UCL = 4.695
   95% Adjusted Gamma KM-UCL = 6.034
   95% GROS Adjusted Gamma UCL = 6.636
3. No (use the information from ProUCL; the suggested UCL > 4.5).
4. Yes (use the information from ProUCL; the calculated LCL = 1.25 > 0.5).

Question 7: Student's *t*-test is probably the most common test for comparing samples. As one of the parametric techniques, what are the underlying assumptions?

Answer 7: Assumption 1, observations must be independent; 2, observations are drawn from normally distributed populations; 3, these populations must have the same variances; 4, the variables involved must have been measured in at least an interval scale.

Question 8: What are the differences between parametric and nonparametric techniques?

Answer 8: For parametric statistics, the four assumptions must be met; however, nonparametric statistical testing is a "distribution-free statistics" and therefore, none of the four assumptions need to be met.

Question 9: Based on the data provided in the following table (as shown in Section 11.6), what is the concentration trend of MTBE for the entire monitored period? Use a Mann-Kendall nonparametric statistics program.

Date	Concentration	Date	Concentration	Date	Concentration
10/26/1995	7.7	10/9/2003	240	4/26/2011	110
4/22/1996	0.01	4/6/2004	180	10/19/2011	150
6/20/1996	19	10/4/2005	170	4/20/2012	260
10/23/1996	83	4/6/2006	170	10/12/2012	170
4/11/1997	143	10/5/2006	160	4/25/2013	280
11/17/1997	150	1/10/2007	200	10/17/2013	110
4/16/1998	130	4/9/2007	120	4/17/2014	160
10/12/1998	220	7/16/2007	180	10/23/2014	130
4/14/1999	620	10/3/2007	170	4/13/2015	91
10/11/1999	750	1/10/2008	81	4/13/2015	91
4/3/2000	1300	4/22/2008	200	10/19/2015	65
10/4/2000	1300	7/28/2008	120	10/19/2015	65
4/12/2001	360	11/11/2008	140	5/4/2016	61
10/10/2001	130	1/14/2009	190	5/4/2016	61
4/16/2002	170	7/16/2009	140	10/18/2016	42
10/11/2002	1000	4/21/2010	200	10/18/2016	42
4/1/2003	400	10/22/2010	210	5/4/2017	38

Answer 9: A decreasing trend. As described in Section 11.5 and shown on Figure 11.2, the MTBE concentrations evolved over time from increasing, to stable with small fluctuations, to decreasing. However, because the period of increasing concentration was short or had many fewer data points as compared with the period of decreasing concentration, the overall decreasing trend dominates the life cycle.

Question 10: What are the nonstatistical methods for comparison between site and background concentrations of contaminants of concern?

Answer 10: Nonstatistical methods include (1) direct comparison of site concentrations with background; (2) comparison of 95% UCL between site and background; (3) weight of evidence demonstration that contaminants are not site related.

Question 11: Why is the determination of background so important in remedial site investigation?

Answer 11: During remedial site investigation, one of the critical tasks is to compare the concentrations of chemicals of concern (COC) with the background concentration, so that the scope of site-specific COC can be established and risk-based remediation levels for soil and groundwater determined.

REFERENCES

Berthouex, P.M. and Brown, L.C., 1994. *Statistics for Environmental Engineers*. Lewis Publishers, Boca Raton, FL.

Chernoff, G., Bosan, W. and Oudiz, D., 2005. Determination of a southern California regional background arsenic concentration in soil, California Department of Toxic Substances Control.

Florida Department of Environmental Protection, 2012. Guidance for comparing background and site chemical concentrations in soil.

Gauthier, T.D., 2002. Chapter 5, Statistical methods. In Murphy, B.L. and Morrison, R.D., (Eds.) *Introduction to Environmental Forensics* (2nd edn). Academic Press, San Diego, CA.

Gilbert, R.O., 1987. *Statistical Methods for Environmental Pollution Monitoring*. Van Nostrand Reinhold, New York, NY.

ITRC, 2013. *Groundwater Statistics and Monitoring Compliance, Statistical Tools for the Project Life Cycle*. GSMC-1. Interstate Technology & Regulatory Council, Groundwater Statistics and Monitoring Compliance Team, Washington, DC. www.itrcweb.org/gsmc-1/.

ITRC, 2016. *Geospatial Analysis for Optimization at Environmental Sites*. GRO-1. Interstate Technology & Regulatory Council, Geostatistics for Remediation Optimization Team, Washington, DC. www.itrcweb.org/gro-1/.

Kendall, M.G., 1975. *Rank Correlation Methods* (4th edn). Charles Griffin, London.

United States Code of Federal Regulations (CFR), 40 CFR Parts 258, 264 and 265.

United States Environmental Protection Agency (EPA), 2006a. Quality system: Guidance on systematic planning using the data quality objectives process (EPA QA/G-4).

United States Environmental Protection Agency (EPA), 2006b. Guidance on choosing a sampling design for environmental data (EPA QA/G-5S).

United States Environmental Protection Agency (EPA), 2006c. Data quality assessment: A reviewer's guide (EPA QA/G-9R).

United States Environmental Protection Agency (EPA), 2006d. Data quality assessment: Statistical methods for practitioners (EPA QA/GS-9. 2006).

U.S. EPA, 1989. Statistical analysis of ground-water monitoring data at RCRA facilities, Section 8 (April). Office of Solid Waste Management Division, Washington DC 20460.

U.S. EPA, 2002. Guidance for comparing background and chemical concentrations in soil for CERCLA sites. Office of Emergency and Remedial Response. EPA 540-R-01-003.

U.S. EPA, 2006. Data quality assessment: Statistical methods for practitioners. EPA QA/G-95. Office of Environmental Information. EPA/240/B-06/003.

U.S. EPA, 2009. Statistical analysis of groundwater monitoring data at RCRA facilities (unified guidance), March 2009. Office of Resource Conservation and Recovery Program Implementation and Information Division. EPA/ 530/R-09/007.

U.S. EPA. Statistical software ProUCL 5.1.00 for environmental applications for data sets with and without nondetect observations. www.epa.gov/land-research/proucl-software

Walpole, R.E., 1974. *Introduction to Statistics*. Macmillan, New York, NY.

Winkler, R.L. and Hays, W.L., 1975. *Statistics: Probability, Inference, and Decision* (2nd edn). Holt, Rinehart and Winston, New York, NY.

12 Laboratory Analytical Methods and Data Interpretation

Yue Rong

CONTENTS

12.1 INTRODUCTION

During environmental site assessment and remediation, sampling results are a very important factor in decision-making. The analytical results indicate the contaminant type, concentration, and concentration distribution and also allow a total mass estimate and choice of remedial methods. As an environmental professional and practitioner, one must learn how to interpret and use the analytical results from the sampling. This chapter will review commonly used analytical methods specified by the U.S. Environmental Protection Agency (USEPA) and present the concept of *detection limit*, which has profound impact on whether the pollutant is detected or not; that is, polluted or not polluted. The presence of a pollutant in samples is based on the detection limit, a conceptually easy but technically complicated term.

Common sample types in site assessment and remediation include soil samples, soil vapor samples, air samples, and groundwater or surface water samples. Since detection limits are analytical method specific, we must first review the concepts and definitions associated with analytical method systems and specifications. In this chapter, the analysis of volatile organic compounds (VOCs) is used as an example to discuss this subject, followed by the definitions and uses of various detection limits in environmental applications. Quality assurance and quality control (QA/QC) procedures used by the analytical laboratory will also be discussed in this chapter.

The purpose of this chapter is to introduce the concept of detection limit and how to use the analytical knowledge for data interpretation. For easier demonstration, discussion is only related to VOC procedures. However, the same principles are also applied to other contaminants, such as semi-volatile organic compounds (SVOCs) and metals.

12.2 ANALYTICAL METHOD REVIEW FOR VOLATILE ORGANIC COMPOUNDS

As discussed in previous chapters, environmental contaminants in site assessment and remediation include primarily VOCs (e.g., tetrachloroethylene [PCE], benzene), SVOCs (e.g., pesticides, PCBs), metals (e.g., lead, arsenic, chromium 6+), and the recently named *emerging compounds* (e.g., constituents in pharmaceutical and cosmetic products). Each analytical method should be relevant to particular contaminant compounds and also specific to sampling media. The USEPA publishes a list of standard analytical methods for solid wastes (e.g., SW-846 methods), water (e.g., drinking/waste water methods), and air (e.g., TO methods) (USEPA, 2014).

To demonstrate the principles of discussion in this chapter, only analytical methods used for analyzing VOCs are discussed. To analyze VOC compounds, gas chromatography (GC), which is a gas separator, is used. Figure 12.1 shows a schematic of the GC process. The volatile compounds are carried by the helium gas (inert gas) and injected into a capillary column where, the lighter compounds (lower molecular weight) go through the column first, and heavier compounds (higher molecular weight) come out the last. In that way, multiple VOCs would be separated. At the end of the column, there is a detector to register which compounds are detected.

There are several types of detector:

1. A flame ionization detector (FID) measures ions by using flame to burn the detected element or compound. This method is good for combustible compounds such as gasoline and methane. However, the detection limit is relatively high compared with the other detectors discussed later.
2. A photo ionization detector (PID) measures ions by using ultraviolet (UV) light to give energy to ionize compounds. This method is good for double bond compounds such as aromatics or PCE. The detection limit is relatively low in comparison with FID.
3. An electron capture detector (ECD) measures electrons by using electron absorption. This method is good for all compounds. The detection limit is relatively low in comparison with FID and PID.
4. An electrolytic conductivity detector (ELCD) (a.k.a. Hall electrolytic conductivity detector [HECD]) measures halogen ions by measuring the conductivity at high temperature. This method is good for halogenated compounds such as PCE or trichloroethylene (TCE). The detection limit is very low for halogenated compounds in comparison with ECD.
5. A mass spectrometer (MS) measures the fragment pattern of the detected element or compound by using energy bombardment. This method is good for all compounds.

Each detector is designed for a certain group of target compounds with different sensitivity of the instrument, and therefore, the comparison of detection limits among individual detectors is all relative and needs to consider other specifications and requirements in each analytical method. Table 12.1 lists the detectors used in common EPA methods applicable to VOC contaminants.

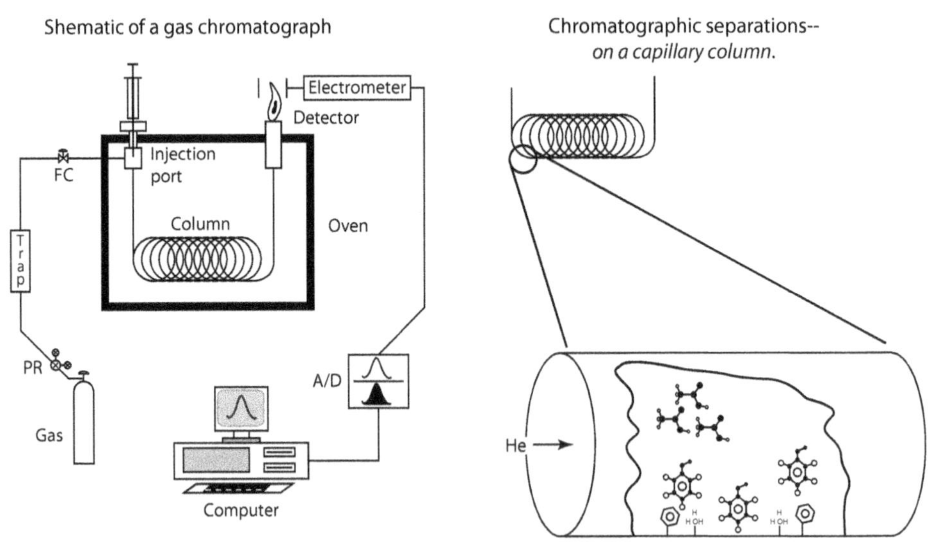

FIGURE 12.1 Schematic of gas chromatography (GC).

TABLE 12.1
Common EPA Methods Applicable to VOC Contaminants

Drinking Water

EPA 502.1 = GC/ELCD(HECD) for halogenated (e.g., PCE, TCE)
EPA 503.1 = GC/PID for aromatics (e.g., benzene, toluene)
EPA 502.2 = GC/PID + ELCD (HECD) for halogenated and aromatics
EPA 524.2 = GC/MS for halogenated and aromatics

Waste Water

EPA 601 = GC/ELCD(HECD) for halogenated (e.g., PCE, TCE)
EPA 602 = GC/PID for aromatics (e.g., benzene, toluene)
EPA 624 = GC/MS for halogenated and aromatics

Solid Waste (Soil)

EPA 8015 = GC/FID for non-halogenated organics
EPA 8021 = GC/PID + ELCD(HECD) for halogenated and aromatics
EPA 8260 = GC/MS for halogenated and aromatics

In summary,

GC/ELCD(HECD) for halogenated compounds: EPA 502.1 (drinking water) and EPA 601 (waste water)
GC/PID or GC/FID for aromatics: EPA 503.1 (drinking water), EPA 602 (waste water), and EPA 8015 (solid)
GC/PID + ELCD (HECD) for halogenated compounds and aromatics: EPA 502.2 (drinking water) and EPA 8021 (solid)
GC/MS for halogenated compounds and aromatics: EPA 524.2 (drinking water), EPA 624 (waste water), EPA 8260 (solid)

12.3 ANALYTICAL TEST AND CONCENTRATION DETERMINATION

The fundamentals of analytical chemistry involve two important aspects: (1) to identify the unknown analytes (also called *qualification*) and (2) to measure the concentrations of the identified compounds (also called *quantification*).

As discussed earlier, in a GC column, the lighter compounds (lower molecular weight) come out faster than the heavier compounds (higher molecular weight). In this principle, retention time is used to identify different compounds from light to heavy. Standard compounds are used to calibrate the instrument so that compounds in samples can be identified according to the retention times of the standards.

Figure 12.2 is an example chromatogram in which different compounds shown in different peaks are separated by their retention times on the horizontal axis.

Quantification is implemented by calibrating the instrument with known concentrations of the standard and then using the instrument response to the known concentrations to quantify unknown levels in samples.

Figure 12.3 shows two calibration curves. Curve (a) is the response, shown as area counts related to known concentrations of the calibration standard. Curve (b) uses the calibrated response to quantify the unknown concentration in samples. In this way, the sample concentration is estimated.

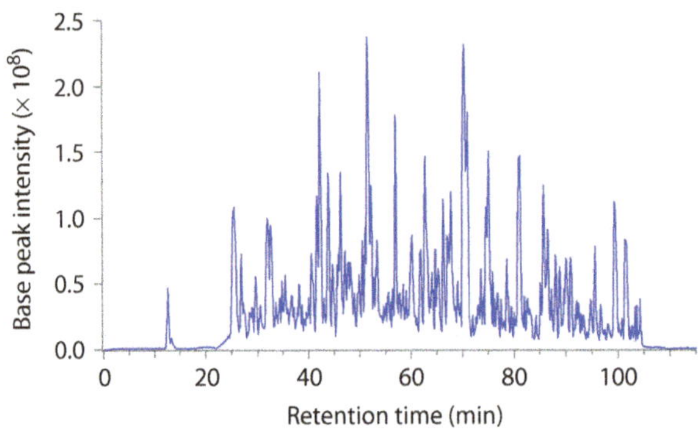

FIGURE 12.2 Example chromatogram.

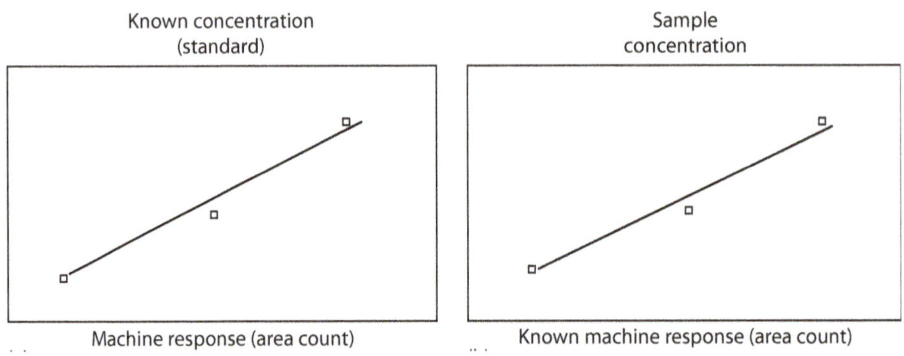

FIGURE 12.3 Calibration curves: (a) standard curve; (b) measurement of sample concentration.

12.4 ANALYTICAL METHOD DETECTION LIMIT

Now, the concept of detection limit needs to be introduced. From the perspective of site assessment, one would like to know what kinds of compounds are present at the site and at what levels. The detection limit indicates the lowest concentration that this analytical method is able to detect. The detection limit is defined as follows.

Each analytical method has its own detection limit, which is called the *method detection limit* (MDL), defined as the minimum concentration of a substance that can be measured and reported with 99% confidence that the analyte concentration is greater than zero.

Therefore,

$$MDL = SD \times t_{0.99} \tag{12.1}$$

where:

$$SD \text{ (standard deviation)} = \left\{ \sum_{i=1}^{n} \left(X_i - X_{\text{avg}}\right)^2 \Big/ (n-1) \right\}^{1/2} \quad n = 7$$

$t_{0.99}$ is value of the student t-distribution at 99% with the degrees of freedom $(n-1)$
X_i is the spiking replicate concentration of sample i
X_{avg} is the average concentration among n samples
n is the number of samples (recommended to be at least 7)

For example, the MTBE method detection limit is determined by spiking known concentrations of seven duplicate samples as follows (usually, the concentrations chosen are close to the machine's detection ability):

0.45 0.46 0.49 0.46 0.45 0.50 0.53

Therefore, mean $X_{avg}=0.48$, $SD=0.0304$

$$t_{0.99}(n=7)=3.143$$

$$MDL = SD \times t_{0.99} = 0.0304 \times 3.143 = 0.0955.$$

Besides MDL, there are other detection limits for different purposes:

- IDL = instrument detection limit (lowest detection above equipment electronic signal noise)
- PQL = practical quantitation limit
- EQL = estimated quantitation limit
- RL = reporting limit (usually specified by the user)

The reason for PQL and EQL is to have confidence in the quantification (i.e., concentration) reported. The RL is much like a PQL or MDL, which is usually specified for a particular need requested by the user. Commonly, analytical laboratories and their clients prefer to push the MDL as low as possible. However, the lower the MDL goes, the more inaccurate the reported concentrations at very low levels may become, due to uncertainties related to the machine response, the operation procedures in a particular laboratory, and the chemist's skill and experience. Thus, the practical quantitation limit (PQL) is set to increase the confidence level in quantification. PQL or EQL is typically defined as 2–10 times above MDL:

$$PQL\ (EQL) = (2-10) \times MDL\ (\text{usually determined by laboratory}) \tag{12.2}$$

By raising MDL by a factor of 2–10, serving as a "safety factor," the commercial laboratories hope to quantify the environmental sample concentrations with a degree of certainty. The degree of the factor (2–10) is decided by the analytical laboratory depending on the skill and experience of the analyst, the quality of the instrument, and the nature of the sample objectives. Therefore, most analytical laboratories report the analytical results above the PQL and may also report results above the MDL, as requested by the client or regulatory agencies.

Concentrations that are detected between MDL and PQL are typically labeled as *J-values*, which originated from footnote J in an EPA document (USEPA, 2014) referring to the laboratory qualifier or flag for estimated values. The J-value is a flag denoting a concentration above MDL and below PQL, which really means that this compound has been positively identified, but the quantitative concentration in this sample is not certain. The concept of J-value is very important to interpret the sample results. *Non-detect*, as referred to in many environmental reports, usually indicates *not detected* per a certain detection limit specified in the laboratory report. The detection limit might be MDL or PQL, as determined by the end-users. Therefore, knowledge of detection limit is necessary to review a laboratory report, and the detection limit must be clearly defined before interpreting data. This will be demonstrated in the case study at the end of this chapter.

In summary, setting an appropriate detection limit is a dilemma in many site assessment and remediation cases. To learn more about the potential pollution, the detection limit needs to be set as low as possible. However, if the detection limit is set too low, the confidence of the reported concentration becomes low, especially at very low levels. On the other hand, if the detection limit is set too high, low-concentration pollutants will not be reported; then, the site assessment may be incomplete. Therefore, the detection limit must be set in a task-specific way to meet the data quality objectives defined by the project-specific needs. References regarding data quality objectives can be seen in USEPA (1994a,b, 2000). For example, the waste water discharge limit should be low enough to meet the permitted limit, while for site investigation, it should be low enough to provide a clue to detect contaminant distribution. The detection limits may be high for testing waste that we know has high concentrations in the first place.

12.5 LABORATORY QUALITY ASSURANCE AND QUALITY CONTROL (QA/QC)

To have relatively high confidence in quantifying the sample results, most EPA methods specify quality assurance and quality control (QA/QC) procedures in the analytical laboratory. Commonly, the QA/QC procedures include initial calibration (initial calibration needs three points or more), daily calibration/continuing calibration, laboratory control spike, internal standards, surrogates, method blank, and field blank. Among them, the calibration procedures (initial and continuing) are most important to quantify the results with consistency and confidence. For the initial calibration, multiple calibration standard samples are needed to calculate the percent relative standard deviation (%RSD). Table 12.2 shows QA/QC standards and acceptable ranges for common EPA methods (seen in USEPA, 2014).

Usually, the QA/QC results are reported along with the sample results in the laboratory report, so that people who use the sample results will have the opportunity to see how much the analytical method used to analyze the field samples is off from the standard used for the calibration. For example, if an analytical method always produces consistently higher results than the spiking concentrations in the standard, although still within the acceptable range, one can conclude that the final reporting results may be at the high end of the analytical quantification.

Sometimes, the samples need to be diluted due to high concentrations that exceed the instrument calibration range. For example, if the calibration range is set for 1–20 ppb, a sample concentration at 40 ppb will be out of range. In this case, the sample will need to be diluted by a factor of two (i.e., 20 ppb) down to the calibration range to continue quantification of the concentration. The dilution factor is used to measure how many times the sample is to be diluted and should be reported in the laboratory reports. It should be noted that a high dilution factor also results in elevated reporting limits for non-detected compounds in a given sample that requires dilution.

Besides the QA/QC procedures and standards performed by the laboratory, QA/QC samples should also be collected for analysis to check the precision and accuracy of field data. Field QA/QC sample types consist primarily of co-located samples, split replicates, travel blanks, equipment blanks, and field blanks. The collected QA/QC samples need to be representative of existing site conditions, and influences due to the sampling and analysis procedures should be minimized. Table 12.3 provides an example of the minimum field QA/QC sample requirements recommended by California regulators (DTSC, 2015).

TABLE 12.2
Common EPA Methods QA/QC Standards and Acceptable Ranges

Initial calibration	%RSD $\leq$ 20% (SD_{n-1}) (SD = standard deviation)
Daily calibration (one point)	Diff $\leq \pm$15%
Laboratory control spike samples	Diff $\leq \pm$15% to 20%
Internal standards	Diff $\leq \pm$15% to 20%
Surrogates	Diff $\leq \pm$25%
Method blank	Non-detect
Field and trip blank	Non-detect

Diff: difference between the spiking concentration and the result concentration.

TABLE 12.3
Example of Minimum Field QA/QC Sample Requirements

Sample Type	Ratios
Co-located replicates	5% of total # of samples
Split replicates	5% of total # of samples
Travel blanks	1 per sample shipment (volatiles)
Equipment blanks	1 per field decontamination event (as needed)
Field blanks	1 per sample day

Source: California Department of Toxic Substances Control (DTSC), Preliminary Endangerment Assessment Guidance Manual (October), 2015.

12.6 CASE STUDY

If a sample concentration is detected between MDL and PQL, the analytical laboratory usually claims that this compound can be identified but its concentration cannot be quantified with confidence, for quantification decided by the laboratory (see Equation 12.2). In this case, the concentration between MDL and PQL is estimated with the indicator J, which is called the *J-value*. Although a J-value may not provide the quantification of a concentration, the result may still be useful. The following example discusses a typical environmental forensics case using J-values reported by the laboratory.

The case is a co-mingled gasoline plume in groundwater impacted by three gasoline stations (denoted A, B, and C) at an intersection in a city. There are three groundwater aquifer layers underlying Station A: the perched shallow groundwater, Upper Silverado aquifer, and Lower Silverado aquifer (see Figure 12.4). The co-mingled plume has been detected at the Lower Silverado aquifer. Groundwater data indicate that the levels of gasoline constituents detected are relatively high at the perched water, only at J-values at the Upper Silverado aquifer, and also relatively high at the Lower Silverado aquifer. Therefore, Station A easily argues that it only contributes contamination to the perched water layer, not the co-mingled plume in the deepest (Lower Silverado) aquifer, since the intermediate (Upper Silverado) aquifer between the two water layers is relatively non-impacted underlying its station. However, further review of the data finds that the gasoline oxygenate additive tertiary amyl methyl ether (TAME) has been detected in J-values in the intermediate aquifer. Furthermore, TAME has been detected in the perched water and the deepest aquifer underlying Station A. This oxygenate compound TAME is known as a unique gasoline additive from a certain brand of gasoline refinery, particularly marketed in the city served, and in this case, sold at Station A. Although the analytical result cannot confidently quantify the concentration of TAME in the intermediate aquifer in this case, the J-value identification can still lead investigation to the linkage of contaminant downward movement from Station A to the perched zone, through the Upper Silverado aquifer, and to the co-mingled plume in the Lower Silverado aquifer. Therefore, the J-value is used as the linkage piece in the investigation.

As we can see, if the concept of the J-value was not understood or used, one might make a different conclusion, based on incomplete information that the TAME concentrations were non-detect (or could not be quantified confidently as meeting the PQL). This case study demonstrates how important it is to understand the concept of detection limit and analytical procedures discussed in this chapter. This knowledge is needed to interpret the analytical data and draw proper conclusions.

12.7 SUMMARY

Most decisions for environmental site assessment and remediation are largely based on the analytical results. Therefore, the analytical methods and procedures should be the fundamental knowledge necessary to interpret the data. This chapter adequately demonstrates how important it is to

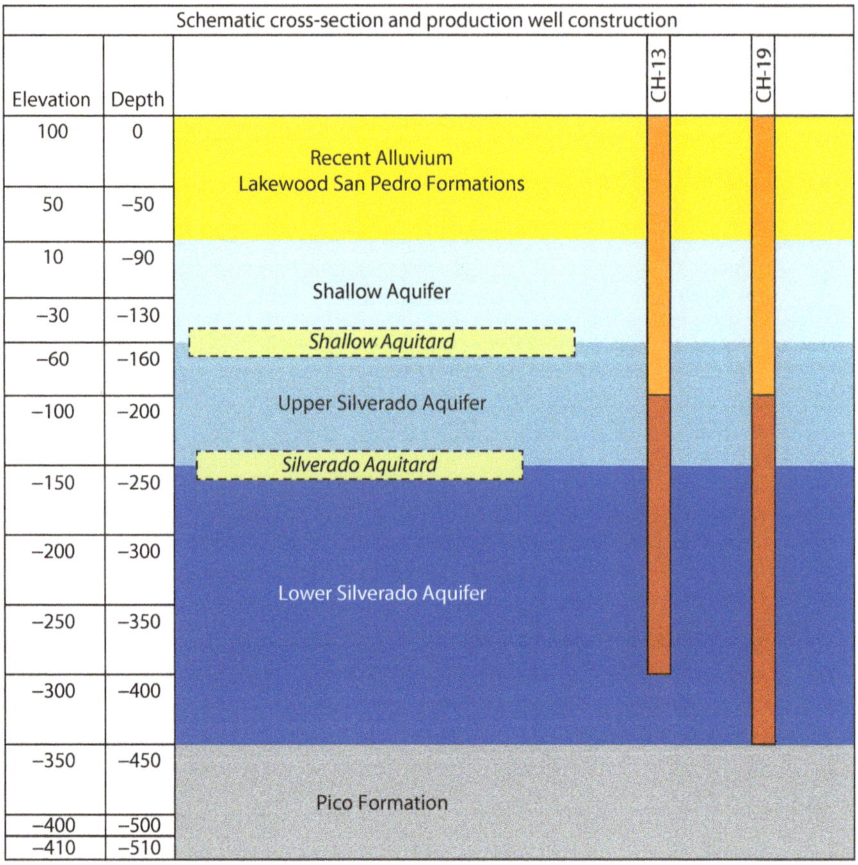

FIGURE 12.4 Case study hydrogeology setting with three groundwater aquifer layers.

understand the concept of detection limit and how the detection limit should be set to serve the purposes of site assessment and remediation. It is recommended for environmental professionals to understand the analytical method prior to evaluating the analytical results. Only doing this can produce a more intelligent and insightful decision based on the analytical data. Environmental professionals are not chemists, but the analytical knowledge is a must to have for decisions in site assessment and remediation.

PRACTICE AND EXERCISES

Question 1: Calculate method detection limit (MDL), given the following seven spiking concentrations in micrograms per liter.

1.45 1.6 1.49 0.96 0.85 0.59 1.53

Therefore, mean $X_{avg} = 1.21$, $SD = 0.401$

$$t_{0.99}(n = 7) = 3.143$$

$$MDL = SD \times t_{0.99} = 0.401 \times 3.143 = 1.26.$$

Question 2: Laboratory report review

Given, the laboratory report provides two limits for benzene: *MDL*=0.11 µg/L and *RL*=2 µg/L. The analytical result shows "non-detect" (ND) for benzene.

(**a**): Do they meet the following regulatory permit limit for benzene at 1 µg/L?

National Pollutant Discharge Elimination System (NPDES) permit limit=1 µg/L for benzene. However, *RL*=2 µg/L is above the permit limit, and therefore, it is considered a non-compliance case.

However, further examination of the report indicates that the laboratory report footnotes stated that ND really means non-detect above its *MDL*.

Therefore, *MDL*=0.11 µg/L is in compliance with the permit limit (1 µg/L).

(**b**): Why give two detection limits?

Answer 2: It allows the laboratory to define the J-value (estimated value) if a detected concentration is between these two limits.

Question 3: Example of baby formula contaminated by melamine

An incident happened in Asia about a decade ago. Baby formula was found to contain melamine, which is toxic to babies. Why was melamine present in baby formula? This was because the routine analytical method for protein testing was too expensive. So, a cheap method was sought to analyze nitrogen to estimate the protein content to check the baby formula quality. Melamine can produce nitrogen and so can "fool" the analytical results that it apparently has a higher protein content. However, melamine has no nutritious value; it is actually toxic to babies and is not a protein "booster."

How do you set a detection limit to test for melamine?

Answer 3:

To set a detection limit requires knowing the regulatory limit. In this incident, we learned the following.

During the detection of melamine (because it is toxic to babies), one regional governmental office in Asia required reporting of whether or not it is detected, but forget to set a detection limit. Therefore, the responsible party may set a self-regulated number, for example, 10 ppm, which means that any concentration below 10 ppm will be reported as ND. Another regional governmental office required a detection limit of 2.5 ppm, but only found out later that the health level limit set by the European Union is 2 ppm. Therefore, the 2.5 ppm detection limit is still considered too high to meet the health goal of 2 ppm.

Question 4: Example of NPDES limit versus laboratory detection limit

NPDES discharge limits (monthly average) are set as follows:

PCE	0.4 µg/L
TCE	1.34 µg/L
1,1-dichloroethane (DCA)	0.03 µg/L

Sometimes, the laboratory detection limits cannot go that low. In this case, we have to set a regulatory "minimum level." After the minimum level is set, the data will be evaluated again.

Using PCE as an example, the laboratory detection limit can only meet the PQL of 0.5 μg/L for GC/MS. How should the newly set regulatory "minimum level" be used to make a decision?

While the NPDES for PCE limit = 0.4 μg/L, the analytical laboratory could not perform to this level, but can meet the PQL = 0.5 μg/L. Therefore, the state minimum level is set at 0.5 μg/L. Based on this level, the regulatory determination will be as follows:

If the sample result < 0.4 μg/L (below MDL or J-value between MDL and PQL), then it is good.

If the sample result is 0.4–0.5 μg/L (J-value between MDL and PQL), then monitoring must be increased.

If the sample result is >0.5 μg/L (above PQL), then it is in violation of the limit.

The same principle may be applied to TCE and 1,1-DCA as well.

REFERENCES

California Department of Toxic Substances Control (DTSC), 2015. Preliminary Endangerment Assessment Guidance Manual (October).

USEPA, 1994a. Guidance for the Data Quality Objective Process II, Office of Research and Development, EPA/600/R-96/055 (September).

USEPA, 1994b. USEPA Contract Laboratory Program National Functional Guidelines for Organic Data Review, EPA/540/R-94/012 (February).

USEPA, 2000. Data Quality Objective Process for Hazardous Waste Site Investigations, EPA/600/R-00/007 (January).

USEPA, 2014. Test Methods for Evaluating Solid Waste: Physical/Chemical Methods (SW-846) (3rd edn) (July). www.epa.gov/hw-sw846/sw-846-compendium.

Appendix

National Drinking Water Regulations (United States)

Contaminant (Microorganism)	MCLG[1] (mg/L)[2]	MCL[1] (mg/L)[2] or TT[1]	Potential Health Effects	Sources of Contaminant
Cryptosporidium	Zero	TT[3]	Gastrointestinal illness (such as diarrhea, vomiting, and cramps)	Human and animal fecal waste
Giardia lamblia	Zero	TT[3]	Gastrointestinal illness (such as diarrhea, vomiting, and cramps)	Human and animal fecal waste
Heterotrophic plate count (HPC)	n/a	TT[3]	HPC has no health effects; it is an analytic method used to measure the variety of bacteria that are common in water	HPC measures a range of bacteria that are naturally present in the environment
Legionella	Zero	TT[3]	Legionnaire's disease, a type of pneumonia	Found naturally in water; multiplies in heating systems
Escherichia coli	Zero	Zero[4]	Disease-causing microbes (pathogens) in these wastes can cause diarrhea, cramps, nausea, headaches, or other symptoms	Human and animal fecal waste
Turbidity	n/a	TT[3]	Turbidity is a measure of the cloudiness of water. It is used to indicate water quality and filtration effectiveness	Soil runoff
Viruses (enteric)	Zero	TT[3]	Gastrointestinal illness (such as diarrhea, vomiting, and cramps)	Human and animal fecal waste

Contaminant (Disinfection By-product)	MCLG[1] (mg/L)[2]	MCL[1] (mg/L)[2]	Potential Health Effects	Sources of Contaminant
Bromate	Zero	0.010	Increased risk of cancer	By-product of drinking water disinfection
Chlorite	0.8	1.0	Anemia; infants and young children: nervous system effects	By-product of drinking water disinfection
Haloacetic acids (HAA5)	n/a[5]	0.060	Increased risk of cancer	By-product of drinking water disinfection
Total trihalomethanes (TTHMs)	n/a[5]	0.080	Liver, kidney or central nervous system problems; increased risk of cancer	By-product of drinking water disinfection

Contaminant (Disinfectant)	MCLG[1] (mg/L)[2]	MRDL[1] (mg/L)[2]	Potential Health Effects	Sources of Contaminant
Chloramines (as Cl_2)	MRDLG=4[1]	MRDL=4.0[1]	Eye/nose irritation; stomach discomfort; anemia	Water additive used to control microbes
Chlorine (as Cl_2)	MRDLG=4[1]	MRDL=4.0[1]	Eye/nose irritation; stomach discomfort	Water additive used to control microbes
Chlorine dioxide (as ClO_2)	MRDLG=0.8[1]	MRDL=0.8[1]	Anemia; infants and young children: nervous system effects	Water additive used to control microbes

Contaminant (Inorganic Chemical))	MCLG[1] (mg/L)[2]	MCL[1] (mg/L)[2]	Potential Health Effects	Sources of Contaminant
Antimony	0.006	0.006	Increase in blood cholesterol; decrease in blood sugar	Discharge from petroleum refineries; fire retardants; ceramics; electronics; solder
Arsenic	0	0.010	Skin damage or problems with circulatory systems, and may have increased risk of getting cancer	Erosion of natural deposits; runoff from orchards; runoff from glass and electronics production wastes
Asbestos (fiber > 10 micrometers)	7 million fibers per liter (MFL)	7 MFL	Increased risk of developing benign intestinal polyps	Decay of asbestos cement in water mains; erosion of natural deposits
Barium	2	2	Increase in blood pressure	Discharge of drilling wastes; discharge from metal refineries; erosion of natural deposits
Beryllium	0.004	0.004	Intestinal lesions	Discharge from metal refineries and coal-burning factories; discharge from electrical, aerospace, and defense industries
Cadmium	0.005	0.005	Kidney damage	Corrosion of galvanized pipes; erosion of natural deposits; discharge from metal refineries; runoff from waste batteries and paints
Chromium (total)	0.1	0.1	Allergic dermatitis	Discharge from steel and pulp mills; erosion of natural deposits
Copper	1.3	TT[6]; action level=1.3	Short-term exposure: gastrointestinal distress; long-term exposure: liver or kidney damage	Corrosion of household plumbing systems; erosion of natural deposits

(Continued)

Appendix: (Continued)

Contaminant (Inorganic Chemical))	MCLG[1] (mg/L)[2]	MCL[1] (mg/L)[2]	Potential Health Effects	Sources of Contaminant
Cyanide (as free cyanide)	0.2	0.2	Nerve damage or thyroid problems	Discharge from steel/metal factories; discharge from plastic and fertilizer factories
Fluoride	4.0	4.0	Bone disease (pain and tenderness of the bones); children may get mottled teeth	Water additive that promotes strong teeth; erosion of natural deposits; discharge from fertilizer and aluminum factories
Lead	Zero	TT[6]; action level=0.015	Infants and children: delays in physical or mental development; children could show slight deficits in attention span and learning abilities Adults: kidney problems; high blood pressure	Corrosion of household plumbing systems; erosion of natural deposits
Mercury (inorganic)	0.002	0.002	Kidney damage	Erosion of natural deposits; discharge from refineries and factories; runoff from landfills and croplands
Nitrate (measured as nitrogen)	10	10	Infants below the age of six months could become seriously ill and, if untreated, may die. Symptoms include shortness of breath and blue-baby syndrome.	Runoff from fertilizer use; leaking from septic tanks, sewage; erosion of natural deposits
Nitrite (measured as nitrogen)	1	1	Infants below the age of six months could become seriously ill and, if untreated, may die. Symptoms include shortness of breath and blue-baby syndrome	Runoff from fertilizer use; leaking from septic tanks, sewage; erosion of natural deposits
Selenium	0.05	0.05	Hair or fingernail loss; numbness in fingers or toes; circulatory problems	Discharge from petroleum refineries; erosion of natural deposits; discharge from mines
Thallium	0.0005	0.002	Hair loss; changes in blood; kidney, intestine, or liver problems	Leaching from ore-processing sites; discharge from electronics, glass, and drug factories

Contaminant (Organic Chemical)	MCLG[1] (mg/L)[2]	MCL[1] (mg/L)[2]	Potential Health Effects	Sources of Contaminant
Acrylamide	Zero	TT[7]	Nervous system or blood problems; increased risk of cancer	Added to water during water/wastewater treatment
Alachlor	Zero	0.002	Eye, liver, kidney, or spleen problems; anemia; increased risk of cancer	Runoff from herbicide used on row crops
Atrazine	0.003	0.003	Cardiovascular system or reproductive problems	Runoff from herbicide used on row crops
Benzene	Zero	0.005	Anemia; decrease in blood platelets; increased risk of cancer	Discharge from factories; leaching from gas storage tanks and landfills
Benzo(a)pyrene (polycyclic aromatic hydrocarbons)	Zero	0.0002	Reproductive difficulties; increased risk of cancer	Leaching from linings of water storage tanks and distribution lines
Carbofuran	0.04	0.04	Problems with blood, nervous system, or reproductive system	Leaching of soil fumigant used on rice and alfalfa
Carbon tetrachloride	Zero	0.005	Liver problems; increased risk of cancer	Discharge from chemical plants and other industrial activities
Chlordane	Zero	0.002	Liver or nervous system problems; increased risk of cancer	Residue of banned termiticide
Chlorobenzene	0.1	0.1	Liver or kidney problems	Discharge from chemical and agricultural chemical factories
2,4-D	0.07	0.07	Kidney, liver, or adrenal gland problems	Runoff from herbicide used on row crops
Dalapon	0.2	0.2	Minor kidney changes	Runoff from herbicide used on rights of way
1,2-Dibromo-3-chloropropane (DBCP)	Zero	0.0002	Reproductive difficulties; increased risk of cancer	Runoff/leaching from soil fumigant used on soybeans, cotton, pineapples, and orchards
o-Dichlorobenzene	0.6	0.6	Liver, kidney, or circulatory system problems	Discharge from industrial chemical factories
p-Dichlorobenzene	0.075	0.075	Anemia; liver, kidney or spleen damage; changes in blood	Discharge from industrial chemical factories
1,2-Dichloroethane	Zero	0.005	Increased risk of cancer	Discharge from industrial chemical factories
1,1-Dichloroethylene	0.007	0.007	Liver problems	Discharge from industrial chemical factories
cis-1,2-Dichloroethylene	0.07	0.07	Liver problems	Discharge from industrial chemical factories
trans-1,2-Dichloroethylene	0.1	0.1	Liver problems	Discharge from industrial chemical factories
Dichloromethane	Zero	0.005	Liver problems; increased risk of cancer	Discharge from drug and chemical factories

(Continued)

Appendix: (Continued)

Contaminant (Organic Chemical)	MCLG[1] (mg/L)[2]	MCL[1] (mg/L)[2]	Potential Health Effects	Sources of Contaminant
1,2-Dichloropropane	Zero	0.005	Increased risk of cancer	Discharge from industrial chemical factories
Di(2-ethylhexyl) adipate	0.4	0.4	Weight loss, liver problems, or possible reproductive difficulties	Discharge from chemical factories
Di(2-ethylhexyl) phthalate	Zero	0.006	Reproductive difficulties; liver problems; increased risk of cancer	Discharge from rubber and chemical factories
Dinoseb	0.007	0.007	Reproductive difficulties	Runoff from herbicide used on soybeans and vegetables
Dioxin (2,3,7,8-TCDD)	Zero	0.00000003	Reproductive difficulties; increased risk of cancer	Emissions from waste incineration and other combustion; discharge from chemical factories
Diquat	0.02	0.02	Cataracts	Runoff from herbicide use
Endothall	0.1	0.1	Stomach and intestinal problems	Runoff from herbicide use
Endrin	0.002	0.002	Liver problems	Residue of banned insecticide
Epichlorohydrin	Zero	TT[7]	Increased cancer risk, and over a long period of time, stomach problems	Discharge from industrial chemical factories; an impurity of some water treatment chemicals
Ethylbenzene	0.7	0.7	Liver or kidney problems	Discharge from petroleum refineries
Ethylene dibromide	Zero	0.00005	Problems with liver, stomach, reproductive system, or kidneys; increased risk of cancer	Discharge from petroleum refineries
Glyphosate	0.7	0.7	Kidney problems; reproductive difficulties	Runoff from herbicide use
Heptachlor	Zero	0.0004	Liver damage; increased risk of cancer	Residue of banned termiticide
Heptachlor epoxide	Zero	0.0002	Liver damage; increased risk of cancer	Breakdown of heptachlor
Hexachlorobenzene	Zero	0.001	Liver or kidney problems; reproductive difficulties; increased risk of cancer	Discharge from metal refineries and agricultural chemical factories
Hexachlorocyclopentadiene	0.05	0.05	Kidney or stomach problems	Discharge from chemical factories
Lindane	0.0002	0.0002	Liver or kidney problems	Runoff/leaching from insecticide used on cattle, lumber, gardens
Methoxychlor	0.04	0.04	Reproductive difficulties	Runoff/leaching from insecticide used on fruits, vegetables, alfalfa, livestock
Oxamyl (Vydate)	0.2	0.2	Slight nervous system effects	Runoff/leaching from insecticide used on apples, potatoes, and tomatoes

(*Continued*)

Appendix: (Continued)

Contaminant (Organic Chemical)	MCLG[1] (mg/L)[2]	MCL[1] (mg/L)[2]	Potential Health Effects	Sources of Contaminant
Polychlorinated biphenyls (PCBs)	Zero	0.0005	Skin changes; thymus gland problems; immune deficiencies; reproductive or nervous system difficulties; increased risk of cancer	Runoff from landfills; discharge of waste chemicals
Pentachlorophenol	Zero	0.001	Liver or kidney problems; increased cancer risk	Discharge from wood preserving factories
Picloram	0.5	0.5	Liver problems	Herbicide runoff
Simazine	0.004	0.004	Problems with blood	Herbicide runoff
Styrene	0.1	0.1	Liver, kidney, or circulatory system problems	Discharge from rubber and plastic factories; leaching from landfills
Tetrachloroethylene	Zero	0.005	Liver problems; increased risk of cancer	Discharge from factories and dry cleaners
Toluene	1	1	Nervous system, kidney, or liver problems	Discharge from petroleum factories
Toxaphene	Zero	0.003	Kidney, liver, or thyroid problems; increased risk of cancer	Runoff/leaching from insecticide used on cotton and cattle
2,4,5-TP (Silvex)	0.05	0.05	Liver problems	Residue of banned herbicide
1,2,4-Trichlorobenzene	0.07	0.07	Changes in adrenal glands	Discharge from textile finishing factories
1,1,1-Trichloroethane	0.20	0.2	Liver, nervous system, or circulatory problems	Discharge from metal degreasing sites and other factories
1,1,2-Trichloroethane	0.003	0.005	Liver, kidney, or immune system problems	Discharge from industrial chemical factories
Trichloroethylene	Zero	0.005	Liver problems; increased risk of cancer	Discharge from metal degreasing sites and other factories
Vinyl chloride	Zero	0.002	Increased risk of cancer	Leaching from PVC pipes; discharge from plastic factories
Xylenes (total)	10	10	Nervous system damage	Discharge from petroleum factories; discharge from chemical factories

Contaminant (Radionuclides)	MCLG[1]	MCL[1]	Potential Health Effects	Sources of Contaminant
Alpha particles	Zero	15 picocuries per liter (pCi/L)	Increased risk of cancer	Erosion of natural deposits of certain minerals that are radioactive and may emit a form of radiation known as alpha radiation
Beta particles and photon emitters	Zero	4 millirems per year	Increased risk of cancer	Decay of natural and manmade deposits of certain minerals that are radioactive and may emit forms of radiation known as photons and beta radiation
Radium 226 and Radium 228 (combined)	Zero	5 pCi/L	Increased risk of cancer	Erosion of natural deposits
Uranium	Zero	30 μg/L	Increased risk of cancer, kidney toxicity	Erosion of natural deposits

NOTES

1 Definitions:

- *Maximum Contaminant Level Goal (MCLG):* The level of a contaminant in drinking water below which there is no known or expected risk to health. MCLGs allow for a margin of safety and are non-enforceable public health goals.
- *Maximum Contaminant Level (MCL):* The highest level of a contaminant that is allowed in drinking water. MCLs are set as close to MCLGs as feasible using the best available treatment technology and taking cost into consideration. MCLs are enforceable standards.
- *Maximum Residual Disinfectant Level Goal (MRDLG):* The level of a drinking water disinfectant below which there is no known or expected risk to health. MRDLGs do not reflect the benefits of the use of disinfectants to control microbial contaminants.
- *Treatment Technique (TT):* A required process intended to reduce the level of a contaminant in drinking water.
- *Maximum Residual Disinfectant Level (MRDL):* The highest level of a disinfectant allowed in drinking water. There is convincing evidence that the addition of a disinfectant is necessary for control of microbial contaminants.

2 Units are in milligrams per liter (mg/L) unless otherwise noted. Milligrams per liter are equivalent to parts per million (PPM).

3 The Environmental Protection Agency (EPA)'s surface water treatment rules require systems using surface water or ground water under the direct influence of surface water to

a. Disinfect their water, and
b. Filter their water, or
c. Meet criteria for avoiding filtration so that the following contaminants are controlled at the following levels:
   - *Cryptosporidium*: Unfiltered systems are required to include *Cryptosporidium* in their existing watershed control provisions.
   - *Giardia lamblia*: 99.9% removal/inactivation.
   - Viruses: 99.99% removal/inactivation.

- *Legionella*: No limit, but EPA believes that if *Giardia* and viruses are removed/inactivated according to the treatment techniques in the Surface Water Treatment Rule, *Legionella* will also be controlled.
- Turbidity: For systems that use conventional or direct filtration, at no time can turbidity (cloudiness of water) be higher than 1 nephelometric turbidity unit (NTU), and samples for turbidity must be less than or equal to 0.3 NTU in at least 95% of the samples in any month. Systems that use filtration other than conventional or direct filtration must follow state limits, which must include turbidity at no time exceeding 5 NTU.
- *Heterotrophic plate count (HPC)*: No more than 500 bacterial colonies per milliliter.
- *Long Term 1 Enhanced Surface Water Treatment Rule*: Surface water systems or groundwater under the direct influence (GWUDI) systems serving fewer than 10,000 people must comply with the applicable Long Term 1 Enhanced Surface Water Treatment Rule provisions (such as turbidity standards, individual filter monitoring, *Cryptosporidium* removal requirements, updated watershed control requirements for unfiltered systems).
- *Long Term 2 Enhanced Surface Water Treatment Rule*: This rule applies to all surface water systems or ground water systems under the direct influence of surface water. The rule targets additional *Cryptosporidium* treatment requirements for higher-risk systems and includes provisions to reduce risks from uncovered finished water storage facilities and to ensure that the systems maintain microbial protection as they take steps to reduce the formation of disinfection by-products.
- *Filter backwash recycling*: This rule requires systems that recycle to return specific recycle flows through all processes of the system's existing conventional or direct filtration system or at an alternate location approved by the state.

4 The maximum contaminant level (MCL) is based on the occurrence of a condition that includes routine and repeat samples.

5 Although there is no collective MCLG for this contaminant group, there are individual MCLGs for some of the individual contaminants:
  - *Trihalomethanes*: Bromodichloromethane (zero); bromoform (zero); dibromochloromethane (0.06 mg/L): chloroform (0.07 mg/L).
  - *Haloacetic acids*: Dichloroacetic acid (zero); trichloroacetic acid (0.02 mg/L); monochloroacetic acid (0.07 mg/L). Bromoacetic acid and dibromoacetic acid are regulated with this group but have no MCLGs.

6 Lead and copper are regulated by a treatment technique that requires systems to control the corrosiveness of their water. If more than 10% of tap water samples exceed the action level, water systems must take additional steps. For copper, the action level is 1.3 mg/L, and for lead, it is 0.015 mg/L.

7 Each water system must certify, in writing, to the state (using third-party or manufacturer's certification) that when acrylamide and epichlorohydrin are used to treat water, the combination (or product) of dose and monomer level does not exceed the levels specified, as follows:
  - Acrylamide = 0.05% dosed at 1 mg/L (or equivalent)
  - Epichlorohydrin = 0.01% dosed at 20 mg/L (or equivalent)

Contaminant (Non-Health)	Secondary MCL	Noticeable Effects
Aluminum	0.05–0.2 mg/L	Colored water
Chloride	250 mg/L	Salty taste
Color	15 color units	Visible tint
Copper	1.0 mg/L	Metallic taste; blue-green staining
Corrosivity	Non-corrosive	Metallic taste; corroded pipes/fixtures staining
Fluoride	2.0 mg/L	Tooth discoloration
Foaming agents	0.5 mg/L	Frothy, cloudy; bitter taste; odor
Iron	0.3 mg/L	Rusty color; sediment; metallic taste; reddish staining
Manganese	0.05 mg/L	Black/brown color; black staining; bitter metallic taste
Odor	3 TON (threshold odor number)	"Rotten-egg", musty, or chemical smell
pH	6.5–8.5	Low pH: bitter, metallic taste; corrosion
		High pH: slippery feel; soda taste; deposits
Silver	0.1 mg/L	Skin discoloration; graying of the white part of the eye
Sulfate	250 mg/L	Salty taste
Total dissolved solids (TDS)	500 mg/L	Hardness; deposits; colored water; staining; salty taste
Zinc	5 mg/L	Metallic taste

Index